STATISTICAL TECHNIQUES FOR TRANSPORTATION ENGINEERING

STATISTICAL TECHNIQUES FOR TRANSPORTATION ENGINEERING

KUMAR MOLUGARAM
Professor, Department of Civil Engineering,
University College of Engineering,
Osmania University, Hyderabad, Telangana, India

G. SHANKER RAO
Formerly Head of the Department,
Department of Mathematics, Govt. Girraj P.G College, and
Asst. Professor(c), University College of Engineering(A),
Osmania University, Hyderabad

BSP **BS Publications**
A Unit of BSP Books Pvt. Ltd., India.

ELSEVIER

Butterworth-Heinemann
An imprint of Elsevier
elsevier.com

Butterworth-Heinemann is an imprint of Elsevier
The Boulevard, Langford Lane, Kidlington, Oxford OX5 1GB, United Kingdom
50 Hampshire Street, 5th Floor, Cambridge, MA 02139, United States

Notices
Knowledge and best practice in this field are constantly changing. As new research and experience broaden our understanding, changes in research methods, professional practices, or medical treatment may become necessary.

Practitioners and researchers must always rely on their own experience and knowledge in evaluating and using any information, methods, compounds, or experiments described herein. In using such information or methods they should be mindful of their own safety and the safety of others, including parties for whom they have a professional responsibility.

To the fullest extent of the law, neither the Publisher nor the authors, contributors, or editors, assume any liability for any injury and/or damage to persons or property as a matter of products liability, negligence or otherwise, or from any use or operation of any methods, products, instructions, or ideas contained in the material herein.

British Library Cataloguing-in-Publication Data
A catalogue record for this book is available from the British Library

Library of Congress Cataloging-in-Publication Data
A catalog record for this book is available from the Library of Congress

ISBN: 978-0-12-811555-8

Publisher: Joe Hayton
Acquisition Editor: Ken McCombs
Editorial Project Manager: Peter Jardim
Production Project Manager: Anusha Sambamoorthy
Designer: Miles Hitchen

CONTENTS

PREFACE

This book "Statistical Techniques for Transportation Engineering" is mainly designed to meet the requirements of engineering students of various Indian Universities. It is intended to be used as a text in basic statistics for the transportation engineering applications of probability and statistics in traffic engineering problems have markedly increased during the past two decades. These techniques are helpful in the study of vehicular traffic. The present text discusses in detail the procedures that have been found useful in treatment of several types of problems encountered by the traffic engineers and it forms the basis of a first course in statistics. It lays foundation necessary to understand the basics of statistical applications in Transportation engineering and other branches of civil engineering.

The topics covered in this book are, Measures of central tendency, Measures of dispersion, Probability, Curve fitting, Correlation and Regression, Sampling, Hypothesis testing, Tests of significance, Index numbers, and Time series. All the concepts have been well presented. The explanations are clear. Numerical examples have been worked out throughout the book to illustrate the concepts and explained the techniques of solving the problems. We earnestly hope that the comprehensive treatment of the various topics covered in the book will be welcomed by both the teachers and the students.

Authors

CHAPTER 1

An Overview of Statistical Applications

1.1 INTRODUCTION

Civil engineering is considered to be one of the oldest engineering disciplines. It deals with planning, analysis, design, construction, and maintenance of the physical and naturally built environment. The subject is grouped into several specialty areas namely, Structural Engineering, Construction Engineering and Management, Transportation Engineering, Water Resource Engineering, Surveying, Environmental Engineering, and Geotechnical Engineering. Transportation engineering involves the collection of huge amount of data for performing all types of traffic and transportation studies. The analysis is carried out based on the collected and observed data. The statistical aspects are also an important element in transportation engineering specifically traffic engineering. Statistics facilitates to resolve how much data will be obligatory, as well as what consequential inferences can confidently be finished based on that observed and collected data. Generally statistics is required whenever it is not possible to directly measure all of the values required. If the traffic engineer needs to know the average speed of all vehicles on a particular section of roadway, not all vehicles could be observed. Even if all speeds of vehicles could be measured over a specified time period, speeds of vehicles arriving before or after the study period or on a different day then the sample day would be unknown. In effect, no matter how many speeds are measured, there are always more that are not known. For all practical and statistical purposes, the number of vehicles using a particular section of roadway over time is infinite. Therefore the traffic engineering often observes and measures the characteristics of a finite sample of vehicles in a population that is effectively infinite. The mathematics of statistics is used to estimate characteristics that cannot be established with absolute certainty and to assess the degree of certainty that exists.

Statistical Techniques for Transportation Engineering.
DOI: http://dx.doi.org/10.1016/B978-0-12-811555-8.00001-5

1.2 PROBABILITY FUNCTIONS AND STATISTICS

Before exploring some of the more complex statistical applications in traffic engineering, some basic principles of probability and statistics that are relevant to transportation and traffic engineering subject are reviewed.

1.2.1 Discrete Versus Continuous Functions

Discrete functions are made up of discrete variables, i.e., they can assume only specific whole values and not any value in between. Continuous functions, made up of continuous variables, on the other hand, can assume any value between two given values. For example, Let $N =$ the number of cars in a family. N can equal 1, 2, 3, etc., but not 1.5, 1.6, 2.3. Therefore it is a discrete variable. Let $H =$ the height of an individual car. H can equal 1.5, 1.75, 2.25, 2.50, and 2.75 m, etc., and therefore, is a continuous variable. Examples of discrete probability functions are the Bernoulli, binomial, and Poisson distributions, which will be discussed in the Chapter 4, Random Variables. Some examples of continuous distribution are the normal, exponential, and chi-square distributions.

1.2.2 Distributions Describing Randomness

Some events are very predictable, or should be predictable. If you add mass to a spring or a force to a beam, you can expect it to deflect a predictable amount. If you depress the gas pedal to a certain amount and you are on level terrain, you expect to be able to predict the speed of the vehicle. On the other hand, some events may be totally random. The emission of the next particle from a radioactive sample is said to be completely random. Some events may have very complex mechanisms and appear to be random for all practical purposes. In some cases, the underlying mechanism cannot be perceived, while in other cases we cannot afford the time or money necessary for the investigation. consider the question of who turns north and who turns south after crossing a bridge. Most of the time, we simply say there is a probability, p, that a vehicle will turn north, and we treat the outcome as a random event. However, if we studied who was driving each car and where each driver worked, we might expect to make the estimate a very predictable event, for each and every car. In fact, if we kept a record of their license plates and their past decisions, we could make very predictable estimates. The events to a large extent are not random. Obviously, it is not worth that trouble because the random assumption serves us well adequate. In fact, a number of things are modeled as random for all practical purposes, given the investment we can afford. Most of the time, these judgments are just fine and are very reasonable but, as with every engineering judgment, they can sometimes cause errors.

1.2.3 Data Organization

In the data collection process, the data is collected for use in traffic studies, the raw data can be looked at as individual pieces of data or grouped into classes of data for easier comprehension. Most of the data will fit into a common distribution. Some of the common distributions found in traffic engineering are the normal distribution, the exponential distribution, the chi-square distribution, the Bernoulli distribution, the binomial distribution, and the Poisson distribution. As part of the process for determining which distribution fits the data, one often summarizes the raw data into classes and creates a frequency distribution table. This makes the data more easily readable and understood. You could put the data into an array format, which means listing the data points in order from either lowest to highest or highest to lowest. This will give you some feeling for the character of the data but it is still not very helpful, particularly when you have a large number of data points.

1.2.4 Common Statistical Estimators

In dealing with a distribution, there are two key characteristics that are of interest. These are discussed in the following subsections.

1.2.4.1 Measures of Central Tendency

Measures of central tendency are measures that describe the center of data in one of several different ways. The Arithmetic mean is the average of all observed data. The true underlying mean of the population, μ, is an exact number that we do not know, but can estimate as:

$$\overline{x} = \frac{1}{N}\sum_{i=1}^{N} x_i \tag{1.1}$$

where $\overline{x}$ = arithmetic average or mean of observed values, x_i = ith individual value of statistic, N = sample size, number of values x_i.

For grouped data, the average value of all observations in a given group is considered to be the midpoint value of the group. The overall average of the entire sample may then be found as:

$$\frac{\bar{x} = \sum_{j} f_j m_j}{N} \tag{1.2}$$

where f_j = number of observation in group j, m_j = middle value of variable in group j, N = total sample size or number of observations.

The median is the middle value of all data when arranged in an array (ascending or descending order). The median divides a distribution in half: half of all observed values are higher than the median, half are lower than the median. For nongrouped data, it is the middle value; e.g., for the set of numbers (3, 4, 5, 5, 6, 7, 7, 7, 8), the median is 6. It is the fifth value (in ascending or descending order) in an array of 9 numbers. For grouped data, the easiest way to get the median is to read the 50 percentile point off a cumulative frequency distribution curve.

The mode is the value that occurs most frequently that is the most common single value. For example, in nongrouped data, for the set of numbers (3, 4, 5, 5, 6, 7, 7, 7, 8) the mode is 7. For the set of numbers (3, 3, 4, 5, 5, 5, 6, 7, 8, 8, 8, 9), both 5 and 8 are modes, and the data is said to be bimodal. For grouped data, the mode is estimated as the peak of the frequency distribution curve. For a perfectly symmetrical distribution, the mean, median, and mode will be the same.

1.2.4.2 Measures of Dispersion

Measures of dispersion are measures that describe how far the data spread from the center. The variance and standard deviation are statistical values that describe the magnitude of variation around the mean, with the variance defined as:

$$S^2 = \frac{\sum (x_i - \bar{x})^2}{N - 1} \tag{1.3}$$

where S^2 = variance of the data, N = sample size, number of observations. All other variables are previously defined.

The standard deviation is the square root of the variance. It can be seen from the equation that what you are measuring is the distance of each data point from the mean. This equation can also be rewritten as:

$$S^2 = \frac{1}{N}\sum_{i=1}^{N} x_i^2 - \left(\frac{N}{N-1}\right)\overline{x}^2 \tag{1.4}$$

For grouped data, the standard deviation is found from:

$$S = \sqrt{\frac{\sum f_m^2 - N(\overline{x})^2}{N-1}} \tag{1.5}$$

where all variables are as previously defined. The standard deviation (STD) may also be estimated as:

$$S_{\text{est}} = \frac{P_{85} - P_{15}}{2} \tag{1.6}$$

where $P_{85} = 85$th percentile value of the distribution (i.e., 85% of all data is at this value or less), $P_{15} = 15$th percentile value of the distribution (i.e., 15% of all data is at this value or less).

The xth percentile is defined as that value below which x% of the outcomes fall. P_{85} is the 85th percentile, often used in traffic speed studies; it is the speed that encompasses 85% of vehicles. P_{50} is the 50th percentile speed or the median.

1.3 APPLICATIONS OF NORMAL DISTRIBUTION

One of the most common statistical distribution is the normal distribution, known by its characteristic bell shaped curve. The normal distribution is a continuous distribution. Probability is indicated by the area under the probability density function $f(x)$ between specified values, such as $P\ (40 < x < 50)$.

The equation for the normal distribution function is:

$$f(x) = \frac{1}{\sigma\sqrt{2\pi}} e^{-\left[\frac{(x-\mu)^2}{2\sigma^2}\right]} \tag{1.7}$$

where $x =$ normally distributed statistic, $\mu =$ true mean of the distribution, $\sigma =$ true standard deviation of the distribution, $\pi = 3.14$.

The probability of any occurrence between values x_1 and x_2 is given by the area under the distribution function between the two values. The area may be found by integration between the two limits. Likewise, the mean, μ, and the variance, α^2, can be found through integration. The normal distribution is the most common distribution, because any process that is the sum of many parts tends to be normally distributed. Speed, travel time, and delay are all commonly described using the normal distribution. The function is completely defined by two parameters: the mean and the variance. All other values in Eq. (1.6) including π, are constants. The notation for a normal distribution is x: $N[\mu, \alpha^2]$, which means that the variable $\underline{x}$ is normally distributed with a mean of μ and a variance of σ^2.

1.3.1 The Standard Normal Distribution

For the normal distribution, the integration cannot be done in closed form due to the complexity of the equation for $f(x)$; thus, tables for a "standard normal" distribution, with zero mean ($\mu = 0$) and unit variance ($\sigma^2 = 1$), are constructed. The standard normal is denoted z: $N[0,1]$. Any value of x on any normal distribution, denoted x: $N[\mu,\sigma^2]$, can be converted to an equivalent value of z on the standard normal distribution. This can also be done in reverse when needed. The translation of an arbitrary normal distribution of values of x to equivalent values of z on the standard normal distribution is accomplished as:

$$z = \frac{x - \mu}{\sigma} \tag{1.8}$$

where z = equivalent statistic on the standard normal distribution, z: $N[0,1]$; x = statistic on any arbitrary normal distribution, x: $N[\mu,\sigma^2]$ other variables as previously defined.

1.3.2 Characteristics of the Normal Distribution Function

The forgoing exercise allow one to compute relevant areas under the normal curve. Some numbers occur frequently in practice, and it is useful to have those in mind. For instance, what is the probability that the next observation will be within one standard deviation of the mean, given that the distribution is normal? That is, what is the probability that x is in the range $(\mu \pm 1.00\sigma)$? For example, the tentative value can find that this probability is 68.3% by adopting some standard method.

The following ranges have frequent use in statistical analysis involving normal distributions:

- 68.3% of the observations are within $\mu \pm 1.00\sigma$
- 95.0% of the observations are within $\mu \pm 1.96\sigma$
- 95.5% of the observations are within $\mu \pm 2.00\sigma$
- 99.7% of the observations are within $\mu \pm 3.00\sigma$

The total probability under the normal curve is 1.00, and the normal curve is symmetric around the mean. It is also useful to note that the normal distribution is asymptotic to the x-axis and extends to values of $\pm \infty$. These critical characteristics will prove to be useful throughout the text.

1.4 CONFIDENCE BOUNDS

What would happen if we asked everyone in class (70 people) to collect 50 samples of speed data and to compute their own estimate of the mean. How many estimates would there be? What distribution would they have? There would be 70 estimates and the histogram of these 70 means would look normally distributed. Thus the "estimate of the mean" is itself a random variable that is normally distributed.

Usually we compute only one estimate of the mean (or any other quantity), but in this class exercise we are confronted with the reality that there is a range of outcomes. We may, therefore, ask how good is our estimate of the mean. How confident are we that our estimate is correct? Consider that:

1. The estimate of the mean quickly tends to be normally distributed.
2. The expected value (the true mean) of this distribution is the unknown fixed mean of the original distribution.
3. The standard deviation of this new distribution of means is the standard deviation of the original distribution divided by the square root of the number of samples, N. (This assumes independent samples and infinite population.)

The standard deviation of this distribution of the means is called the standard error of the mean (E):

$$E = \sigma / \sqrt{N} \tag{1.9}$$

where the sample standard deviation, s, is used to estimate σ, and all variables are as previously defined. The same characteristics of any normal distribution apply to this distribution of means as well. In other words, the single value of the estimate of the mean, $\overline{x}_n$, approximates the true population, μ, as follows:

$\mu = \overline{x} \pm E$, with 68.3% confidence

$\mu = \overline{x} \pm 1.96\ E$, with 95% confidence

$\mu = \overline{x} \pm 3.00\ E$, with 99.7% confidence

The $\pm$ term (E, 1.96E, or 3.00E, depending upon the confidence level) in the above equation is also called the tolerance and is given the symbol e.

Consider the following: 54 speeds are observed, and the mean is computed as 47.8 km/h, with a standard deviation of 7.80 km/h. What are the 95% confidence bounds?

$$P\left[47.8 - 1.96 \times (7.80/\sqrt{54})\right] \leq \mu \leq \left[47.8 + 1.96 \times 7.80/\sqrt{54})\right] = 0.95 \text{ or}$$
$$P(45.7 \leq \mu \leq 49.9) = 0.95$$

Thus it is said that there is a 95% chance that the true mean lies between 45.7 and 49.9 km/h. Further, while not proven here, any random variable consisting of sample means tends to be normally distributed for reasonably large n, regardless of the original distribution of individual values.

1.5 DETERMINATION OF SAMPLE SIZE

We can rewrite the equation for confidence bounds to solve for N, given that we want to achieve a specified tolerance and confidence. Resolving the 95% confidence bound equation for N gives:

$$N \geq \frac{1.96^2 x^2}{e^2} \tag{1.10}$$

where 1.96^2 is used only for 95% confidence. If 99.7% confidence is desired, then the 1.96^2 would be replaced by 3^2.

Consider another example: With 99.7% and 95% confidence, estimate the true mean of the speed on a highway, ± 1 mph. We know from previous work that the standard deviation is 7.2 km/h. How many samples do we need to collect?

$$N = \frac{3^2 \times 7.2^2}{1^2} \approx 467 \text{ samples for 99.7\% confidence, and}$$

$$N = \frac{1.96^2 \times 7.2^2}{1^2} \approx 200 \text{ samples for 95\% confidence}$$

Consider further that a spot speed study is needed at a location with unknown speed characteristics. A tolerance of ± 0.2 km/h and a confidence of 95% is desired. What sample size is required? Since the speed characteristics are unknown, a standard deviation of 5 km/h (a most common result in speed studies) is assumed. Then for 95% confidence, $N = (1.96^2 \times 5^2)/0.2^2 = 2401$ samples. This number is unreasonably high. It would be too expensive to collect such a large amount of data. Thus the choices are to either reduce the confidence or increase the tolerance. A 95% confidence level is considered the minimum that is acceptable: thus in this case, the tolerance would be increased. With a tolerance of 0.5 mph:

$$N = \frac{1.96^2 \times 5^2}{0.5^2} = 384 \text{ vehicles}$$

Thus the increase of just 0.3 mph in tolerance resulted in a decrease of 2017 samples required. Note that the sample size required is dependent on s, which was assumed at the beginning. After the study is completed and the mean and standard deviation are computed, N should be rechecked. If N is greater (i.e., the actual s is greater than the assumed s) then more samples may need to be taken.

Another example: An arterial is to be studied, and it is desired to estimate the mean travel time to a tolerance of ± 5 seconds with 95%, confidence. Based on prior knowledge and experience, it is estimated that the standard deviation of the travel times is about 15 seconds. How many samples are required?

Based on an application of Eq. 13.10, $N = 1.96^2(15^2)/(5^2) = 34.6$, which is rounded to 35 samples.

As the data is collected, the s computed is 22 seconds, not 15 seconds. If the sample size is kept at $n = 35$, the confidence bounds will be $\pm 1.96 (22)/\sqrt{35}$ or about ± 7.3 seconds. If the confidence bounds must be kept at ± 5 seconds, then the sample size must be increased so that $N \geq 1.96^2(22^2)/(5^2) = 74.4$ or 75 samples. Additional data will have to be collected to meet the desired tolerance and confidence level.

1.6 RANDOM VARIABLES SUMMATION

One of the most common occurrences in probability and statistics is the summation of random variables, often in the form $Y = a_1X_1 + a_2X_2$ or in the more general form:

$$Y = \sum a_i X_i \tag{1.11}$$

where the summation is over i, usually from 1 to n.

It is relatively straightforward to prove that the expected value (or mean) μ_Y of the random variable Y is given by:

$$\mu_Y = \sum a_i \mu_{Xi} \tag{1.12}$$

and that if the random variables x_i are independent of each other, the variance σ_Y^2 of the random variable Y is given by:

$$\sigma_Y^2 = \sum a_i^2 \sigma_{x_i}^2 \tag{1.13}$$

The fact that the coefficients, $a_{i,}$ are multiplied has great practical significance for us in all our statistical work.

1.6.1 The Central Limit Theorem

One of the most impressive and useful theorems in probability is that the slim of n similarly distributed random variables tends to the normal distribution, no matter what the initial, underlying distribution is. That is, the random variable $Y = \Sigma X_i$, where the X_i have the same distribution, tends to the normal distribution.

The words "tends to" can be read as "tends to look like" the normal distribution. In mathematical terms, the actual distribution of the random variable Yapproaches the normal distribution asymptotically.

1.6.1.1 Sum of Travel Times

Consider a trip made up of 15 components, all with the same underlying distribution, each with a mean of 10 minutes and standard deviation of 3.5 minutes. The underlying distribution is unknown. What can you say about the total travel time?

While there might be an odd situation to contradict this, $n = 15$ should be quite sufficient to say that the distribution of total travel times tends to look normal. From the standard equation, the mean of the distribution of total travel times is found by adding 15 terms (a_i μ_i) where

$a_i = 1$ and $\mu_i = 10$ minutes, or

$\mu_y = 15 \times (1 \times 10) = 150$ minutes

The variance of the distribution of total travel times is found from Eq. (1.13) by adding 15 terms (a_i^2 σ_i^2) where a_i is again 1, and σ_i is 3.5 minutes. Then:

$$\sigma_y^2 = 15 \times (1 \times 3.5^2) = 183.75 \text{ minutes}^2$$

The standard deviation, σ_y is, therefore, 13.6 minutes.

If the total travel times are taken to be normally distributed, 95% of all observations of total travel time will lie between the mean (150 minutes) ±1.96 standard deviations (13.6 minutes), or:

$$X_y = 150 \pm 1.96\ (13.6)$$

Thus 95% of all total travel times would be expected to fall within the range of 123–177 minutes (values rounded to the nearest minute).

1.6.1.2 Hourly Volumes

Five-minute counts are taken, and they tend to look rather smoothly distributed but with some skewness (asymmetry). Based on many observations, the mean tends to be 45 vehicles in the 5-minute count, with a standard deviation of seven vehicles. What can be said of the hourly volume?

The hourly volume is the sum of 12 five-minute distributions, which should logically be basically the same if traffic levels are stable. Thus the hourly volume will tend to look normal, and will have a mean computed using Eq. (1.12), with $a_i = 1$, $\mu_i = 45$ vehicles (veh), and $n = 12$, or $12 \times (1 \times 45) = 40$ veh/h. The variance is computed using Eq. (1.13), with $a_i = 1$, $\sigma_i = 7$, and $n = 12$, or $12 \times (1^2 \times 7^2) = 588$ $(\text{veh/h})^2$. The standard deviation is 24.2 veh/h. Based on the assumption of normality, 95% of hourly volumes would be between $540 \pm 1.96(24.2) = 540 \pm 47$ veh/h (rounded to the nearest whole vehicle).

Note that the summation has had an interesting effect. The σ/μ ratio for the 5-minute count distribution was $7/45 = 0.156$, but for the hourly volumes it was $47/540 = 0.087$. This is due to the summation, which tends to remove extremes by canceling "highs" with "lows" and thereby introduces stability. The mean of the sum grows in proportion n, but the standard deviation grows in proportion to the square root of n.

1.6.1.3 Sum of Normal Distributions

Although not proven here, it is true that the sum of any two normal distributions is itself normally distributed. By extension, if one normal is formed by n_1 summations of one underlying distribution and another normal is formed by n_2 summations of another underlying distribution, the sum of the total also tends to the normal.

Thus in the foregoing travel time example, not all of the elements had to have exactly the same distribution as long as subgroupings each tended to the normal.

1.7 THE BINOMIAL DISTRIBUTIONS

1.7.1 Bernoulli and the Binomial Distribution

The Bernoulli distribution is the simplest discrete distribution, consisting or only two possible outcomes: yes or no, heads or tails, one or zero, etc. The first occurs with probability p, and therefore the second occurs with probability $(1-p=q)$. This is modeled as:

$$P(X=1)=p$$
$$P(X=0)=I-p=q$$

In traffic engineering, it represents any basic choice—to park or not to park; to take this route or that; to take auto or transit (for one individual). It is obviously more useful to look at more than one individual, however, which leads us to the binomial distribution. The binomial distribution can be thought of in two common ways:

1. In N outcomes of the Bernoulli distribution, make a record of the number of events that have the outcome "1," and report that number as the outcome X.
2. The binomial distribution is characterized by the following properties:
   a. There are N events, each with the same probability p of a positive outcome and $(1-p)$ of a negative outcome.
   b. The outcomes are independent of each other.
   c. The quantity of interest is the total number X of positive outcomes, which may logically vary between 0 and N.
   d. N is a finite number.

   The two ways are equivalent, for most purposes.

   Consider a situation in which people may choose "transit" or "auto" where each person has the same probability $p=0.25$ of choosing transit, and each person's decision is independent of that of all other persons. Defining "transit" as the positive choice for the purpose of this example and choosing $N=8$, note that:

1. Each person is characterized by the Bernoulli distribution, with $p=0.25$.
2. There are $2^8=256$ possible combinations of choices, and some of the combinations not only yield the same value of X but also have the same probability of occurring. For instance, the value of $X=2$ occurs for both.

   TTAAAAAA

   and

   TATAAAAA

and several other combinations, each with probability of $p^2(1 - P)^6$, for a total of 2^8 such combinations.

Stated without proof is the result that the probability $P(X = x)$ is given by

$$P(X = x) = \frac{N!}{(N - x)!x!} p^x (1 - p)^{N-x} \tag{1.14}$$

with a mean of Np and a variance of Npq where $q = 1 - p$. The derivation may be found in any standard probability text.

There is an important concept that the reader should master, in order to use statistics effectively throughout this text. Even though on average two out of eight people will choose transit, there is absolutely no guarantee what the next eight randomly selected people will choose, even if they follow the rules (same p, independent decisions, etc.). In fact, the number could range anywhere from $X = 0$ to 8. And we can expect that tile result $X = 1$ will occur 10.0% of the time, $X = 4$ will occur 8.7% of the time, and $X = 2$ will occur only 31.1% of the time.

This is the crux of the variability in survey results. If there were 200 people in the senior class and each student surveyed eight people from the subject population, we would get different results. Likewise, if we average our results, the result would probably be close to 2.00, but would almost surely not be identical to it.

1.7.2 Asking People Questions Survey Results

Consider that the commuting population has two choices $X = 0$ for auto and $X = 1$ for public transit. The probability p is generally unknown, and it is usually of great interest. Assuming the probability is the same for all people (to our ability to discern, at least). then each person is characterized by the Bernoulli distribution.

If we ask $n = 50$ people for their value of X, the resulting distribution of the random variable Y is binomial and may tend to look like the normal. Applying some "quick facts" and noting that the expected value (that is the mean) is 12.5 (50×0.25), the variance is 9.375 ($50 \times 0.25 \times 0.75$), and the standard deviation is 3.06, one can expect 95%, of the results 10 fall in the range 12.5 ± 6.0 or between 6.5 and 18.5.

If $n = 200$ had been selected. then the mean of Y would have been 50 when $p = 0.25$ and the standard deviation would have been 6.1, so that 95% of the results would have fallen in the range of 38–62.

1.7.3 The Binomial and the Normal Distributions

The central limit theorem informs us that the sum of Bernoulli distributions (i.e., the binomial distribution) tends to the normal distribution. The only question is: How Fast? A number of practitioners in different fields use a rule of thumb that says "for large n and small p" the normal approximation can be used without restriction. This is incorrect and can lead to serious errors.

The most notable case in which the error occurs is when rare events are being described, such as auto accidents per million miles traveled or aircraft accidents. Which is an exact rendering of the actual binomial distribution for $p = 0.7(10)^{-6}$ and two values of n namely $n = 10^6$ and $n = 2$ $(10)^{-6}$, respectively. Certainly p is small and n is large in these cases, and, just as clearly, they do not have the characteristic symmetric shape of the normal distribution.

It can be shown that in order for there to be some chance of symmetry, i.e., in order for the normal distribution to approximate the binomial distribution, the condition that $np/(1-p) \geq 9$ is necessary.

1.8 THE POISSON DISTRIBUTION

The Poisson distribution is known in traffic engineering as the "counting" distribution. It has the clear physical meaning or a number of events X occurring in a specified counting interval of duration T and is a one-parameter distribution with:

$$P(X = x) = e^{-m}\frac{m^x}{x!} \tag{1.15}$$

with mean $\mu = m$ and variance $\sigma^2 = m$.

The fact that one parameter m specifies both the mean and the variance is a limitation, in that if we encounter field data where the variance and mean are clearly different, the Poisson does not apply.

The Poisson distribution often applies to observations per unit of time, such as the arrivals per 5-minute period at a toll booth. When headway times are exponentially distributed with mean $\mu = 1/\lambda$, the number of arrivals in an interval of duration T is Poisson distributed with mean $\mu = m = \lambda T$.

Applying the Poisson distribution is done the same way we applied the Binomial distribution earlier. For example, say there is an average of five accidents per day on the Florida freeways.

1.9 TESTING OF HYPOTHESIS

Very often traffic engineers must make a decision based on sample information. For example, is a traffic control effective or not? To test this, we formulate a hypothesis, H_0, called the null hypothesis and then try to disprove it. The null hypothesis is formulated so that there is no difference or no change, and then the opposite hypothesis is called the alternative hypothesis, H_1.

When testing a hypothesis, it is possible to make two types of errors: (1) We could reject a hypothesis that should be accepted (e.g., say an effective control is not effective). This is called a Type I error. The probability of making a Type I error is given the variable name, α. (2) We could accept a false hypothesis (e.g., say an ineffective control is effective). This is called a Type II error. A Type II error is given the variable name β.

Consider this example: An auto inspection program is going to be applied to 100,000 vehicles, of which 10,000 are "unsafe" and the rest are "safe," "Of course, we do not know which cars are safe and which are unsafe."

We have a test procedure, but it is not perfect, due to the mechanic and test equipment used. We know that 15% of the unsafe vehicles are determined to be safe, and 5% of the safe vehicles are determined to be unsafe.

We would define: H_0: The vehicle being tested is "safe," and H_1: the vehicle being tested is "unsafe." The Type I error, rejecting a true null hypothesis (false negative), is labeling a safe vehicle as "unsafe." The probability of this is called the level of significance, α, and in this case $\alpha = 0.05$. The Type II error. failing to reject a false null hypothesis (false positive), is labeling an unsafe vehicle as "safe." The probability of this, β is 0.15. In general for a given test procedure, one can reduce Type I error only by living with a higher Type II error, or vice versa.

1.9.1 Before-and-After Tests With Two Distinct Choices

In a number of situations, there are two clear and distinct choices, and the hypotheses seem almost self-defining:

- Auto inspection (acceptable, not acceptable)
- Disease (have the disease, do not)
- Speed reduction of 5 mph (it happened, it did not)
- Accident reduction of 10% (it happened, it did not)
- Mode shift by 5% points (it happened, it did not)

Of course, there is the distinction between the real truth (reality, unknown to us) and the decision we make, as already discussed and related to Type I and Type II errors. That is, we can decide that some cars in good working order need repairing and we can decide that some unsafe cars do not need repairing.

There is also the distinction that people may not want to reduce the issue to a binary choice or might not the able to do so. For instance, if an engineer expects a 10% decrease in the accident rate, should we test "H_0: no change" against "H_1: 10% decrease" and not allow the possibility of a 5% change? Such cases are addressed in the next section. In the present section, we will concentrate on binary choices.

1.9.1.1 Application: Travel Time Decrease

Consider a situation in which the existing travel time on a given route is known to average 60 minutes, and experience has shown the standard deviation to be about 8 minutes. An "improvement" is recommended that is expected to reduce the true mean travel time to 55 minutes.

This is a rather standard problem, with what is now a fairly standard solution. The logical development of the solution follows. The first question we might ask ourselves is whether we can consider the mean and standard deviation of the initial distribution to be truly known or whether they must be estimated. Actually, we will avoid this question simply by focusing on whether the after situation has a true mean of 60 minutes or 55 minutes. Note that we do not know the shape of the travel time distribution, but the central limit theorem tells us: a new random variable Y, formed by averaging several travel time observations, will tend to the normal distribution if enough observations are taken. The shape of Y for two different hypotheses, which we now form:

H_0: The true mean of Y is 60 minutes

H_1: The true mean of Y is 55 minutes

A logical decision rule: if the actual observation Y falls to the right of a certain point, Y^*, then accept H_0; if the observation falls to the left of that point, then accept H_1.

Note that:

1. The n travel time observations are all used to produce the one estimate of Y.
2. If the point Y^* is fixed, then the only way the Type I and Type II errors can be changed is to increase n, so that the shapes of the two

distributions become narrower because the standard deviation of Y involves the square root of n in its denominator.

3. If the point Y^* is moved, the probabilities of Type I and Type II errors vary, with one increasing while the other decreases.

To complete the definition of the test procedure, the point Y^* must be selected and the Type I and Type II errors determined. It is common to require that the Type I error (also known as the level of significance, α) be set at 0.05, so that there is only a 5% chance of rejecting a true null hypothesis. In the case of two alternative hypotheses, it is common to set both the Type I and Type II errors to 0.05, unless there is very good reason to imbalance them (both represent risks, and the two risks-repairing some cars needlessly versus having unsafe cars on the road, for instance may not be equal).

Y^* will be set at 57.5 minutes in order to equalize the two probabilities. The only way these errors can be equal is if the value of Y^* is set at exactly half the distance between 55 and 60 minutes. The symmetry of the assumed normal distribution requires that the decision point be equally distant from both 55 and 60 minutes, assuming that the standard deviation of both distributions (before and after) remains 8 minutes.

To ensure that both errors are not only equal but have an equal value of 0.05, Y^* must be 1.645 standard deviations away from 60 minutes, based on the standard normal table. Therefore $n \geq (1.645^2)\ (8^2)/2.5^2$ or 2^8 observations, where $8 =$ the standard deviation, $2.5 =$ the tolerance (57.5 mph is 2.5 mph away from both 55 and 60 mph), and 1.645 corresponds to the z statistic on the standard normal distribution for a beta value of 0.05 (which corresponds to a probability of $z \leq 95\%$).

The test has now been established with a decision point of 57.5 minutes. If the "after" study results in an average travel lime of under 57.5 minutes, we will accept the hypothesis that the true average travel time has been reduced to 55 minutes. If the result of the "after" study is an average travel time of more than 57.5 minutes, the null hypothesis— that the true average travel time has stayed at 60 minutes is accepted.

Was all this analysis necessary to make the commonsense judgment to set the decision time at 57.5 minutes half way between the existing average travel time of 60 minutes and the desired average travel time of 55 minutes? The answer is in two forms: the analysis provides the logical basis for making such a decision. This is useful. The analysis also provided the minimum sample size required for the "after" study to restrict both alpha and beta errors to 0.05. This is the most critical result of the analysis.

1.9.1.2 Application: Focus on the Travel Time Difference

The preceding illustration assumed that we would focus on whether the underlying true mean or the "after" situation was either 60 minutes or 55 minutes. What are some of the practical objections that people could raise?

Certainly one objection is that we implicitly accepted at face value that the "before" condition truly had an underlying true mean or 60 minutes. Suppose, to overcome that, we focus on the difference between before and after observations.

The n_1 "before" observations can be averaged to yield a random variable Y_1 with a certain mean μ_1 and a variance of σ_1^2/n_1. Likewise, the n_2 "after" observations can be averaged to yield a random variable Y_2 with a (different?) certain mean μ_2 and a variance of σ_2^2/n_2. Another random variable can be formed as $Y = (Y_2 - Y_1)$, which is itself normally distributed and that has an underlying mean of $(\mu_2 - \mu_1)$ and variance $\sigma^2 = \sigma_2^2/n_2 + \sigma_1^2/n_1$. This is often referred to as the normal approximation.

What is the difference between this example and the previous illustration? The focus is directly on the difference and does not implicitly assume that we know the initial mean. As a result, "before" samples are required. Also there is more uncertainty, as reflected in the larger variance. There are a number of practical observations stemming from using this result: it is common that the "before" and "after" variances are equal, in which case the total number of observations can be minimized if $n_1 = n_2$. If the variances are not known, the estimators s_i^2 are used in their place.

If the "before" data was already taken in the past and n_1 is therefore fixed, it may not be possible to reduce the total variance enough (by just using n_2) to achieve a desired level of significance, such as $\alpha = 0.05$. Comparing with the previous problem, note that if both variances are 8^2 and $n_1 = n_2$ is specified, then $n_1 \geq 2 \times (1.645^2)\ (8^2)/(2.5^2)$ or 55 and $n_2 \geq 55$. The total required is 110 observations. The fourfold increase is a direct result of focusing on the difference of −5 mph rather than the two binary choices (60 or 55 minutes).

1.9.2 Before-and-After Tests With Generalized Alternative Hypothesis

It is also common to encounter situations in which the engineer states the situation as "there was a decrease" or "there was a change" versus "there was not," but does not state or claim the magnitude of the change. In these cases, it is standard practice to set up a null hypothesis of "there was no change" ($\mu_1 = \mu_2$) and an alternative hypothesis of "there was a change" ($\mu_1 \neq \mu_2$). In such cases, a level of significance of 0.05 is generally used.

The first case implicitly considers that if there were a change, it would be negative—i.e., either there was no change or there was a decrease in the mean. The second case does not have any sense (or suspicion) about the direction of the change, if it exists, Note that:

1. The first is used when physical reasoning leads one to suspect that if there were a change, it would be a decrease. In such cases, the Type I error probability is concentrated where the error is most likely to occur, in one tail.
2. The second is used when physical reasoning leads one to simply assert "there was a change" without any sense of its direction. In such cases, the Type I error probability is spread equally in the two tails.

In using the second case often we might hope that there was no change, and really do not want to reject the null hypothesis. That is, not rejecting the null hypothesis in this case is a measure of success. There are, however, other cases in which we wish to prove that there is a difference. The same logic can he used, but in such cases, rejecting the null hypothesis is "success."

1.9.2.1 An Application: Travel Time Differences

Let us assume, we have made some improvements and suspect that there is a decrease in the true underlying mean travel time. Using information from the previous illustration, let us specify that we wish a level of significance, $a = 0.05$. The decision point depends upon the variances and the n_i. If the variances are as stated in the prior illustration and $n_1 = n_2 = 55$, then the decision point $Y^* = -2.5$ mph, as before.

Let us now go one step further. The data is collected, and $Y = -3.11$ results. The decision is clear: reject the null hypothesis of "there is no decrease." But what risk did we take?

Consider the following:

- Under the stated terms, had the null hypothesis been valid, we were taking a 5% chance of rejecting the truth. The odds favor (by 19 to 1, in case you are inclined to wager with us) not rejecting the truth in this case.
- At the same time, there is no stated risk of accepting a false hypothesis H_1, for the simple reason that no such hypothesis was stated.

- The null hypothesis was rejected because the value of Y was higher than the decision value of 2.5 mph. Since the actual value of -3.11 is considerably higher than the decision value, one could ask about the confidence level associated with the rejection. The point $Y = -3.11$ is 2.033 standard deviations away from the zero point, as can be seen from:

$$\sigma_Y = \sqrt{\frac{\sigma_1^2}{n_1} + \frac{\sigma_2^2}{n_2}}$$

$$= \sqrt{\frac{8^2}{55} + \frac{8^2}{55}} = 1.53$$

and $z = 3.11/1.53 = 2.033$ standard deviations. Entering the standard normal distribution table with $z = 2.033$ yields a probability of 0.9790. This means that if we had been willing to take only a 2% chance of rejecting a valid H_0. we still would have rejected the null hypothesis: we are 98% confident that our rejection of the null hypothesis is correct. This test is called the Normal Approximation and is only valid when $n_1 \geq 30$ and $n_2 \geq 30$.

Since this reasoning is a little tricky, let us state it again: If the null hypothesis had been valid, you were initially willing to take a 5% chance of rejecting it. The data indicated a rejection. Had you been willing to take only a 2% chance, you still would have rejected it.

1.9.2.2 One-Sided Versus Two-Sided Tests

The material just discussed appears in the statistics literature as "one-sided" tests, for the obvious reason that the probability is concentrated in one tail (we were considering only a decrease in the mean). If there is no rationale for this, a "two-sided" test should be executed, with the probability split between the tails. As a practical matter, this means that one does not use the probability tables with a significance level of 0.05, but rather with $0.05/2 = 0.025$.

1.9.3 Other Useful Statistical Tests

1.9.3.1 The t-Test

For small sample sizes ($N < 30$), the normal approximation is no longer valid. It can be shown that if x_1 and x_2 are from the same population, the statistic t is distributed according to the tabulated, distribution, where:

$$t = \frac{x_1 - x_2}{S_p\sqrt{1/n_1 + 1/n_2}} \tag{1.16}$$

and S_p is a pooled standard deviation which equals:

$$S_p = \sqrt{\frac{(n_1 - 1)S_1^2 + (n_2 - 1)S_2^2}{n_1 + n_2 - 2}} \tag{1.17}$$

The t distribution depends upon the degrees of freedom, f, which refers to the number of independent pieces of data that form the distribution. For the t distribution, the nonindependent pieces of data are the two means, x_1 and x_2. Thus:

$$f = n_1 + n_2 - 2 \tag{1.18}$$

Once the t statistic is determined, the tabulated values, yield the probability of a t value being greater than the computed value. In order to limit the probability of a Type I error to 0.05, the difference in the means will be considered significant only if the probability is less than or equal to 0.05, i.e., if the calculated t value falls in the 5% area of the tail, or in other words, if there is less than a 5% chance that such a difference could be found in the same population. If the probability is greater than 5% that such a difference in means could be found in the same population, then the difference would be considered not significant.

1.9.3.2 The F-Test

In using the t-test, and in other areas as well, there is an implicit assumption made that the $\sigma_1 = \sigma_2$. This may be tested with the F distribution, where:

$$F = \frac{S_1^2}{S_2^2} \tag{1.19}$$

(by definition the larger s is always on top).

It can be proven that this F value is distributed according to the F distribution. The F distribution is tabulated according to the degrees of freedom in each sample, thus $f_1 = n_1 - 1$ and $f_2 = n_2 - 1$. Since the f distribution in the shaded area in the tail, like the t distribution, the decision rules are as follows:

- If [Prob $F \geq F$] ≤ 0.05, then the difference is significant.
- If [Prob $F \geq F$] > 0.05, then the difference is not significant.

The F distribution is tabulated for various probabilities, as follows, based on the given degrees of freedom:

when

$p = 0.10$;	$F = 1.94$
$p = 0.05$;	$F = 2.35$
$p = 0.025$;	$F = 2.77$

The F values are increasing, and the probability $[F \geq 1.56]$ must be greater than 0.10 given this trend; thus, the difference in the standard deviations is not significant. The assumption that the standard deviations are equal, therefore, is valid.

1.9.3.3 Chi-Square Test: Hypotheses or an Underlying Distribution f(x)

One of the early problems stated was a desire to "determine" the underlying distribution. such as in a speed study. The most common rest to accomplish this is the Chi-square (χ^2) goodness-of-fit test.

In actual fact, the underlying distribution will not be determined. Rather, a hypothesis such as "H_0: The underlying distribution is normal" will be made, and we test that it is not rejected, so we may then act as if the distribution were in fact normal.

The procedure is best illustrated by an example. Consider data on the height of 100 people. To simplify the example, we will test the hypothesis that this data is uniformly distributed (i.e., there are equal numbers of observations in each group).

In order to test this hypothesis, the goodness-or-fit test is done by following these steps:

1. Compute the theoretical frequencies, f_i, for each group. Since a uniform distribution is assumed and there are 10 categories with 100 total observations, $f_i = 100/10 = 10$ for all groups.
2. Compute the quantity:

$$\chi^2 = \sum_{i=1}^{N} \frac{(n_i - f_i)^2}{f_i} \tag{1.20}$$

3. As shown in any standard statistical text, the quantity χ^2 is chi-squared distributed, and we expect low values if our hypothesis is correct. (If the observed samples exactly equal the expected, then the quantity is zero.) Therefore refer to a table of the chi-square distribution and look up the number that we would not exceed more than 5% of the time (i.e., $\alpha = 0.05$). To do this, we must also have the number of degrees of freedom, designated df, and defined as $df = N - 1 - g$, where N is the number of categories and g is the number of things we estimated from the data in defining the hypothesized distribution. For the uniform distribution, only the sample size was needed to estimate the theoretical frequencies, thus "0" parameters were used in computing $\chi^2(g = 0)$. Therefore for this case, $df = 10 - 1 - 0 = 9$.

4. Now entered with $\alpha = 0.05$ and $df = 9$. A decision value of $\chi^2 = 16.92$ is found. As the value obtained is $43.20 > 16.92$, the hypothesis that the underlying distribution is uniform must be rejected.

In this case, a rough perusal of the data would have led one to believe that the data were certainly not uniform, and the analysis has confirmed this. The distribution actually appears to be closer to normal, and this hypothesis can be tested. Determining the theoretical frequencies for a normal distribution is much more complicated, however. An example applying the χ^2 test to a normal hypothesis is discussed in detail in Chapter 9, Distribution. Note that in conducting goodness-of-fit tests, it is possible to show that several different distributions could represent the data. How could this happen? Remember that the test does not prove what the underlying distribution is in fact. At best, it does not oppose the desire to assume a certain distribution. Thus it is possible that different hypothesized distributions for the same set of data may be found to be statistically acceptable. A final point on this hypothesis testing: the test is not directly on the hypothesized distribution, but rather on the expected versus the observed number of samples. The computations involve the expected and observed number of samples; the actual distribution is important only to the extent that it influences the probability of category, i.e., the actual test is between two histograms. It is therefore important that the categories be defined in such a way that the "theoretic histogram" truly reflects the essential features and detail of the hypothesized distribution. That is one reason why categories of different sizes are sometimes used: the categories should each match the fast-changing details of the underlying distribution.

1.10 SUMMARY

In transportation engineering the specifically traffic studies cannot be done without using some basic statistics. While using statistics, the engineer is often faced with the need to act despite the lack of certainty, which is why statistical analysis is employed. This chapter is meant to review basic statistics for the reader to be able to understand and perform everyday traffic studies before entering to the main and detailed statistics applications to traffic and transportation engineering.

There are a number of very important statistical techniques presented in this book in further chapters that are useful for traffic engineers and transportation engineers.

CHAPTER 2

Preliminaries

2.1 INTRODUCTION

Statistics is a branch of applied mathematics concerned with the collection and interpretation of quantitative data and the use of probability theory to estimate population parameters. The word *statistics* is derived from the Latin word "Status." It deals with the collection, analysis, and analysis of numerical facts of data and is regarded as one of the best tools for taking decisions. In early stages, the word statistics meant the data in relation to the activities of a state collected for official purposes. But the subject gradually gained broader meaning and word statistics is now means the numerical statements as well as statistical methodology. The word statistics is used in two senses viz. singular and plural. In narrow sense it denotes numerical data and in wider sense it denotes statistical methods. Different authors gave different definitions of the word "statistics" from time to time but no definition is satisfactory. Some of the definitions are given below:

> *By statistics we mean quantitative data affected to a marked extent by multiplicity of cause.*
>
> —***Yule and Kendall***

> *Statistics are numerical statements of facts in any department of enquiry placed in relation to each other.*
>
> —***Bowley***

The above definitions are incomplete, since they do not possess all the characteristics of statistics. It refers only to the numerical data affected by a multiplicity of causes.

We now give a comprehensive and exhaustive definition that is considered as the best, as follows:

> *By statistics we mean aggregates of facts affected to a marked extent by multiplicity of causes, numerically expressed, enumerated or estimated according to reasonable standard of accuracy, collected in a systematic manner for a predetermined purpose and placed in relation to each other.*

Statistical Techniques for Transportation Engineering.
DOI: http://dx.doi.org/10.1016/B978-0-12-811555-8.00002-7

According to the above definition, statistics should possess the following characteristics:

1. Statistics are aggregate of facts; which means that statistics deals with groups but not individual items.
2. Statistics are affected to a marked extent by a multiplicity of causes.
3. Statistics are expressed numerically.
4. Statistics should be enumerated or estimated: When the field of enquiry is not large, the numerical data pertaining to the field collected by actual counting, i.e., by enumeration and when the field of enquiry is very large the data can be estimated.
5. Statistics should be collected with reasonable standard of accuracy.
6. Statistics should be collected in a systematic manner for a predetermined purpose.
7. Statistics should be placed in relation to each other.

These definitions describe statistics as numerical data. The following are some definitions, which describe statistics as a method:

> *Statistics is a body of methods for making definitions in the face of uncertainty.*
>
> —***Wallis and Robert***

> *Statistics may be called the science of counting.*
>
> —***A.L. Bowley***

> *Statistics is the science of estimates and probabilities.*
>
> —***Boddington***

> *Statistics may be define as the collection, presentation analysis and interpretation of numerical data.*
>
> —***Croxton and Cowden***

The definition given by Croxton and Cowden is a simple definition. According this definition there are four stages in statistics, namely Collection of data, Presentation of data, Analysis of data, and Interpretation of data.

Broadly speaking, there are five stages:

1. *Collection of data*: The first step in any investigation is the collection of data. Special care must be taken by the investigator in collecting the data. If the data is faulty, the results will be faulty. There are different methods of collection of data.
2. *Organization of data*: A large collection of data must be edited very carefully and organized. After editing the data must be classified and tabulated. Classification refers to the determination of various classes, categories of or group heads in which the whole data shall be

distributed. The tabulation of data refers to actual sorting and placing of the data in well designed and systematic tables.

3. *Presentation of data*: After organizing the data, it must be presented systematically in the form of tables or diagrams or graphs in order to facilitate further analysis.
4. *Analysis of data*: The next step is to analyze the presented data, which includes condensation, summarization, etc.
5. *Interpretation of data*: The last stage is correct interpretation that leads to a valid conclusion.

The real purpose of statistical methods is to make sense out of facts and figures, to prove the unknown, and cast light upon the situation. Statistics helps us in drawing conclusions from facts affected by a multiplicity of causes in any department of enquiry. It is a body of methods for making wise decisions in the face of enquiry. Statistics is essential for a country. It is an indispensable tool in all aspects of economic study and business. The subject of statistics is widely used in education, accounting, astronomy, and in all sciences.

Characteristics of statistics can be outlined as follows:

1. Statistics are the aggregate of facts.
2. Statistics are expressed numerically.
3. Statistics are affected by multiplicity of causes.
4. Statistics must be related to the field of investigation.
5. Reasonable accuracy should be maintained in statistics.

Limitations of Statistics:

1. Statistics is deals with quantitative data only.
2. Statistics does not deal with individual items.
3. Statistical laws may mislead to wrong conclusions in the absence of details.
4. The person who has an expert knowledge can only handle statistical data.
5. Statistical results are ascertained by samples. If the selection of samples is biased the results obtained may not be reliable.

Statistical procedures are divided into two major categories: (1) Descriptive statistics and (2) Inferential statistics.

Descriptive statistics: Descriptive statistics serve as devices for organizing data and bringing into focus their essential characteristics for the purpose of reaching conclusions at a later stage. It is the discipline of quantitatively describing main features of a collection of data. The primary function of descriptive statistics is to provide meaningful and convenient techniques for describing features of data that are of interest. The subject provides simple summaries about the sample through the use of graphic, tabular, or numerical devices, the abstraction of various properties of sets of observations. Descriptive statistics allows us to present the data in a meaningful way and interpret the data in a simpler way.

Inferential statistics: Inferential statistics are used to make generalizations from a sample to population. It always involves the process of sampling.

2.2 BASIC CONCEPTS

In this section we briefly introduce some basic concepts, which are extensively used in this book.

2.2.1 Characteristics

The quality possessed by an individual is called a characteristic. Characteristics are of two types: Nonmeasurable characteristics and measurable characteristics.

2.2.2 Attributes

The nonmeasurable characteristics that cannot be numerically expressed in terms of units are known as attributes. Nationality, religion, etc., are examples of attributes.

2.2.3 Variables

The measurable characteristics that can be numerically expressed in terms of some units are called variables. A variable is any characteristic, number, or quantity that can be measured or counted. They may be the conditions or characteristics that an experiment manipulates, controls, or observes. In general variables are measurable characteristics, which can be numerically expressed in terms of some units. Age, income, marks scored, eye color are examples of variables.

There are different ways variables can be described accordingly they can be studied and presented.

2.2.4 Numeric Variables

Numeric variables have values that describe a measurable quantity. They are quantitative variables. Numeric variables may be further described as either continuous or discrete.

Continuous variables can take a value based on measurement at any point along a continuum.

Examples of continuous variables include age, height, and temperature.

Discrete variables can take a value based on a count from a set of distinct whole values. A discrete variable cannot take the value of a fraction between one value and the next closest value.

Examples of discrete variables include number of children in a family, number of wells in a village, and number of tractors sold all of which are measured as whole units.

2.2.5 Categorical Variables

Categorical variables have values that describe a "quality" "or" characteristic of data. These variables are qualitative variables and tend to be represented by a nonnumerical value. Categorical variables may be further described as ordinal or nominal.

Ordinal variables take on values that can be logically ordered or ranked. Examples of ordinal values include academic grades (i.e., A, B, C, D) and attitudes (i.e., strongly agree, agree, disagree, and strongly disagree).

Nominal variables take on values that are not able to be organized in a logical sequence. Examples of nominal categorical variables include religion, eye color, sex the data collected for a numerical variables are quantitative data whereas the data collected for a categorical variable are qualitative data.

2.2.6 Data

Data is a collection of observations expressed in numerical figures. Data are facts and is the search for answers to questions. Some examples of data are:

1. The items of raw materials required for a product line.
2. The type and frequency of breakdowns occurred in a particular brand of cars.

Data provides facts and figures required for constructing measurement scales and tables, which are analyzed by statistical techniques. The scientific process of measurements, analysis, testing, and inferences depend on the availability of relevant data and their accuracy.

The choice of methods of data collection depends upon, the nature of the study of subject matter, the unit of enquiry, the type and depth of information to be collected, the availability of skilled and trained man power, the rate of accuracy and representative nature of the data required, and the size and the spread of the sample.

Statistical data may be classified as primary and secondary.

2.2.6.1 Primary Data

The data collected by the investigator himself to study any particular problem is called primary data. They are collected for the first time and are original in character.

Primary data can be collected by any one of the following methods:

1. Direct personal observation
2. Indirect oral interview
3. Information through agencies
4. Mailed questionnaire
5. Schedules through enumerator.

Primary data provides with detailed information. It is free from transcribing errors and estimation of errors. Primary data often include the methods of procuring the data.

2.2.6.2 Secondary Data

The data which are already collected by someone for some purpose and are available for the present study is called secondary data. Those data which have been already collected and analyzed by some earlier agency for its own use, and later the same data are used by a different agency are considered as secondary data.

Secondary data includes both the published and unpublished records. Census reports, statistical statements, reports of government departments, publications of international organizations such as IMF, UNO, WHO, World Bank, trade and financial journals are published sources for secondary data. Various records and registers maintained by firms and organizations are unpublished sources for the secondary data. They include minutes of meetings, personal files, and accounting records, etc.

Secondary data sources are not limited in time and space. The data if available can be secured cheaply and quickly, but the available data may not be accurate as desired. In some cases the available data may not meet our specific needs and the data may be inappropriate, inadequate, and obsolete. The units of measurements may not match and time periods may be different.

In application of statistical treatments, two types of data are recognized, namely *parametric data* and *nonparametric data*. Parametric data are measured data and nonparametric data are either counted or ranked.

2.2.7 Classification and Tabulation

In any statistical investigation the data collected are known as raw data, which may not be fit for further analysis. Hence the collected data should be organized in a systematic manner. The first step in the analysis and interpretation of data is classification and tabulation. Classification is the process of arranging the raw data into homogeneous groups or classes according to resemblances and similarities. According to Stockton and Clark "The process of grouping a large number of individual facts or observations on the basis of similarity among the items," called classification.

The main objective of classification is to condense the mass of the data and present the facts in a simple form. The classification aims at eliminating unnecessary details and bringing out clearly the points of similarity and dissimilarity. There are four important types of classification. They are:

1. Geographical classification
2. Chronological classification
3. Qualitative classification
4. Quantitative classification

Geographical classification is also known as spatial classification. The basis of classification is geographical or local differences between various items in the statistical data. Area-wise sales of TV sets is an example of geographic classification. In *chronological classification* data is classified according to the time of its occurrence. Time series is an example of chronological classification. When the data is classified according some quantity or attribute, the classification is called *qualitative classification.*

There may be two types of qualitative classification: (1) Simple classification and (2) Manifold classification. In simple classification the items are classified according to one attribute only. In main fold classification the population (universe) is classified. The population on the basis of more than one attribute at a time. Simple classification is usually twofold. When the data are classified according to some characteristics, which is capable of quantitative measurements like income, sales, loss, profit, production, etc., the classification is called *quantitative classification.*

The first step in organizing the statistical data is the preparation of an *Ordered array*. An ordered array is a listing of collected values in the order of magnitude from the smallest value to the largest value.

When items or attributes are arranged according to some logical order they are said to form statistical series.

Tally mark: A tally mark is an upward slanted stroke (/ or |), which is put against each occurrence of a value. When a value occurs more than four times the fifth occurrence of the value is denoted by a cross (/) tally mark, running diagonally, across the four tally marks.

Frequency: The total count of tally against each item or value is called its frequency.

Statistical data are presented in three different forms namely:

1. Textual presentation
2. Tabular presentation
3. Graphical presentation

When data are presented in descriptive form it is called textual presentation of the data. When numerical data is presented in a logical and systematic manner it is called tabular presentation. The presentation of data by graphs and charts is known as graphical presentation of the data.

2.3 TABULATION OF DATA

Tabulation is the process of presenting data in tables. It is the logical and systematic presentation of numerical data in rows and columns designed to simplify the presentation and facilitate comparisons. It helps in understanding complex numerical data and make them in simple and clear way that their similar and dissimilar facts are separated. The columns are vertical arrangements and the rows are horizontal arrangements. The main objective of a statistical table is to simplify the complex data and to facilitate comparison and another objective is to present facts in minimum space facilitating statistical processing. Tabulation helps us in detecting errors and omissions.

There are two types of tabulation

1. Simple tabulation
2. Complex tabulation.

In simple tabulation the data are classified according to only one characteristic. In complex tabulation two or more characteristics are considered.

Example 2.1: Simple tabulation (Table 2.1).

Table 2.1 Distributions of marks

Class marks	Number of students
0–25	18
25–50	53
50–75	16
75–100	8

Example 2.2: Complex tabulation (Table 2.2).

Table 2.2

S.No.	Name of the college	Number of students			
		BA	BSc	BCom	Total
1.	Gautami degree college	52	64	68	184
2.	Nalanda degree college	48	56	74	178
3.	St. Ann's degree college	60	80	142	282

A good statistical table should be simple, compact, and self-explanatory. It should suit the purpose of investigation. The table should not contain ditto marks that may be mistaken. It should be free from statistical errors. The different parts of a table are:

1. *Table number*: A good statistical table should always be numbered for its identification and future reference.
2. *Title*: The title briefly describes the contents of the table and is at the top of the table. It should be clear and precise. It explains the contents of the table and is the main heading.
3. *Stubs (row captions)*: A stub is the extreme left part of the table where the description of the rows are shown. The horizontal headings and subheadings of the row are called row captions and the space where the row headings is called stub.
4. *Box head (column captions)*: The captions are headings for vertical columns. Captions must be brief and self-explanatory. Only first letter of the box head is in capital letters and the remaining must be written in small letters.
5. *Body of the table*: It is the most important part of the table and it contains the numerical information classified with respect to row and column captions.
6. *Footnote*: It appears immediately below the body of the table where the source of the data is and any explanation is shown.
7. *Source notes*: Source notes are given at the end of the table indicating the source from where the information is taken.

2.4 FREQUENCY DISTRIBUTION

Frequency distribution is the tabulation of raw data obtained by dividing the data into classes of some size or magnitude and computing the number of observations falling within each pair of class boundaries. It is the list of values that a variable takes in a sample.

There are two types of frequency distribution:

1. Simple frequency distribution
2. Grouped frequency distribution.

2.4.1 Simple Frequency Distribution

Involves the ordering of observations to lowest or lowest to highest and counting the number of times each observation occurs. Fundamentally there are two columns in the table. One to list the values or observation and the other is to indicate the frequency of the occurrence. The distribution individually shows the values of the variable in groups or intervals. The following is an example of simple frequency distribution. In simple frequency distribution the data are classified according to only one characteristic.

Example 2.3: See Table 2.3.

Table 2.3 Simple frequency distribution

Number of accidents	Number of drivers
1	3
2	7
3	12
4	8
5	5

2.4.2 Grouped Frequency Distribution

Grouped frequency distribution is used when the original scale of measurement consists of more categories than can be listed in an original table. To construct grouped distribution the data are sorted and separated into groups called classes. Usually 5–20 classes are used. The frequency of the data belonging to each class is then recorded in a table of frequencies called the frequency table. The classes must be mutually exclusive, i.e., nonoverlapping and continuous. They must all be inclusive or exhaustive. They must be of equal width. The class width must be an odd number. The class limits of a grouped distribution are the smallest and the largest values of the class. The difference between the lower limit and the upper limit of the class is known as the class interval. If L denotes the largest value and S denotes the smallest value, then the formula to find the class interval C is

$$C = \frac{L - S}{K}$$

where L = largest value, S = smallest value, and K = number of classes.

There are two methods of forming the class intervals:

1. Exclusive method.
2. Inclusive method.

In the *Exclusive method* the upper limit of one class interval is the lower limit of the next class interval.

Example 2.4: See Table 2.4.

Table 2.4 Exclusive internal table

Marks	Number of students
10–20	15
20–30	25
30–40	12
40–50	18
50–60	24

In the *Inclusive method* the upper limit of one class interval is included in that class itself.

Example 2.5: See Table 2.5.

Table 2.5 Inclusive internal table

Marks	Number of students
0–19	10
20–39	26
40–59	19
60–79	11
80–99	7

The magnitude of class intervals is determined by the formula

$$K = 1 + 3.322 \log N$$

where K = number of classes and N = the total number of observations.

The midpoint of a class interval is computed by the formula:

$$x_i = \frac{\text{upper limit of the class} + \text{lower limit of the class}}{2}$$

2.4.2.1 Solved Examples

Example 2.6: The marks scored by 24 students are given below:

5, 0, 4, 7, 9, 2, 7, 5, 3, 9, 0, 6, 7, 2, 5, 8, 1, 3, 1, 3, 6, 2, 4, 9

Construct a frequency distribution table.

Solution: A simple frequency table can be constructed as follows (Table 2.6):

Table 2.6 Simple frequency table

Marks	Tally marks	Frequency
0	||	2
1	||	2
2	|||	3
3	|||	3
4	||	2
5	|||	3
6	||	2
7	|||	3
8	|	1
9	|||	3

Example 2.7: The heights of 40 students in centimeters are given below:

102, 104, 101, 103, 106, 103, 105, 107, 102, 101, 108, 105, 108, 104, 109, 112, 121, 114, 123, 118, 111, 120, 119, 108, 117, 115, 123, 103, 108, 121, 115, 119, 124, 101, 105, 116, 115, 120, 116, 122

Construct a grouped frequency distribution table, by inclusive method for the class intervals (Table 2.7).

Table 2.7 Frequency distribution table

Heights (cm)	Number of students
101–105	13
106–110	7s
111–115	6
116–120	8
121–125	6
Total	40

Solution: Taking the length of the class interval as 5, the frequency distribution can be formed as follows:

2.5 CUMULATIVE FREQUENCY TABLE

The total frequency of all classes less than the upper class limit of a given class is called the cumulative frequency of that class.

A frequency table showing the cumulative frequencies is called cumulative frequency distribution.

The cumulative frequencies of a given distribution are of two types:

1. Less than cumulative frequency.
2. More than cumulative frequencies.

In less than cumulative frequency distribution the cumulative frequency of class is obtained by adding successively the frequencies of all the previous classes including the one which it is written.

In more than cumulative frequency distribution, the frequency of a class is obtained by finding the cumulative total of frequencies starting from the highest to lowest class.

The less than and more than frequency tables of the examples are given below:

2.5.1 Less Than Cumulative Frequency Table

See Table 2.8.

Table 2.8 Less than cumulative frequency table

Upper limits of class intervals	Less than cumulative frequency
105.5	13
110.5	20
115.5	26
120.5	34
125.5	40

2.5.2 More Than Cumulative Frequency Table

See Table 2.9.

Table 2.9 More than cumulative frequency table

Lower limits of class intervals	More than cumulative frequency
100.5	40
105.5	27
110.5	20
115.5	14
120.5	6

2.6 MEASURES OF CENTRAL TENDENCY

Measure of central tendency is a typical value of the data. It is general term describing the central value of the given series and is also known as the average of the given series. It is a single statistical expression (measure), which represents the entire data.

Definitions (Average):

An average is a single number describing some features of a set of data
—***Wallis and Roberts***

Average is an attempt to find one single figure to describe whole of figures
—***Clark and Sekkade***

The purpose of a measure of central tendency is to represent a group of individual values in a simple and concise manner. It helps to obtain the picture of a complete population by means of a single figure. Averages are valuable in setting standards, estimation, and decision-making.

Characteristics of an Average (a measure of central tendency):

1. It should be rigidly defined. "An average should be properly defined and it should not depend on the personal prejudice and bias of the investigator."
2. It should be based on all items.
3. It should be easily understood.
4. It should be easy to interpret.
5. The average should not be unduly affected by the extreme values.
6. It should be least affected by fluctuations in sampling.
7. It should be subjected to further mathematical calculations.

The common measures of central tendency are:

1. Arithmetic mean (AM)
2. Median
3. Mode
4. Geometric mean (GM)
5. Harmonic mean (HM)

2.7 ARITHMETIC MEAN

It is a common type and mostly used measure of central tendency. It is simply called the mean of the data. The mean of a given series is the figure obtained by dividing the total, i.e., sum of the various values by their number.

Arithmetic average is of two types:

1. Simple arithmetic average.
2. Weighted arithmetic average.

2.7.1 Simple Arithmetic Average

Arithmetic mean of ungrouped data (Individual observations) by direct method:

Let $x_1, x_2, \ldots, x_n$ denote n observations. Then the AM $\overline{x}$ of the n observations is defined as:

$$\overline{x} = \frac{\sum_{i=1}^{n} x_i}{n}$$

where $\sum_{i=1}^{n} x_i$ = Sum of the observations (given values), n = Number of observations, $\overline{x}$ = Arithmetic mean.

Example 2.8: Find the AM of the data: 45, 25, 60, 30, 20

Solution: We have $x_1 = 45$, $x_2 = 25$, $x_3 = 60$, $x_4 = 30$, $x_5 = 20$

$$\sum_{i=1}^{n} x_i = x_1 + x_2 + x_3 + x_4 + x_5 = 45 + 25 + 60 + 30 + 20 = 180$$

$$\overline{x} = \frac{\sum_{i=1}^{n} x_i}{n} = \frac{180}{5} = 36$$

2.7.1.1 Shortcut Method (Method of Deviations)

The AM can also be computed by shortcut method. In this method we assume any one value as the AM of the given data. The arbitrary number (i.e., Assumed mean) is denoted by A. We find the difference, i.e., deviation d_i of each value from A and add all deviations. To find the AM we then apply the formula:

$$\overline{x} = A + \overline{d}$$

where

$$\overline{d} = \frac{\sum_{i=1}^{n} d_i}{n}$$

A = Assumed mean, n = Number of observations (i.e., items in the data), $\overline{x}$ = Arithmetic mean.

Example 2.9: Consider the data 45, 25, 60, 30, 20. Let us assume that the assumed mean, A, of the data is 30. The deviations from A are:

$d_1 = 45-30 = 15$, $d_2 = 25-30 = -5$, $d_3 = 60-30 = 30$, $d_4 = 30 - 30 = 0$, $d_5 = 20 - 30 = -10$

The sum of deviations

$$\sum_{i=1}^{n} d_i = 15 - 5 + 30 + 0 - 10 = 30$$

$$\therefore \quad \overline{d} = \frac{\sum_{i=1}^{n} d_i}{n} = \frac{30}{5} = 6$$

$$\overline{x} = A + \overline{d} = 30 + 6 = 36$$

The assumed mean can be taken as any number, but the value of the AM remains the same. It is observed that the sum of all deviations of the given values from the AM $\overline{x}$ is zero. If all the observations of a given data are increased by a constant k, then the AM of the data is also increased by the number k. Simple AM gives equal importance to all the items of the data. When the importance of the items varies we allocate weights to the items and calculate AM known as weighted AM.

2.7.2 Weighted Arithmetic Mean

If the w_i are weights, w_i are assigned to the values x_i then the weighted AM of the distribution is defined as follows:

$$\text{Weighted Arithmetic mean} = \frac{w_1x_1 + w_2x_2 + \cdots + w_nx_n}{w_1 + w_2 + \cdots + w_n} = \frac{\sum_{i=1}^{n} w_ix_i}{\sum_{1}^{n} w_i}$$

2.7.2.1 Combined Arithmetic Mean

If $\overline{x}_1$ and $\overline{x}_2$ denote the AM of two series with sizes n_1 and n_2, respectively then the combined AM $\overline{X}$ is defined as:

$$\overline{X} = \frac{n_1\overline{x}_1 + n_2\overline{x}_2}{n_1 + n_2}$$

2.7.2.1.1 Solved Examples

Example 2.10: A man traveled by car for 3 days. He covered 500 km each day. He drove the first day for 12 hours at 50 km/h, the second day for 10 hours at 60 km/h, and the third day for 8 hours at 70 km/h. Find his average speed?

Solution:

Speed, x_i (km/h)	Hour, w_i	$w_i\, x_i$
50	12	600
60	10	600
70	8	560
	30	1760

$$\text{Weighted Arithmetic mean} = \frac{w_i x_i + w_2 x_2 + w_n x_n}{w_1 + w_2 + \cdots + w_n} = \frac{w_1 x_1 + w_2 x_2 + w_3 x_3}{w_1 + w_2 + w_3}$$

$$= \frac{1760}{30} = 58.67 \text{ km/h}$$

Example 2.11: There are two branches of a transport company employing 250 and 150 persons, respectively. If the AMs of the monthly salaries paid by the two branches of the company are Rs. 1500 and Rs. 1800 respectively, find the arithmetic of the salaries of the companies as a whole?

Solution: We have $n_1 = 250$, $\overline{x}_1 = \text{Rs. } 1500$, $n_2 = 150$, $\overline{x}_2 = Rs.\ 1800$

The combined AM is

$$\overline{X} = \frac{n_1\overline{x}_1 + n_2\overline{x}_2}{n_1 + n_2} = \frac{250 \times 1500 + 150 \times 1800}{250 + 150} = \frac{375000 + 270000}{400}$$

$$= \frac{645000}{400} = \frac{6450}{4} = 1612.50$$

Therefore the AM of the salaries as a whole = Rs. 1612.50

Arithmetic mean of discrete series

The mean of the discrete data is given by the formula:

$$\overline{X} = \frac{\sum_{i=1}^{n} f_i x_i}{\sum_{i=1}^{n} f_i}$$

where f_i = frequency of the ith item, x_i = value of the ith item.

Shortcut method

In the shortcut method the AM is calculated by using the formula:

$$\overline{X} = A + \frac{\sum_{i=1}^{n} f_i d_i}{\sum_{i=1}^{n} f_i}$$

where $\overline{X}$ = Arithmetic mean, A = Assumed mean, $\sum_{i=1}^{n} f_i d_i$ = Total sum of deviations, $\sum_{1}^{n} f_i$ = Total frequency.

Arithmetic mean of continuous series

If x_i denotes the midpoints of class intervals and f_i denote the corresponding frequencies of class intervals, the AM of continuous series is defined by the formula:

$$\overline{X} = \frac{\sum_{i=1}^{n} f_i x_i}{\sum_{i=1}^{n} f_i}$$

where $\overline{X}$ = Mean.

Shortcut method

In the shortcut method the mean is calculated by applying the formula:

$$\overline{X} = A + \frac{\sum_{i=1}^{n} f_i d_i}{\sum_{1}^{n} f_i}$$

where $\overline{X}$ = Arithmetic mean, A = Assumed mean, f_i = The frequencies of class intervals, d_i = Deviations of the midpoints from assumed mean A.

Step-deviation method

In this method we choose one of the mid values of class intervals as the assumed mean A and apply the formula:

$$\overline{X} = A + \frac{\sum_{i=1}^{n} f_i d_i}{\sum_{1}^{n} f_i} \times C$$

where $\overline{X}$ = Arithmetic mean, A = Assumed Mean, f_i = The frequencies of class intervals, x_i = Mid values of class intervals, C = Common factor (usually the uniform length of class intervals), and $d_i = (x_i - A)/C$.

2.7.3 Merits of Arithmetic Mean

AM is a simple measurement. It is easily understood. It is easy to calculate. It is rigidly defined and is based on every value of the data. AM can be used for further analysis and algebraic treatment. It is an ideal average and it provides a good basis for comparison.

2.7.4 Demerits of Arithmetic Mean

AM cannot be located by inspection or by graphic method. It is unrealistic. It is unduly affected by extreme values. It may lead to false conclusions. It gives greater importance to bigger items of the data. AM is not useful for the qualities like character, honesty, and intelligence.

2.7.5 Properties of Mean

The sum of the deviations of the values in a distribution from their mean is zero, and the sum of squares of deviations of the values of the distribution is minimum when taken about mean.

2.7.5.1 Solved Examples

Example 2.12: Find the Arithmetic mean (AM) of the data given below:

x	12	14	16	18	20	22	24
f	6	12	18	20	16	10	8

Solution:

$$\text{Arithmetic mean} = \overline{x} = \frac{\sum_{i=1}^{n} f_i x_i}{\sum_{i=1}^{n} f_i}$$

$$= \frac{12 \times 6 + 14 \times 12 + 16 \times 18 + 18 \times 20 + 20 \times 16 + 22 \times 10 + 24 \times 8}{6 + 12 + 18 + 20 + 16 + 10 + 4}$$

$$= \frac{1620}{90} = 18$$

Example 2.13: Find the Arithmetic mean (AM) for the following data:

Marks	0–10	10–20	20–30	30–40	40–50
Number of students	5	10	15	10	10

Solution:

Marks (class intervals)	Number of students, f_i (frequency)	Midpoints of class intervals, x_i	$f_i x_i$
0–10	5	5	25
10–20	10	15	150
20–30	15	25	375
30–40	10	35	350
40–50	10	45	450
Total	$\sum f_i = 50$		$\sum f_i x_i = 1350$

$$\text{Arithmetic mean} = \frac{\sum_{i=1}^{n} f_i x_i}{\sum_{i=1}^{n} f_i} = \frac{1350}{50} = 27$$

Example 2.14: Find the Arithmetic mean (AM) of following data by (1) shortcut method (2) direct method:

Wages (Rs.)	10–20	20–30	30–40	40–50	50–60	60–70
Number of workers	5	15	25	35	12	8

Solution:

1. Shortcut method:

   Let the assumed mean be 45, i.e., $A = 45$

   We have the following table:

Class interval	Midpoint, x_i	Number of workers, f_i	$d_i = x_i - A$	$f_i d_i$
10–20	15	5	15–45 = –30	–150
20–30	25	15	25–45 = –20	–300
30–40	35	25	35–45 = –10	–250
40–50	45	35	45–45 = 0	0
50–60	55	12	55–45 = 10	120
60–70	65	8	65–45 = 20	160
Total		$N = 100$		–420

$$\text{Mean, } \overline{x} = A + \frac{\sum_{i=1}^{n} f_i d_i}{\sum_{1}^{n} f_i} = 45 + \left(\frac{-420}{100}\right) = 40.8$$

2. Direct method

Class interval	Midpoint, x_i	Number of workers, f_i	f_ix_i
10–20	15	5	75
20–30	25	15	375
30–40	35	25	875
40–50	45	35	1575
50–60	55	12	660
60–70	65	8	520
Total		$N = \sum f_i = 100$	$\sum f_ix_i = 4080$

$$\text{Arithmetic mean} = \frac{\sum_{i=1}^{n} f_ix_i}{\sum_{i=1}^{n} f_i} = \frac{4080}{100} = 40.80$$

Example 2.15: Find the Arithmetic mean (AM) of following data by step-deviation method:

Height (in.)	60–62	63–65	66–68	69–71	71–73
Number of students	15	54	126	81	24

Solution: Let the assumed mean be $A = 67$

Height (in.)	Midpoints of class intervals, x_i	Frequency, f_i	$X_i–A$	$d_i = \frac{1}{C}(x_i - A)$	f_id_i
60–62	61	15	−6	−2	−30
63–65	64	54	−3	−1	−54
66–68	67	126	0	0	0
69–71	70	81	3	1	81
72–74	73	24	6	2	48
		300			45

$$\text{Arithmetic mean} = \overline{X} = A + \frac{\sum_{i=1}^{n} f_id_i}{\sum_{1}^{n} f_i} \times C = 67 + \frac{45}{300} \times 3 = 67 + 0.45$$

$$= 67.45 \text{ in.}$$

Exercises 2.1

1. Define "Mean," and "Weighted mean."
2. Find the mean height of 8 students whose heights in centimeters are 59, 71, 65, 67, 61, 63, 69, 73.

   Ans: 66 cm
3. Find the Arithmetic mean (AM) of the following data:

Class intervals	0–10	10–20	20–30	30–40	40–50	50–60
Frequency	42	44	58	35	26	15

   Ans: 25.2
4. A man traveled by motor car for 3 days. He covered 960 km each day. He drove the first day 10 hours at 96 km/h, the second day 12 hours at 80 km/h, and the third day 15 hours at 64 km/h. What was his average speed?

   Ans: 77.8 km/h
5. The mean wage of 100 laborers working in a factory, running two shifts of 60 and 40 workers respectively, is Rs. 38. The mean wage of 60 laborers working in the first shift is Rs. 40. Find the mean wage of 40 laborers working in the second shift.

   Ans: Rs. 35
6. The average monthly production of certain factory for the first 9 months is 2, 584 units, and for the remaining three months it is 2416 units. Calculate the average monthly production for the year.

   Ans: 2542 units
7. The mean of the following data is 50. Find the missing value?

x	10	12	60	70	40
f	3	7	–	2	5

8. The average height of 25 male workers in a transportation company is 61 cm and the average height of female workers in the same company is 58 cm. Calculate the combined average height of the workers.

   Ans: 59.25 cm
9. Compute the mean of the following distribution:

Class intervals	1–7	8–14	15–21	22–28	29–35
Frequency	3	17	12	11	7

   Ans: 18.28

10. Calculate the weighted Arithmetic mean (AM) for the following data:

Components	A	B	C	D	E
Price index (Rs.)	250	200	135	325	400
Weights	42	14	22	30	16

Ans: Rs. 261.45

11. Compute the Arithmetic mean (AM) of the data given below:

x	20	30	40	50	60	85
f	8	12	20	10	6	4

Ans: 42

12. The following data is related to the distance traveled by 520 villagers to buy their weekly requirements:

Miles traveled	2	4	6	8	10	12	14	16	18	20
Number of villagers	38	104	140	78	48	42	28	24	16	2

Calculate the arithmetic average.

Ans: 7.78

13. A railway train runs for 30 minutes at a speed of 40 mph and then because of repairs of the track, the train runs for 10 minutes at a speed of 8 mph, after which it resumes its previous speed and runs for 20 minutes except for a period of 2 minutes when it had to run over the bridge with a speed of 30 mph. What is the average speed?

Ans: 34.33

14. An average rainfall of a city from Monday to Saturday is 0.3 in. Due to heavy rainfall on Sunday, the average rainfall for the week increased to 0.5 in. What was the rainfall on Sunday?

Ans: 1.7 in.

15. A cyclist covers his first 3 miles at an average speed of 8 mph; another 2 miles at 3 mph and the last 2 miles at 2 mph. find the average speed for the entire journey?

Ans: 3.4 mph

16. A man traveled by car for 3 days. He covered 960 km each day. He drove the first day 10 hours at 96 km/h, the second day 12 hours at 80 km/h, and the third day 15 hours at 64 km/h. Find his average speed?

Ans: 77.8 km/h

17. Given the speed's data in kilometers per hour, estimate the true, underlying mean of the data:

53	43	63	61	54
41	57	38	43	58
63	46	37	31	34
52	47	44	46	54
41	45	49	48	47
39	48	58	37	62
55	51	47	42	48
34	44	37	39	54
43	46	47	51	47
57	53	32	32	36
47	52	54	47	50
37	36	63	62	43
58	57	53	59	48

Ans: 47.7 km/h

18. Find the Arithmetic mean (AM) of the data on oxygen percentages in respirometer: 11, 11, 12, 12, 13, 14, 13, 13, 13, 15,16, 16, 15,17, 17, 15, and 18.

Ans: 14.17

19. A piece of equipment will function only when all the three components A, B, and C are working. The probability of A failing during 1 year is 0.15, that of B failing is 0.10. What is the probability that the equipment will fail before the end of the day?

Ans: 0.273 [Hint: E denote the event that all components work. Then $P(E)=(1-0.15)\ (1-0.05)(1-0.10)=0.727$, Therefore, $P(E)=1-P(E)=0.273$]

20. An average monthly production of a certain factory for the first 9 months is 2584 units and for the remaining 3 months is 2416 units. Calculate the average monthly production for the year.

Ans: 2542 units

21. A man traveled by motor car for 3 days. He covered 960 km each day. He drove the first day 10 h at 96 km/h, the second day 12 hours at 80 km/h and the third day 15 hours at 64 km/h. What was his average speed?

Ans: 77.8 km/h

22. A train travels first 300 km at an average of 30 km/h, and further travels the same distance at an average rate of 40 km/h. What is the average speed over the whole distance?

Ans: 34.28 km/h

23. The consumption of petrol by a motor car was a gallon for 20 km while going up from plains to hill station and a gallon for 24 km while coming down. What particular average would you consider appropriate for finding the average consumption in kilometers per gallon for their up and down journey and why?

    Ans: 21.82 km/gallon

24. There are 2000 scooter drivers, 4000 car drivers, and 6000 truck drivers. The probabilities of an accident involving a scooter, a car, and a truck are 0.01, 0.03, and 0.15, respectively. One of the insured driver met with an accident. What is the probability he/she is a scooter driver?

    Ans: $\frac{1}{52}$

2.7.6 Statistical Applications to Transportation Engineering

We know that each distribution has expected value. Which is known as the mean of the distribution. Mean of the distribution is an exact number. Estimate of this number is given by $\overline{x} = \frac{1}{n}\sum x_i$ (which is not true mean).

The most common distributions have one or two parameters (fixed quantities or numbers) that define the exact shape of the distribution. One parameter sets mean and the other sets the spread. The variance is the second moment about the measure of central tendency [i.e., mean or center of gravity]. The most common estimator of for the variance is given by $S^2 = (1/(n-1))\sum_1^n (x_i - \overline{x})^2$ [unbiased estimator of the variance].

For very large samples $\overline{x}$ yields μ, the true mean.

S^2 yields σ, the true variance

The square root of variance is called standard deviation (SD), variance and SD are common measures of the spread of distribution.

Distribution of hourly traffic volume, on a proposed highway to find hourly volume to be used for the design.

1. *Average Travel Time and Average Travel Speed*: Speed is a principal parameter describing the state of a given traffic stream. It is defined as the rate of motion, in distance per unit time. It is given by

$$S = \frac{d}{t}$$

where S = Speed in mph or fps; d = distance traversed in miles or ft; t = time to traverse the distance d.

The average or mean speed can be computed in two different ways.

Mathematically, the time mean speed is an average of the individual vehicle speeds and the Space mean speed is the HM of the individual speeds.

Annual Average Daily Traffic (AADT): Traffic data on all traffic volume maps is represented as AADT, a theoretical estimate of the total number of vehicles using a specific segment of roadway (in both directions) on any given day of the year.

AADT estimates are subject to many sources of variability. Therefore it is suggested that historical AADT's be referenced in addition to the most currently available information. Construction effects are unavoidable when collecting traffic data. If possible, traffic counts are scheduled before a project starts or after it is completed. It is important to remember that construction affects traffic patterns on the entire road network, another reason why it is valuable to reference historical traffic volumes.

AADT is defined as the average 24 hour traffic volume at a given location over a full 365 days/year.

$$\text{i.e., } AADT = \frac{\text{Total number of vechicles passing the site in a year}}{365}$$

Average daily traffic (ADT) is the average 24 hour traffic volume at a given location for some period of time less than a year (6 months or a season, a month or, a week or some days).

2. *Time Mean Speed (TMS)*: TMS is the average speed of all vehicles passing a point on a highway over some specified time period.

   It can also be defined as the AM of the speeds of vehicles passing a point on a highway during an interval of time.

   It is computed by using the formula:

$$TMS = \frac{\sum \frac{d}{t_i}}{n}$$

where TMS = Time mean speed, d = distance traversed, n = number of travel times observed, and t_1 = travel time of ith vehicle.

3. *Space Mean Speed (SMS)*: SMS is defined as the average speed of all vehicles occupying a given section of a highway over some specified time period.

It is the HM of the speeds of vehicles passing a point on a highway during an interval of time. It is computed by the formula:

$$\text{SMS} = \frac{d}{\sum \frac{t_i}{n}} = \frac{n}{\sum t_i}$$

The time mean speed is always higher than the Space mean speed.

2.8 MEDIAN

The median of a given series is the middle item in a series, when the items are arranged according to the order of their magnitudes. It is defined by L.R. Conor as follows:

> *The Median is that value of the variable which divides the data into two equal parts, one part comprising all the values greater, and the other all the values less than the median.*

1. *Median of ungrouped data (Simple series)*:
   a. When the number of observations n is odd, the data is arranged either in ascending order or descending order and the value of the $(n + 1)\text{th}/2$ item, i.e., the middle—most item is taken as median.

   Example 2.16: The wages of persons working in a transportation company are given as follows (Rs.): 5000, 4500, 3800, 7000, 6500, 5500, 6800. Find the median?

   Solution: Arranging the data in ascending order we have (Rs.): 3800, 4500, 5000, 5500, 6500, 6800, 7000

   Here we have $n = 7$ (odd)

$$\frac{n+1}{2} = \frac{7+1}{2} = 4$$

   The value of the fourth item in ascending order is Rs. 5500. Hence the median is Rs. 5500.

   b. When the number of observations n is even, the data is arranged either in ascending order or descending order. In this case the average value of the $n\text{th}/2$ and the $\left((n/2) + 1\right)\text{th}$ items is taken as the median.
2. *Median of discrete series*: In this case we arrange the data in ascending or descending order and find the cumulative frequencies. The median, M, is taken as the size of $\left((n + 1)/2\right)\text{th}$ item.
   where $N = \sum f_i$ (Sum of all frequencies is odd);

when $N = \sum f_i$ (Sum of all frequencies is even), the median in the average of Nth/2 and the $((N/2) + 1)$th items.

Example 2.17: The wages of 10 persons are given as follows:

(Rs.) 5200, 4800, 3600, 7000, 6000, 5500, 6500, 4800, 6700, 5800.

Calculate the median wage.

Solution: We arrange the data in the ascending order as follows:

(Rs.) 3600, 4800, 4800, 5200, 5500, 5800, 6000, 6500, 6700, 7000

Since $n = 10$ is even, the average value of the nth/2 and the $((n/2) + 1)$th items taken as the median. We have

$$\frac{n}{2} = \frac{10}{2} = 5; \quad \frac{n}{2} + 1 = 5 + 1 = 6$$

the fifth item in ascending order is Rs. 5500, and the sixth item is Rs. 5800. Therefore the median is $(5500+5800)/2=(11,300)/2=5650$

3. *Median of continuous series*: When the data is given in the form of frequency table, the median is calculated by applying the formula:

$$\text{Median}(M) = L + \frac{\frac{N}{2} - m}{f} \times c$$

where $L =$ lower limit of the median class (i.e., the class in which the median lies) $N = \sum f_i$; $m =$ cumulative frequency of the class preceding median class; $f =$ frequency of the median class; $c =$ length (width) of median class; $M =$ median.

2.8.1 Merits of Median

Median is easily understood. It can easily be computed. It can be located graphically. It is the best measure for qualitative data such as beauty, intelligence, etc. Median is not affected by extreme values. It can be determined by inspection in some cases.

2.8.2 Demerits of Median

Median cannot be computed when the distribution of items is irregular. When the number of items is large median cannot easily be computed. It ignores extreme values. It is not amenable for algebraic treatment. To compute median the data must be arranged either in ascending order or descending order.

Remarks:

The cross-section of a highway refers to the lateral design of elements within the highway right of way and covers the elements (1) The travel way, (2) The road side, and (3) The median.

Median refers to the recovery area between directions of traffic flow and to barriers separating directional flows.

Many types of facilities have no median at all. Where medians are provided, they may act as a recovery area for vehicles, a pedestrian refuge and/or a location for barriers of various types. All medians narrower than 20 ft on freeways or rural divided facilities should have barriers separating directional traffic flows.

The purpose of every median barrier is to guide a vehicle, hitting it back to the direction of primary flow, absorbing much of the collision shock, and safely bringing the vehicle to a stop. The most effective type is the concrete median barrier. It is virtually impossible to pass or bounce over it during accidents.

For a cross-section of a highway the median refers to the recovery area between directions of traffic flow and to barriers separating directional flows.

On surface streets, the median is often used as a pedestrian refuge, should be at least 4 ft wide for such use.

The median speed equally divides the distribution. That is, half of all vehicles travel faster than this speed and half travel slower. It is found by entering the cumulative frequency distribution curve at 50% and is also referred to as the 50% speed P_{50} which is a measure of center of the distribution.

2.8.2.1 Solved Examples

Example 2.18: Find the median of the data given below:

x_i	16	20	25	40	50	70
f_i	3	7	16	7	9	1

Solution:

x_i	f_i	Cumulative frequency (C′)
16	3	3
20	7	10
25	16	26

(Continued)

x_i	f_i	Cumulative frequency (C′)
40	7	33
50	9	42
70	1	43
	$N = \sum f_i = 43$	

We have $N = 43$, $N + 1 = 44$, $(N + 1)/2 = 22$, $44/2 = 22$

From the above table it is clear that all the items from 11th to 26th item have their value 25. The 22nd item's value is 25. Hence the median is 25.

Example 2.19: Compute the median of the following distribution:

Class intervals	51–55	56–60	61–64	65–69	70–74
Frequency	12	25	45	30	8

Solution:

Class intervals	True limits of class intervals	Frequency, f_i	Cumulative frequency
51–55	50.5–55.5	12	12
56–60	55.5–60.5	25	37
61–64	60.5–64.5	45	82
65–69	64.5–69.5	30	112
70–74	69.5–74.5	8	120

We have $N = \sum f_i = 120, \quad \frac{N}{2} = \frac{120}{2} = 60$

60th item lies in the class interval 61–64, which is the median class.

Therefore, we have $L = 60.5; f = 45; m = 37, C = 5$.

$$\text{Median}(M) = L + \frac{\frac{N}{2} - m}{f} \times C = 60.5 + \frac{60 - 37}{45} \times 5 = 60.5 + 2.56 = 63.06$$

Exercise 2.2

1. Define the term "median."
2. Find the median of the following distribution:

Class intervals	15–25	25–35	35–45	45–55	55–65	65–75
Frequency	4	11	19	14	0	2

   Ans: 40.263
3. The median of the observations 8, 11, 13, 15, $x + 1$, $x + 3$, 30, 35, 40, 43 arranged in ascending order is 22. Find x.

   Ans: 20

4. Find the median of the following distribution:

Class intervals	0–10	10–20	20–30	30–40	40–50	50–60
Frequency	5	7	8	10	5	15

Ans: 35

5. Find the median of the following distribution:

Weekly income (Rs.) in locality x	Number of families
Below 100	50
100–200	500
200–300	555
300–400	100
400–500	3
500 and above	2

Ans: Rs. 209.9

6. Find the median of the following distribution:

Marks	Number of students
Below 10	3
Below 20	8
Below 30	17
Below 40	20
Below 50	22

Ans: Rs. 22.83

2.9 MODE

Mode is that value of the variate for which the frequency is maximum. It is the most common item of the given series. It is also known as the modal value of the data. According to Croxton and Cowden,

> *The Mode of a distribution is the value at the point around which the items tend to be most heavily concentrated.*

Modal speed is the most frequently occurring value in the distribution. It is defined as the peak of the frequency distribution.

Mode can often be found out by inspection. The given distribution may or may not have a mode. In some cases there may be more than one mode. If the given series has one mode then it said to be Uni-modal, if it has two modes then it is said to be bi-modal, if it has three modes then is said to be tri-modal, and if there are many modes then it is called multimodal.

1. *Mode of ungrouped data*: In the case of ungrouped the mode is determined by identifying the value which occurs maximum number of times. For example consider the data 4, 6, 5, 2, 1, 6, 7, 8, 5, 8, 7, 5, 4, 3, 5, 6, 3, 3, 4, 2, 8, 9, 4, 3, 1, 4, 3, 4, 2, 4

   We prepare a frequency table for the given data as follows

Value	Number of times (frequency)
1	2
2	3
3	5
4	7
5	4
6	3
7	2
8	3
9	1

   The maximum frequency occurs at 4.
   Hence the mode of the given data is 4.

2. *Mode of grouped data*

   Discrete series: The method of finding mode of discrete series by inspection may not give exact value of the mode. To find the mode of discrete series we prepare a grouping table and analyze the table. Follow the steps given below:

   a. Prepare grouping table with six columns.
   b. Write the size of each item in the margin and mark the maximum frequency.
   c. Write the frequencies against the respective items in Column 1 and mark the maximum frequency.
   d. In Column 2, group the frequencies in twos and mark the maximum frequency.
   e. Leaving the first frequency group the frequencies of Column 3, in twos and mark the maximum frequency.
   f. Group the frequencies of Column 4 in threes, and mark the maximum frequency.
   g. Leaving the first frequency, group the frequencies of Column 5 in threes, and mark the maximum frequency.
   h. Leaving the first two frequencies, group the frequencies of Column 6 in threes, and mark the maximum frequency.
   i. Analyze the table prepared, and identify the item with maximum frequency. Which is the mode.

Remark:
The maximum frequencies are marked by bold letters or by circles.

Example 2.20: Calculate the mode of the following frequency distribution:

Height	150	155	156	157	158	159	160
Number of persons	6	18	20	38	10	8	6

Solution: We construct the grouping table as follows:

Height	(1)	(2)	(3)	(4)	(5)	(6)
150	6	24				
155	18		38	44	**76**	
156	20	**58**				**68**
157	**38**		**48**	**56**		
158	10	18			24	
159	8		14			
160	6	6				

Analysis table

Column number	150	155	156	157	158	159	160
1.				1			
2.		1	1	1			
3.				1	1		
4.				1	1	1	
5.		1	1	1			
6.			1	1	1		

Analyzing the table we observe that the item 157 is repeated many times with maximum frequency. Hence the mode of the data is 157.

3. *Mode of continuous series*: When all the class intervals of a continuous series with equal class intervals is calculated by applying the formula:

$$\text{Mode} = L + \frac{f - f_1}{2f - f_1 - f_2} \times C$$

$$= L + \frac{f - f_1}{(f - f_1) + (f - f_2)} \times C$$

where L = Lower limit (boundary) of the modal class; f = frequency of the modal class; f_1 = frequency of the class preceding the modal class; f_2 = frequency of the class succeeding the modal class; C = width of modal class; Modal class = the class with maximum frequency.

When the class intervals of the distribution are not continuous, they are transformed into class boundaries. We can also find the mode of a continuous series by inspection or by constructing a grouped table and analyzing the table. If the lengths of the class intervals are unequal, they are transformed into class intervals of equal length to compute the mode. In some cases the modal class obtained by inspection may be different from the modal class obtained analyzing the table. In such cases we compute the mode by applying the formula:

$$\text{Mode} = L + \frac{f_2}{f_1 + f_2} \times C$$

When the distribution is asymmetrical, we have the following relation:

$$\text{Mean} - \text{Median} = \frac{1}{3}(\text{Mean} - \text{Mode})$$

$$\text{or Mode} = 3\ \text{Median} - 2\ \text{Mean}$$

$$\text{or Mean} - \text{Mode} = 3(\text{Mean} - \text{Median})$$

$$\text{or Median} = \frac{1}{3}(2\ \text{Mean} + \text{Mode})$$

The mode can be calculated by using any one of the formulae mentioned above.

2.9.1 Merits of Mode

It is easy to understand and easy to calculate mode. In some cases mode can also be located by inspection. It can be calculated from open-end classes. It is not affected by extreme values. Mode can also be determined by graphic method.

2.9.2 Demerits of Mode

Mode is not a rigidly defined measure. It is not suitable for further algebraic treatment. It may not give weight to extreme values. Mode cannot easily be calculated, when the data contains both positive and negative items, and when the data contains one or more zeros. It is not easy to calculate mode, when there are two Modal classes in the distribution.

2.9.2.1 Solved Examples

Example 2.21: Compute the following distribution of tracheal ventilation scores (milliliter per minute) of a sample of beetle:

Class interval	61–65	66–70	71–75	76–80	81–85
Frequency	12	25	45	30	8

Solution:

We have the following table:

Class intervals	True limits of class intervals	Frequencies
61–65	60.5–65.5	12
66–70	65.5–70.5	25—f_1
71–75	70.5–75.5	45—f
76–80	75.5–80.5	30—f_2
81–85	80.5–85.5	8

From the above table we have 71–75 as the Modal class.

Maximum frequency = 45, Modal class limits are 70.5 and 75.5, $f = 45, f_1 = 25, f_2 = 30, L = 70.5, C = 5$ therefore,

$$\text{Mode} = L + \frac{f - f_1}{2f - f_1 - f_2} \times C = 70.5 + \frac{45 - 25}{2 \times 45 - 30 - 25} \times 5$$

$$= 70.5 + 2.86 = 73.36$$

Remark:

The mode of a distribution by constructing a grouping table and using the formula:

$$Mode = L_1 + \frac{f - f_1}{2f - f_1 - f_2} \times (L_2 - L_1)$$

where L_1 = Lower limit (boundary) of the modal class; L_2 = upper limit (boundary) of the modal class; f = frequency of the modal class; f_1 = frequency of the class preceding the modal class; f_2 = frequency of the class succeeding the modal class; $L_2 - L_1 = C$ = Width of modal class; Modal class = the class with maximum frequency.

The method is explained with the help of an example as follows:

Example 2.22: Compute the mode of the data given below:

Class interval	10–20	20–30	30–40	40–50	50–60	60–70
Frequency	4	6	20	10	7	3

Solution:

Class interval	Frequency (1)	(2)	(3)	(4)	(5)	(6)
10–20	4	10				
20–30	6			30		
30–40	20	30	26		36	
40–50	10					37
50–60	7	10	17	20		
60–70	3					

Analysis table

Column number	10–20	20–30	30–40	40–50	50–60	60–70
1			1			
2			1	1		
3		1	1			
4	1	1	1			
5		1	1	1		
6			1	1	1	
	1	3	6	3	1	

Analyzing the table we observe that the interval 30–40 appears maximum number of times with maximum frequency. Hence the Modal class is 30–40.

We have $L_1 = 30$, $L_2 = 40$, f = frequency of the modal class = 20; f_1 = frequency of the class preceding the modal class = 6; f_2 = frequency of the class succeeding the modal class = 10; $L_2 - L_1 = C$ = Width of modal class $= 40 - 30 = 10$.

Modal class = the class with maximum frequency = 30–40.

$$Mode = L_1 + \frac{f - f_1}{2f - f_1 + f_2} \times (L_2 - L_1) = 30 + \frac{20 - 6}{2 \times 20 - 6 - 10} \times 10$$

$$= 30 + 5.83 = 35.83$$

Example 2.23: The mean and mode of a moderately asymmetrical distribution are 35 and 36, respectively. Find the mode of the distribution?

Solution: We have mean = 35, Median = 36.

$\therefore$ Mode = 3 Median − 2 Mean $= 3 \times 36 - 2 \times 35 = 108 - 70 = 38$

Example 2.24: In a moderately asymmetrical distribution the values of mean and mode are 35.4 and 32.1, respectively. Find the Median of the distribution?

Solution: We have Mean = 35.4, Mode = 32.1.

$$\therefore \quad \text{Median} = \frac{1}{3}(2\,Mean + Mode) = \frac{1}{3}(2 \times 35.4 + 32.1) = 34.3$$

Exercise 2.3

1. Define the term "Mode."
2. If the mean and median of a moderately asymmetrical series are 26.8 and 27.9, respectively, what would be it most probable mode?

   Ans: 30.1
3. The values of mode and median for a moderately skewed distribution are 64.2 and 68.6, respectively. Find the value of the mode?

   Ans: 70.8
4. The values of mode and median for a distribution are 28 and 25, respectively. Find the value of the mode?

   Ans: 26
5. If the mode and median of a moderately asymmetrical series are 20 and 24, respectively. Find the value of the mean?

   Ans: 26
6. Find the mode of the following data:

Class interval	0–10	10–20	20–30	30–40	40–50	50–60
Frequency	5	7	8	10	5	15

   Ans: 34.6

7. Calculate the mode of the following distribution:

Class interval	1–5	6–10	11–15	16–20	21–25
Frequency	7	10	16	32	24

   Ans: 18.83

8. If the values of mode and mean are 60 and 66, respectively. Find the value of median?

   Ans: 64

Percentile

The *n*th percentile of a set of data is the point where *n*% of the data is below it.

A percentile score tells us what percent of other scores are less than the data point we are investigating. The median, first quartile and third quartile can all be stated in terms of percentiles. Since half of the data is less than the median, and one half is equal to 50%, we could call the median the 50th percentile. One-fourth is equal to 25%, and so the first quartile the 25th percentile. Similarly the third quartile is the same as the

75th percentile. A decile serves as a demarcation of 10% of a set of data. This means that the first decile is the 10th percentile. The second decile is the 20th percentile. Deciles provide a way to split a dataset into more pieces than quartiles without splitting it into 100 pieces as with percentiles. Percentile scores have a variety of uses. Anytime that a set of data needs to be broken into digestible chunks, percentiles are helpful. Median speed, percentile speeds often used to describe speed distribution. The median speed equally divides the distribution. The 50th percentile speed P_{50} is a measure of the center of the distribution. The 85th percentile speed is used as a measure of upper limit of reasonable speeds and 15th percentile speed is the lower limit. One common application of percentiles is for use with tests, such as the SAT, to serve as a basis of comparison for those who took the test.

2.10 GEOMETRIC MEAN

The GM of n observations is defined as the nth root of the product of the n observations.

1. *GM of individual series*: If $x_1, x_2, \ldots, x_n$ denote n observations then the GM of the n observations is:

$$G = \sqrt[n]{x_1 x_2 \ldots x_n} \quad (x_i > 0,\ i = 1, 2, \ldots, n)$$

i.e., $G = (x_1 x_2 \ldots x_n)^{\frac{1}{n}}$

where G denotes the Geometric mean. It is also denoted by GM.

When n is large, we make use of logarithms and reduce the formula to the form:

$$\text{Geometric mean} = \text{antilog}\left[\frac{\log x_1 + \log x_2 + \cdots + \log x_n}{n}\right]$$

$$= \text{antilog}\left[\frac{\sum_1^n f \log x_i}{n}\right]$$

2. *GM of discrete series*: If $x_1, x_2, \ldots, x_n$ denote n observations with frequencies $f_1, f_2, \ldots, f_n$ respectively then, the GM of the series is defined by

$$G = \text{antilog}\left[\frac{f_1 \log x_1 + f_2 \log x_2 + \cdots + f_n \log x_n}{n}\right] = \text{antilog}\left[\frac{\sum_1^n f_i \log x_i}{n}\right]$$

where $n = \sum f_i$

3. *GM of continuous series*: If x_1, x_2, . . ., x_n denote the midpoints of class intervals and f_1, f_2, . . ., f_n denote the corresponding frequencies of the class intervals, then, the GM of the continuous series is given by

$$G = \text{antilog}\left[\frac{f_1 \log x_1 + f_2 \log x_2 + \cdots + f_n \log x_n}{n}\right] = \text{antilog}\left[\frac{\sum_{1}^{n} f_i \log x_i}{n}\right]$$

where $n = \sum f_i$

Note: It is not possible to compute the GM, if the data contains negative observations or if there are zero observations.

4. *Weighted GM*: The weighted GM of the data is computed by the applying the formula:

$$G = \text{antilog}\left[\frac{w_1 \log x_1 + w_2 \log x_2 + \cdots + w_n \log x_n}{w_1 + w_2 + \cdots + w_n}\right] = \text{antilog}\left[\frac{\sum_{1}^{n} w_i \log x_i}{\sum w_i}\right]$$

where w_1, w_2, . . ., w_n denote the weights.

2.10.1 Merits of Geometric Mean

The GM is rigidly defined. It is based on all observations. It is capable of further algebraic treatment. It is less affected by extreme values.

2.10.2 Demerits of Geometric Mean

It is difficult to understand. It is difficult to understand and compute GM if the data contains both positive and negative values it is impossible to compute GM.

2.10.2.1 Solved Examples

Example 2.25: Find the Geometric mean (GM) of the data given below:

0.0009, 0.005, 0.08, 0.8, 5, 75, 475, 2574

Solution:

x	$\log x$
0.0009	$\bar{4}.9542$
0.005	$\bar{3}.6990$

(Continued)

x	$\log x$
0.08	$\bar{2}.9031$
0.8	$\bar{1}.9031$
5	0.698
75	1.875
475	2.676
2574	3.410
	$\sum \log x_i = 2.1208$

We have $n = 8$, $\sum \log x_i = 2.1208$, therefore

$$G = \text{antilog}\left(\frac{\sum_{1}^{n} \log x_i}{n}\right) = \text{antilog}\left(\frac{2.1208}{8}\right) = \text{antilog}(0.2651) = 1.841$$

Example 2.26: The depreciation on a building for the first year is fixed at 5%, in the second year it is fixed at 10%, in the third year the depreciation is fixed at 15%, and in the next 2 years the depreciation rate is fixed at 20% per year. Find the average depreciation rate in these 5 years?

Solution: Since we are to find the rate of change, GM is the appropriate average. We have the following table:

Year	Depreciation	Value after depreciation (x_i)	$\log x_i$
1	5	95	1.9777
2	10	90	1.9542
3	15	85	1.9294
4	20	80	1.9031
5	20	80	1.9031
$n = 5$			$\sum \log x_i = 9.6675$

$$GM = \text{antilog}\left(\frac{\sum_{1}^{n} \log x_i}{n}\right) = \text{antilog}\left(\frac{9.6675}{5}\right) = \text{antilog}(1.9225) = 85.8$$

Therefore the average depreciation rate is 100−85.80 = 14.20

Example 2.27: Find the Geometric mean (GM) of the following distribution:

Class interval	10−20	20−30	30−40	40−50	50−60
Frequency (f)	5	10	15	7	4

Solution: Taking logarithms, we have the following table:

Class interval	f_i	Midpoints of class intervals, x_i	$\log x_i$	$f_i \log x_i$
10–20	5	15	1.1761	5.8805
20–30	10	25	1.3979	13.9790
30–40	15	35	1.5441	23.1615
40–50	7	45	1.6532	11.5724
50–60	4	55	1.7404	6.9616
	$\sum f_i = 41$			$\sum f_i \log x_i = 61.5550$

We have $\sum f_i = 41, \quad \sum f_i \log x_i = 61.5550$

$$GM = \text{antilog}\left(\frac{\sum_1^n f_i \log x_i}{n}\right) = \text{antilog}\left(\frac{61.5550}{41}\right) = \text{antilog}(1.5013) = 31.72$$

Exercise 2.4

1. Compute Geometric mean (GM) of the following data:
10, 37, 50, 52, 60, 80, 110, 120
Ans: 52.84

2. Calculate the Geometric mean (GM) of the following data:
50, 54, 72, 82, 93
Ans: 68.26

3. Find the Geometric mean (GM) of the following data:

x	350	150	200	150	225
f	10	2	2	2	4

Ans: 255.8

4. Compute the weighted Geometric mean (GM) of the following data:

Index number	8	10	52	25	37
Weight	5	3	4	2	1

Ans: 17.74

5. Find the Geometric mean (GM) of the following data:

Observation	20	40	60	80	100
Weight	14	8	7	12	5

Ans: 45.6

6. Compute the weighted Geometric mean (GM) of the following data:

Index number	350	180	200	150	220
Weight	15	2	4	5	2

Ans: 67.441

2.11 HARMONIC MEAN

HM if a set of observations is defined as the reciprocal of the arithmetic average of the reciprocals of the observations.

1. *HM of individual series*: If $x_1, x_2, \ldots, x_n$ denote n various values in the observations, then the HM of the n observations is

$$HM = \frac{n}{\frac{1}{x_1} + \frac{1}{x_2} + \cdots + \frac{1}{x_n}} = \frac{n}{\sum \frac{1}{x_i}}$$

2. *HM of discrete series*: If $x_1, x_2, \ldots, x_n$ denote n items which occur with frequencies $f_1, f_2, \ldots, f_n$, respectively, then the HM of these items is

$$HM = \frac{\sum f_i}{\frac{f_1}{x_1} + \frac{f_2}{x_2} + \cdots + \frac{f_n}{x_n}} = \frac{N}{\sum \frac{f_i}{x_i}}$$

where $N = \sum f_i$.

3. *HM of continuous series*: If $x_1, x_2, \ldots, x_n$ denote the midpoints of class intervals and frequencies $f_1, f_2, \ldots, f_n$ denote the corresponding frequencies of a frequency distribution, the HM of the distribution is

$$HM = \frac{\sum f_i}{\frac{f_1}{x_1} + \frac{f_2}{x_2} + \cdots + \frac{f_n}{x_n}} = \frac{N}{\sum \frac{f_i}{x_i}}$$

where $N = \sum f_i$.

4. *Weighted HM*: The weighted HM is computed by the formula

$$HM_w = \frac{\sum w_i}{\frac{w_1}{x_1} + \frac{w_2}{x_2} + \cdots + \frac{w_n}{x_n}}$$

2.11.1 Merits of Harmonic Mean

HM is rigidly defined. It is easy to calculate. It is based on all items and is suitable for algebraic treatment. HM is useful in finding the averages involving time rate and price. It gives less weight to large items and greater weight to smaller items.

2.11.2 Demerits of Harmonic Mean

HM is difficult to calculate. Its value is difficult to calculate when there are both positive and negative values in the data.

2.11.3 Relation Between AM, GM, and HM

If AM denotes the Arithmetic mean, GM denotes the Geometric mean, and HM denotes the HM of a distribution then we have

AM $\geq$ GM $\geq$ HM

2.11.3.1 Solved Examples

Example 2.28: The consumption of petrol by a motor car was a gallon for 20 miles while going up from plains to a hill station and a gallon for 24 miles while coming down. What is the average consumption in miles per gallon for the up and down journey?

Solution: We have $x_1 = 20$, $x_2 = 24$, $n = 2$

The average consumption is the HM, which is given by

$$HM = \frac{n}{\frac{1}{x_1} + \frac{1}{x_2} + \cdots + \frac{1}{x_n}} = \frac{n}{\sum \frac{1}{x_i}} = \frac{2 \times 120}{\frac{1}{20} + \frac{1}{24}} = \frac{2 \times 120}{6 + 5}$$

$$= 21.82 \text{ miles/gallon}$$

Example 2.29: Find the HM of the following data:

Size of the item	6	7	8	9	10	11
Frequency	4	6	9	5	2	8

Solution: We have the following calculation table:

x_i	f_i	$\frac{1}{x_i}$	$\frac{f_i}{x_i}$
6	4	0,1667	0.6668
7	6	0.1429	0.8574
8	9	0.1250	1.1250
9	5	0.1111	0.5555
10	2	0.1000	0.2000
11	8	0.0909	0.7272
	$N = \sum f_i = 34$		$\sum \frac{f_i}{x_i} = 4.1319$

$$HM = \frac{N}{\sum \frac{f_i}{x_i}} = \frac{34}{4.1319} = 8.23$$

Example 2.30: An aeroplane traveled a distance of 800 miles, of which the plane covered the first lap consisting a distance of 200 miles at a velocity of 100 mph, and the second lap of 200 miles at a velocity of 200 mph, the third distance of 200 miles at a velocity of 300 mph, and the remaining distance of 200 miles at a velocity of 400 mph. Find the average velocity of the aeroplane?

Solution: To find the average velocity of the aeroplane compute the Harmonic mean (HM), which is an appropriate measure. We have $n = 4$, $x_1 = 100$ mph, $x_2 = 200$ mph, $x_3 = 300$ mph, $x_4 = 400$ mph

$$HM = \frac{n}{\frac{1}{x_1} + \frac{1}{x_2} + \cdots + \frac{1}{x_n}} = \frac{n}{\sum \frac{1}{x_i}} = \frac{4}{\frac{1}{100} + \frac{1}{200} + \frac{1}{300} + \frac{1}{400}}$$

$$= 192 \text{ mph}$$

Example 2.31: A train travels a distance of 16 km at a speed of 20 km/h, the next distance of 20 km at a speed of 40 km/h, and the remaining distance of 10 km at a speed of 15 km/h. Find the average speed of the train?

Solution: Since the distances traveled and speeds are different we apply weighted Harmonic mean (HM). Here we have $w_1 = 16$, $w_2 = 20$, $w_3 = 10$; $x_1 = 20$, $x_2 = 40$, $x_3 = 15$

$$HM_w = \frac{\sum w_i}{\frac{w_1}{x_1} + \frac{w_2}{x_2} + \cdots + \frac{w_n}{x_n}} = \frac{16 + 20 + 10}{\frac{16}{20} + \frac{20}{40} + \frac{10}{15}} = \frac{46}{1.97} = 23.35 \text{ km/h}$$

Exercise 2.5

1. Define "Harmonic mean" and write the merits and demerits of Harmonic mean (HM).
2. Find the average rate of motion. In the case of a person who rides the first kilometer at 10 Km/h, the next kilometer at 8 Km/h, and the third kilometer at 6 Km/h?

   Ans: 7.6 km/h
3. A train travels first 300 km at an average of 39 Km/h and further travels the same distance at an average rate of 40 Km/h. What is the average speed over the whole distance.

   Ans: $34\frac{2}{7}$ km/h

4. Find the HM on the following frequency distribution:

Class interval	0–30	30–60	60–90	90–120	120–150
Frequency	2	5	11	4	2

*Ans***:** 54.06

5. Find the HM of the following weighted distribution:

Item	60	25	350	25
Weight	900	3000	4000	55

*Ans***:** 137.03

6. A person travels first 900 km in a train at a speed of 60 km/h and travels the distance of 300 km in a ship at an average rate of 25 km/h, covers a distance of 400 km in an aeroplane at a speed of 350 km/h, and further travels a distance of 15 km in a taxi at a speed of 20 km/h. What is the average speed over the whole distance?

*Ans***:** 55.27 km/h

7. A taxi driver traveled in a motor 100 km at a speed of 30 km/h while going up from plains to a hill station. While coming down he traveled at a speed of 2 km/h. What is the average speed for the up and down journey?

*Ans***:** 24 km/h

8. An aeroplane flies, along the four sides of a square at speeds of 100, 200, 300, and 400 km/h, respectively. What is the average speed of the plane in its flight around the square?

*Ans***:** 192 km/h

9. The following is the distribution of 100 accidents in New Delhi during 7 days of a week, of a given month. During that month there were 5 Mondays, 5 Tuesdays, and 5 Wednesdays and four of each of the other days. Calculate the number of accidents per day.

*Ans***:** $14.13 \cong 14$

10. The following distribution gives the number of accidents met by 160 workers in a factory during a month:

Number of accidents	0	1	2	3	4
Number of workers	70	52	34	3	1

Find mean of the data?

*Ans***:** $0.83 \cong 1$ (approx.)

11. A Train travels first 300 km at an average speed of 30 km/h, and further travels the same distance at an average speed of 40 km/h. What is the average speed of the train over the whole distance?
 Ans: 34.29 km/h

12. A Scooterist purchased petrol at the rate of Rs. 24, 29.50, and 36.85 per liter during three successive years. Calculate the average price of the petrol:
 a. if he purchased 150, 180, 195 liters of petrol in the respective years, and
 b. if he spent Rs. 3850, 4675, and 5825 in the 3 years.
 Ans: Rs. 30.66, 30.09

13. The following table gives the distribution of total household expenditure in rupees of manual workers in a city:

Expenditure (Rs.)	100–150	150–200	200–250	250–300	300–350	350–400	400–450	450–500
Frequency	24	40	33	28	30	22	16	7

 Find the average expenditure per household?
 Ans: Rs. 266.25

14. You take a trip which entails traveling 900 miles by train at an average speed of 60 mph, 300 miles by boat at an average speed of 25 mph, 400 miles by plane at 350 mph and finally by taxi at 25 mph. What is your average speed for the entire distance?
 Ans: 56.19 mph

15. An aeroplane flies along the four sides of a square which measures 100 km each at speeds of 100, 200, 300, and 400 km/h, respectively. Find the average speed of the flight around the square?
 Ans: 192 km/h

16. A cyclist pedals from his house to his college at a speed of 10 km/h and back from the college to his house at 15 km/h. Find his average speed?
 Ans: 12 km/h

17. A cyclist covers first 3 km at an average speed of 8 km/h and another 2 km at 3 km/h, and the last 2 km at 2 km/h. Find the average speed of the entire journey?
 Ans: 3.43 km/h

18. A person living in Hyderabad started for a village at a 6-km distance. He traveled in his car at a speed of 40 km/h. After traveling for 4 km the car stopped running. He then traveled in a rickshaw at a speed of

10 km/h. After traveling a distance of 1.5 miles he left rickshaw and covered the remaining distance on foot at a speed of 4 km/h. What is his average speed?

***Ans*:** 16 km/h

19. You take a trip which entails traveling 90 km by train at an average speed of 60 km/h an average of 25 km/h. What is your average speed?

***Ans*:** 31.49 km/h

2.12 PARTITION VALUES (QUARTILES, DECILES, AND PERCENTILES)

The values which divide a given series into a number of equal parts are called partition values.

2.12.1 Quartiles

A quartile is a measure which divides the given distribution (set of observations or data) into four equal parts. We have three quartiles:

1. First quartile (Lower quartile)
2. Second quartile (Middle quartile)
3. Third quartile (Upper quartile).

The first quartile is denoted by Q_1. It has 25% of the observations below it and 75% of the observations above it. The second quartile is denoted by Q_2. It has 50% of the items of the distribution below it and 50% of the items above it. Q_2 is called Median. The upper quartile is denoted by Q_3. It contains 75% of the items less than it and 25% of the items above it.

1. *Quartiles of individual series*: If N denotes the number of observations (items) the first quartile $Q_1 =$ value of the $(N+1)\text{th}/4$ item provided the items are arranged either in the ascending or descending order of their magnitudes.

   If N is the number of observations, the third quartile $Q_3 =$ value of the $3(N+1)\text{th}/4$ item provided the items are arranged either in the ascending or descending order of their magnitudes.
2. *Quartiles of discrete series*: If N denotes the sum of all frequencies of the series the first quartile $Q_1 =$ size of the $(N+1)\text{th}/4$ item provided the cumulative frequencies are calculated if N denotes the sum of all frequencies of the series the third quartile $Q_3 =$ size of the $3(N+1)\text{th}/4$ item provided the cumulative frequencies are calculated.

3. *Quartiles of continues series*: We first calculate the cumulative frequencies and the values of $N/4$ and $3N/4$ where N denotes the sum of all frequencies (i.e., total frequency). The class interval containing the $N^{th}/4$ item is taken as the first quartile Q_1 class, and the class interval containing $3N/4$ item is taken as the third quartile Q_3 class. The values of Q_1 and Q_3 are computed by using the formulae given below:

$$Q_1 = L_1 + \frac{\frac{N}{4} - m_1}{f_1} \times C_1$$

where L_1 = lower limit of the Q_1 class, $N = \sum f_i$, m_1 = cumulative frequency of the class preceding Q_1 class, f_1 = frequency of the Q_1 class, C_1 = length (width) of Q_1 class.
and

$$Q_3 = L_3 + \frac{\frac{3N}{4} - m_3}{f_3} \times C_3$$

where L_3 = lower limit of the Q_3 class, $N = \sum f_i$, m_3 = cumulative frequency of the class preceding Q_3 class, f_3 = frequency of the Q_3 class, c_3 = length (width) of Q_3 class.

2.12.2 Deciles

The values which divide the given series into 10 equal parts are called deciles. We have 9 deciles which divide the data into 10 equal parts. The 9 deciles are denoted by the symbols $D_1, D_2, \ldots, D_9$.

1. *Deciles of individual and discrete series*: The deciles of the individual and discrete series are computed by using the formula

$$D_i = \text{size of } \frac{i_N}{10}^{th} \text{ item}$$

where $i = 1, 2, 3, \ldots, 9$.

2. *Deciles of continuous series*: The deciles of continuous series are calculated by applying the formula

$$D_i = L_i + \frac{\frac{i_N}{10} - m_i}{f_i} \times c_i$$

where $i=1, 2, \ldots, 9$; $N=\sum f_i$; $m_i=$ cumulative frequency of the class preceding D_i class; $f_i=$ frequency of the D_i class; $c_i=$ length (width) of D_i class.

2.12.3 Percentiles

The values which divide the data into 100 equal parts are called percentiles. A percentile is denoted by the letter P. We have 99 percentiles denoted by the symbols $P_1, P_2, \ldots, P_{99}$.

1. *Percentiles of individual and discrete series*: The percentiles of the individual and discrete series are computed by using the formula

$$P_i = \text{size of } \frac{iN}{100}^{th} \text{ item}$$

where $i=1, 2, 3, \ldots, 99$.

2. *Percentiles of continuous series*: The percentiles of continuous series are calculated by applying the formula

$$P_i = L_i + \frac{\frac{iN}{100} - m_i}{f_i} \times C_i$$

where $i=1, 2, \ldots, 99$; $N=\sum f_i$; $m_i=$ cumulative frequency of the class preceding P_i class; $f_i=$ frequency of the P_i class; $C_i=$ length (width) of P_i class.

A measure of central tendency (i.e., average) is the representative the given distribution but the partitioned values are the averages of parts of the distribution. The partition values are used to study the values, which are used to study the dispersion of the items in relation to the median. The relation between the partition values are given below:

$$P_{25} = Q_1, \quad P_{50} = Q_2 = \text{Median} = D_5, \quad P_{75} = Q_3$$

Example 2.32: Compute the values of Q_1, Q_3, D_6, and P_{20} for the following data:

41, 45, 49, 47, 51, 42, 53, 54, 55

Solution: Arranging the given values in the ascending order we have

41, 42, 45, 47, 49, 51, 53, 54, 55

$N=9$, $N+1=10$, $3(N+1)=30$

$$Q_1 = \text{size of the } ((N+1)/4)\text{th item} = \frac{10}{4} = \text{value of 2.5th item}$$

Since it is 2.5, we take the average of 2nd and 3rd items as the value of Q_1.

$$Q_1 = \frac{42 + 45}{2} = 43.5$$

Q_3 = value of the $(3(N+1)/4)$th item = the value of $(30/4)$th item $= 30/4 = 7.5$.

Since it is 7.5 we take the mean of 7th and 8th items as the value of Q_3.

$$\therefore \quad Q_3 = \frac{53 + 54}{2} = 53.5$$

D_6 = The value of $(6(N+1)/10)$th item = the value of $(6(9+1))/10 = 60/10 = 6$th item

The 6th item is 51. Therefore, $D_6 = 51$.

P_{20} = The size of $((20(N+1))/100)$th item = the size of $(20(9+1))/100 = 2$nd item

The 2nd item is 45, therefore $P_{20} = 45$.

2.13 MEASURES OF DISPERSION

A measure of central tendency of a distribution is a single significant value which is used to describe a distribution. The averages like mean, median, and mode tell something about the general level of magnitude of a distribution but they fail to show further about the distribution. According to George Simpson and Fritz Kafka "An average does not tell the full story. It is hardly fully representative of a mass, unless we know the manner in which the individual items scatter around it. A further description of the series is necessary if we are to gauge how representative the average is." In order to describe a frequency distribution adequately it is necessary not only to know the average of the distribution but to have some idea of the variability in the measurements. In this section we introduce measures of dispersion (variability) that are designed to state numerically the extent to which individual observations vary on the average. Averages will be meaningful, when they are studied through dispersion. The term dispersion is defined as follows:

Dispersion is a measure of extent to which the individual items vary

—L.R. Connor

Dispersion is the measure of the variation of the items

—A.L. Bowley

The degree to which numerical data tend to spread about an average value is called the variation or dispersion of the data

—***Spiega***

The measures of dispersion are also known as the measures of scatteredness or the measures of spread of the data. They are used to test the reliability of an average. When the dispersion (or scatter) is small it is reliable and when the scatter is large, the average is less reliable. Measures of dispersion are indispensable to determine the nature and find the causes of variation. They are used to compare two or more series with regard to their variability. The purpose of measuring variation is to facilitate as a basis for further statistical analysis.

2.13.1 Characteristics of an Ideal Measure of Dispersion

A measure of dispersion should be simple to understand and easy to calculate. It should be rigidly defined. It should be based on all observations. A good measure of dispersion should not be duly affected by extreme values and it should have sampling stability. It should be amenable to further algebraic treatment.

2.13.2 Types of Measures of Dispersion

There are two types of measures of dispersion, namely:

1. Absolute measures of dispersion
2. Relative measures of dispersion

Absolute measures of dispersion are expressed in terms of the original data. These measures may be used to compare the variation in two sets of observations provided, the variables are expressed in the same units and the same average size. Absolute measures of dispersion cannot be used for comparison when expressed in different units. The most commonly used absolute measures of dispersion are:

1. Range
2. Quartile deviation
3. Mean deviation
4. Standard deviation

Relative measures of dispersion are independent of the units of measurement. For comparing the variability, even if the distributions are expressed in the same units, the relative measure of dispersion is computed. The relative measure is the ratio of absolute dispersion to an appropriate measure of central value and is expressed as a pure number.

These measures are also used to compare the relative accuracy of the data. The most commonly used relative measures of dispersion are:

1. Coefficient of variation
2. Coefficient of quartile deviation
3. Coefficient of mean deviation

We now briefly introduce these measures of dispersion.

2.14 RANGE

Range is a simple measure of dispersion. The value of range depends upon the extreme values of the data.

If L denotes the largest value of the distribution and S denotes the smallest value of the distribution, then the range of the distribution is defined as the difference between largest value (L) and the smallest value (S), thus

$$\text{Range}(R) = L - S$$

2.14.1 Coefficient of Range

It is relative measure corresponding to range and is defined as follows:

$$\text{Coefficient of Range} = \frac{L - S}{L + S}$$

2.14.2 Merits of Range

Range is simple to understand and easy to compute. The units of range are the same as the units of the variable being measured. It takes minimum time to calculate. It gives a rough answer.

2.14.3 Demerits of Range

The value of range does not depend on all values of the distribution. The measure is highly affected by extreme values. It imparts limited information about the data. The range of a distribution with open-end classes cannot be calculated. It is not reliable and has limited use.

2.14.4 Uses of Range

In quality control of manufactured products, range is used to study the variation in the quality of the units manufactured. Range is used when variations in the variable are not much. It is a measure of dispersion that is used to study variations in money rates, share values, gold prices, etc. It is also used in weather forecasting by the meteorological department.

Example 2.33: Find the range and the coefficient of range for the following data:

180, 210, 208, 225, 260, 190.

Solution: Arranging the data in ascending order we have

180, 190, 208, 210, 225, 260.

$L = 260,\ S = 180$

$$\text{Range}(R) = L - S = 260 - 180 = 80$$

$$\text{Coefficient of Range} = \frac{L - S}{L + S} = \frac{260 - 180}{260 + 180} = \frac{80}{440} = 0.1818$$

Example 2.34: Find the range of the following distribution:

x	10	20	30	40	50	60
f	8	12	15	6	20	9

Solution: We have

$L = 60,\ S = 10$

Range $(R) = L - S = 60 - 10 = 50$

Note: To calculate the range of a distribution we consider only the values of x, not the frequencies.

Example 2.35: Find the range of the following distribution:

Class intervals	0–5	5–10	10–15	15–20	20–25	25–30
Frequencies	8	4	5	12	6	3

Solution: Here we consider the class intervals to find the range

We have $L = 30,\ S = 0,\ Range = L - S = 30 - 0 = 30$

2.15 INTERQUARTILE RANGE

Definition:

The interquartile range (IQR) is defined as the difference between the third and first quartiles. Symbolically

$$IQR = Q_3 - Q_1$$

The interquartile range is a measure of variation. It is used in building Box plot. When the data is skewed it represents the spread better than other measures of dispersion.

2.16 QUARTILE DEVIATION

Quartile deviation is an absolute measure of dispersion. It is defined as follows:

Definition:
The quartile deviation of a distribution is defined as half the difference between the third and first quartiles. Symbolically

$$\text{Quartile deviation} = QD = \frac{Q_3 - Q_1}{2}$$

Quartile deviation is also known as semiinterquartile range.

2.16.1 Coefficient of Quartile Deviation

If Q_3 and Q_1 denote third and first quartiles respectively, the coefficient of quartile deviation is calculated by the formula:

$$\text{Coefficient of quartile deviation} = \frac{\frac{Q_3 - Q_1}{2}}{\frac{Q_3 + Q_1}{2}} = \frac{Q_3 - Q_1}{Q_3 + Q_1}$$

Example 2.36: Find the quartile deviation and the coefficient of quartile deviation of the following data:

41, 34, 52, 42, 37, 61, 58, 71, 64, 69, 74

Solution: Arranging the data in ascending order we have

34, 37, 41, 42, 52, 58, 61, 64, 69, 71, 74.

Since $n = 11$, Q_1 = Size of $((n+1)/4)$th item = Size of $((11+1)/4)$th item = size of 3rd item.

Therefore, $Q_1 = 41$

Q_3 = Size of $(3(n+1)/4)$th item = Size of $(3(11+1)/4)$th item = size of 9th item.

Therefore, $Q_3 = 69$

$$\text{Quartile deviation} = QD = \frac{Q_3 - Q_1}{2} = \frac{69 - 41}{2} = 14$$

$$\text{Coefficient of quartile deviation} = \frac{\frac{Q_3 - Q_1}{2}}{\frac{Q_3 + Q_1}{2}} = \frac{Q_3 - Q_1}{Q_3 + Q_1} = \frac{69 - 41}{69 + 41}$$

$$= \frac{28}{110} = 0.255$$

Example 2.37: Find the interquartile range (IQR) and quartile deviation of the data given below:

Marks	20	30	40	50	60	70	80
Number of students	5	9	16	12	4	3	1

Solution: We form the cumulative frequency table of the given distribution as follows:

Marks	**Frequency**	**Cumulative frequency**
20	5	5
$30 - Q_1$	9	14
40	15	29
$50 - Q_3$	120	41
60	4	45
70	3	48
80	1	49

We have $N - 49$, $N + 1 = 50$, $\frac{N+1}{4} = \frac{50}{4} = 12.5$, $3\left(\frac{N+1}{4}\right) = 37.5$

The size of 12.5th item is less than the size of the 13th item.
The value of the item against the cumulative frequency 13 is 30. Therefore $Q_1 = 30$ 37.5 item is less than the 41st item, the size of the item against the cumulative frequency 41 is Q_3. Therefore $Q_3 = 50$.

$$\therefore \quad \text{Interquartile range} = Q_3 - Q_1 = 50 - 30 = 20$$

$$\text{Coefficient of quartile deviation} = \frac{Q_3 - Q_1}{Q_3 + Q_1} = \frac{50 - 30}{50 + 30} = \frac{20}{80} = 0.25$$

Example 2.38: Calculate the quartile deviation from the following:

Wages (in Rs.)	30–32	32–34	34–36	36–38	38–40	40–42	42–44
Number of laborers	12	18	16	14	12	8	6

Solution: Cumulative frequency table

Class intervals (wages)	Frequency (*f*)	Cumulative frequency
30–32	12	12
32–34 – Q_1 class	18	30
34–36	16	46
36–38	14	60
38–40 – Q_3 class	12	72
40–42	8	80
42–44	6	86
	$N = 86$	

We have $N = 86$, $N = 86/4 = 21.5$, $3N/4 = 64.5$

21.5th item lies in the interval against the cumulative frequency 30, i.e., in the interval 32–34.

64.5th item lies in the interval against the cumulative frequency 72, i.e., in the interval 38–40.

Where L_1 = lower limit of the Q_1 = class = 32, $N = \sum f_i = 86$, m_1 = cumulative frequency of the class preceding Q_1 class = 12, f_1 = frequency of the Q_1 class = 18, C_1 = length (width) of Q_1 class = 2.

$$Q_1 = L_1 + \frac{\frac{N}{2} - m_1}{f_1} \times C_1 = 32 + \frac{21.5 - 12}{18} \times 2 = 32 + \frac{19}{18}$$

$$= 32 + 1.06 = 33.06$$

and

L_3 = lower limit of the Q_3 class = 38, $N = \sum f_i = 86$, m_3 = cumulative frequency of the class preceding Q_3 class = 60, f_3 = frequency of the Q_3 class = 12, C_3 = length (width) of Q_3 class = 2.

$$Q_3 = L_3 + \frac{\frac{3N}{4} - m_3}{f_3} \times C_3 = 38 + \frac{64.5 - 60}{12} \times 2 = 38 + \frac{4.5}{12} \times 2$$

$$= 38 + 0.75 = 38.75$$

$$\text{Quartile deviation} = QD = \frac{Q_3 - Q_1}{2} = \frac{38.75 - 33.06}{2} = \frac{5.69}{2} = 2.85$$

2.16.1.1 Merits of Quartile Deviation

Quartile deviation is a simple measure of dispersion, which is easy to understand and easy to compute. The measure is not influenced by extreme values. It can be found by extreme values. It can be found with open-end distribution. It is not affected by presence of extreme values.

2.16.1.2 Demerits of Quartile Deviation

Quartile deviation is a positional average; therefore it is not amenable to further algebraic treatment. It ignores the first 25% of the items and the last 25% of the items. The value of quartile deviation is affected by sampling fluctuations.

2.17 MEAN DEVIATION

Mean deviation is an absolute figure. It is defined as the AM of the deviations of a series computed from any measure of central tendency provided all the deviations are taken as positive.

Mean deviation is denoted by the letter δ.

We usually compute mean deviation about any one of the three averages mean, median, or mode. In some cases mode may be ill defined. Hence we calculate mean deviation either from mean or median.

1. *Mean deviation of ungrouped data (individual series)*: If $x_1, x_2, \ldots, x_n$ denote n observations (items), then the mean deviation of the observations about mean $\overline{x}$ is

$$\frac{\sum |x_i - \overline{x}|}{n}$$

and the mean deviation about the median M_e is

$$\frac{\sum |x_i - \overline{x}|}{N}$$

Similarly we can also define the mean deviation about mode.

2. *Mean deviation of grouped data*
   a. *Mean deviation of discrete series*: If $x_1, x_2, \ldots, x_n$ denote N observations (items), with frequencies $f_1, f_2, \ldots, f_n$ respectively, then the mean deviation of the observations about mean $\overline{x}$ is

$$\frac{\sum f_i|x_i - \overline{x}|}{N}$$

   and the mean deviation about the median M_e is

$$\frac{\sum f_i|x_i - M_e|}{N}$$

   where $N = \sum f_i$.

   Similarly we can also define the mean deviation about mode.

   b. *Mean deviation of continuous series*: If $x_1, x_2, \ldots, x_n$ denote the mid-points of class intervals, with frequencies $f_1, f_2, \ldots, f_n$ respectively, then the mean deviation of the observations about mean $\overline{x}$ is

$$\frac{\sum f_i|x_i - \overline{x}|}{N}$$

   And the mean deviation about the median M_e is

$$\frac{\sum f_i|x_i - M_e|}{N}$$

2.17.1 Coefficient of Mean Deviation

$$\text{Coefficient of mean deviation} = \frac{\text{Mean deviation}}{\overline{x}}$$

Example 2.39: Find the mean deviation of the following series about mean:

34, 36, 37, 38, 39, 40, 41, 43, 44, 48

Solution: Mean of the given series is

$$\overline{x} = \frac{34 + 36 + 37 + 38 + 39 + 40 + 41 + 43 + 44 + 48}{10} = \frac{400}{10} = 40$$

We have the following table:

Item, x_i	$\lvert x_i - \overline{x}\rvert$
34	$\lvert -6\rvert = 6$
36	$\lvert -4\rvert = 4$

(*Continued*)

Item, x_i	$\lvert x_i - \overline{x} \rvert$
37	$\lvert -3 \rvert = 3$
38	$\lvert -2 \rvert = 2$
39	$\lvert 1 \rvert = 1$
40	$\lvert 0 \rvert = 0$
41	$\lvert 1 \rvert = 1$
43	$\lvert 3 \rvert = 3$
44	$\lvert 4 \rvert = 4$
48	$\lvert 8 \rvert = 8$
$\sum x = 400$	$\sum \lvert x_i - \overline{x} \rvert = 32$

Therefore the mean deviation of the given observations about mean $\overline{x}$ is

$$\frac{\sum |x_i - \overline{x}|}{N} = \frac{32}{10} = 3.2$$

Example 2.40: Compute the mean deviation of the data given below:

x	2	4	6	8	10
f	1	4	6	4	1

Solution:

x_i	Frequency, f_i	$f_i x_i$	$\lvert x_i - \overline{x} \rvert$	$f_i \lvert x_i - \overline{x} \rvert$
2	1	2	4	4
4	4	16	2	8
6	6	36	0	0
8	4	32	2	8
10	1	10	4	4
	$N - \sum f_i - 16$	$\sum f_i x_i - 96$		$\sum f_i \lvert x_i - \overline{x} \rvert = 24$

$$\text{we have mean of the data } \overline{x} = \frac{\sum_{i=1}^{n} f_i x_i}{\sum_{i=1}^{n} f_i} = \frac{96}{16} = 6$$

$$\text{Mean deviation} = \frac{\sum f_i |x_i - \overline{x}|}{N} = \frac{24}{16} = 1.5$$

Example 2.41: Find the mean deviation of the following data from Arithmetic mean (AM):

Class intervals	0–5	5–10	10–15	15–20	20–25	25–30	30–35
Frequency	2	7	10	12	9	6	4

Solution:

We have the following table:

Class intervals	Frequency, f_i	Midpoints, x_i	$f_i x_i$	$\lvert x_i - \overline{x} \rvert$	$f_i \lvert x_i - \overline{x} \rvert$
0–5	2		5	15.3	30.6 72.1
05–10	7	7.5	52.5	10.3	53
10–15	10	12.5	125.0	5.3	3.6
15–20	12	17.5	210.0	0.3	42.3
20–25	9	22.5	202.5	4.7	58.2
25–30	6	27.5	165.0	9.7	58.8
30–35	4	32.5	130.0	14.7	
	$N = \sum f_i = 50$		$\sum f_i x_i = 890$		$\sum f_i \lvert x_i - \overline{x} \rvert = 318.6$

We have

$$\text{Arithmetic mean} = \overline{X} = \frac{\sum_{i=1}^{n} f_i x_i}{\sum_{i=1}^{n} f_i} = \frac{890}{50} = 17.8$$

$$\text{Mean deviation} = \frac{\sum f_i \lvert x_i - \overline{x} \rvert}{N} = \frac{318.6}{50} = 6.372$$

$$\text{Coefficient of mean deviation} = \text{Mean deviation } \overline{x} = \frac{6.372}{17.8} = 0.36$$

2.17.2 Merits of Mean Deviation

Mean deviation is a simple measure of dispersion. It is easy to understand and easy to calculate. The measure is less affected by fluctuations of sampling. It is based on all items of data and it gives weight according to their size. Mean deviation is a rigidly defined measure of variation. It can be used for comparison. It is less affected by the extreme values of the data.

2.17.3 Demerits of Mean Deviation

Mean deviation is not an accurate measure of dispersion. It is not suitable for further mathematical calculations. Mean deviation calculated from median yields better results though the variability of median is too high. While computing mean deviation the signs of all deviations are taken as positive.

2.17.4 Uses of Mean Deviation

Mean deviation is useful in marketing problems. It is used in Social Sciences. The measure is useful in forecasting business cycles. It is used in studying problems such as income and wealth.

Exercise 2.6

1. Find the mean deviation of 10, 15, 20, 17, 19, 21 13, 12, 12, 9, and 17?

   Ans: 3.45

2. Find the mean deviation of 33, 49, 27, 61, 76, 104, and 126?

   Ans: 29.14

3. Find the mean deviation and coefficient of mean deviation for the following data:

   34, 48, 41, 36, 44, 37, 43, 39, 40, and 38.

   Ans: 3.2, 0.08

4. Compute mean deviation from mean for the following data:

x	2	4	6	8	10
f	1	4	6	4	1

   Ans: 1.5

5. Calculate mean deviation and the coefficient of mean deviation of the following data:

x	17	21	25	29	33	37	41
f	6	5	1	2	4	2	3

   Ans: $MD = 8.36$, Coefficient of $MD = 0.19$

6. Find the mean deviation of distribution given below:

Class interval	2–4	4–6	6–8	8–10
Frequency	3	4	2	1

   Ans: 1.48

2.18 STANDARD DEVIATION

SD is defined as the positive square root of the AM of the squares of deviations of the given items (Observations) from their AM. It is also called the Root Mean Square Deviation. It is denoted by the letter σ.

There are two methods for calculating SD. In the first method we consider the deviations from actual mean while computing the SD. In the second method we calculate the SD by taking deviations from the assumed mean. In general we define the symbolic form of SD as follows:

1. *SD of individual series*: Let $x_1, x_2, \ldots, x_n$ denote the n values of taken by the random variable x, then the SD of these n observations is given by

$$\sigma = \sqrt{\sum \frac{(x_i - \overline{x})^2}{n}}$$

where $\overline{x}$ is the AM of the n observations $x_1, x_2, \ldots, x_n$.

Alternatively the above formula for finding the SD can be written as

$$\sigma = \sqrt{\frac{\sum x_i^2}{n} - (\overline{x})^2}$$

If A denotes the assumed mean and $d_i = x_i - A$, then the SD of the n observations is

$$\sigma = \sqrt{\frac{\sum d_i^2}{n} - (\overline{d})^2}$$

where $\overline{d} = \dfrac{\sum d_i}{n}$

2. *SD of discrete data*: If $\overline{x}$ is the actual mean of the series, then the SD is calculated by the formula:

$$\sigma = \frac{\sum f_i d_i^2}{\sum f_i}$$

where $d_i = x_i - \overline{x}$, (x_i is the discrete data). Hence we have

where $\sigma = \sqrt{\dfrac{\sum f_i (x_i - \overline{x})^2}{\sum f_i}}$ i.e., $\sigma = \sqrt{\dfrac{\sum f_i (x_i - \overline{x})^2}{N}}$

where $N = \sum f_i$.

If A denotes the assumed mean of the series then SD is given by

$$\sigma = \sqrt{\frac{\sum f_i d_i^2}{\sum f_i} - \left(\frac{\sum f_i d_i}{\sum f_i}\right)^2}$$

where $d_i = x_i - A$

where x_i denotes the discrete series.

In the step-deviation method the SD is computed by the formula

$$\sigma = \sqrt{\frac{\sum f_i d_i^2}{\sum f_i} - \left(\frac{\sum f_i d_i}{\sum f_i}\right)^2} \times C$$

where $d_i = \dfrac{x_i - A}{C}$, where C is the common factor.

3. *SD of continuous series*: If $\overline{x}$ is the actual mean of the series, then the SD is calculated by the formula

$$\sigma = \sqrt{\frac{\sum f_i d_i^2}{\sum f_i}}$$

where $d_i = x_i - \overline{x}$ (x_i denote the midpoints of the class intervals).
If A denotes the assumed mean of the series then SD is given by

$$\sigma = \sqrt{\frac{\sum f_i d_i^2}{\sum f_i} - \left(\frac{\sum f_i d_i}{\sum f_i}\right)^2}$$

where $d_i = x_i - A$
where x_i denotes the midpoints of the class intervals.
In the step-deviation method, the SD is computed by the formula

$$\sigma = \sqrt{\frac{\sum f_i d_i^2}{\sum f_i} - \left(\frac{\sum f_i d_i}{\sum f_i}\right)^2} \times C$$

where $d_i = \dfrac{x_i - A}{C}$, where C is the common factor which is usually the size of the class intervals.

2.18.1 Coefficient of Standard Deviation

Coefficient of SD is defined as the value of $\sigma \overline{x}$.

Variance: Variance is the square of SD. It is denoted by Var or V or σ^2

Coefficient of variance: Coefficient of variance is defined by the formula

$$\sigma \overline{x} \times 100$$

Standard error of mean (SE): The standard error of mean is defined as the SD of the sample divided by the square root of the number observations of the sample and given by $\sigma / \sqrt{N}$.

2.18.2 Merits of Standard Deviation

The SD is rigidly defined. It is based on all items. It possesses almost all the requisite characteristics of a good measure of dispersion. It is a stable measure since the measure is less affected by fluctuations in sampling. It is amenable to algebraic treatment.

2.18.3 Demerits of Standard Deviation

It is difficult to calculate SD. It is not easily understood. The measure gives more weight to extreme items. It is an absolute measure of variability; hence it cannot be used for the purpose of comparison.

2.18.3.1 Uses

SD is widely used in statistics. The measure helps in finding the standard error to determine the difference between means of two similar samples is by chance or real. SD is useful in assessing the degree of dispersion around its mean.

Example 2.42: Find the standard deviation (SD) of the data given below:

3. 2, 5, 3, 8, and 7.

Solution: We have $n = 5$

x	x^2
2	4
5	25
3	9
8	64
7	49
$\sum x_i = 25$	$\sum x_i^2 = 151$

$$\text{Arithmetic mean} = \overline{x} = \frac{\sum x_I}{n} = \frac{25}{5} = 5$$

$$\sigma^2 = \frac{\sum x_i^2}{n} - (\overline{x})^2 = \frac{151}{5} - 5^2 = 5.2$$

Standard deviation $= \sqrt{5.2} = 2.2803$

Example 2.43: Compute standard for the following data:

x	10	12	15	18	20
f	5	10	6	5	8

Solution: We have $N = \sum f_i = 34$

x_i	f_i	$f_i x_i$	$(x_i - \overline{x})$	$(x_i - \overline{x})^2$	$f_i(x_i - \overline{x})^2$
10	5	50	−5	25	125
12	10	120	−3	3	90
15	6	90	0	0	0

(Continued)

x_i	f_i	f_ix_i	$(x_i - \overline{x})$	$(x_i - \overline{x})^2$	$f_i(x_i - \overline{x})^2$
18	5	90	3	9	45
20	8	160	5	5	200
	$\sum f_i = 34$	$\sum f_ix_i = 510$			$\sum f_i(x_i - \overline{x}) = 460$

$$AM = \overline{X} = \frac{\sum f_ix_i}{34} = \frac{510}{34} = 15$$

$$\sigma = \frac{\sum f_i(x_i - \overline{x})^2}{N} = \sqrt{\frac{460}{34}} = 3.678$$

Example 2.44: Calculate the standard deviation (SD) of the following data:

Class intervals	95–105	105–115	115–125	125–135	135–145
Frequency	19	23	36	70	32

Solution: We have $C =$ length of class interval $= 10$

Let the assumed mean $= A = 130$, then taking $\frac{x_i - A}{C} = d_i$ we have the following table:

Class intervals	Frequency, f_i	Midpoints of class intervals, x_i	$\frac{x_i - A}{C} = d_i$	$f_i d_i$	$f_id_i^2$
95–105	19	100	−3	−57	171
105–115	23	110	−2	−46	92
115–125	36	120	−1	−36	36
125–135	70	130	0	0	0
135–145	52	140	1	52	52
	$N = \sum f_i = 200$			$\sum f_id_i = -87$	$\sum f_id_i^2 = 351$

$$\text{Standard deviation} = \sigma = \sqrt{\frac{\sum f_id_i^2}{\sum f_i} - \left(\frac{\sum f_id_i}{\sum f_i}\right)^2} \times C$$

$$= \sqrt{\frac{351}{200} - \left(\frac{-87}{200}\right)^2} \times 10 = 12.89$$

Example 2.45: Compute the coefficient of variance of the data of 20 items whose mean is 5.12 and standard deviation (SD) is 2.81.

Solution: We have $\overline{x} = 5.12$, Standard deviation $= \sigma = 2.81$

Compute the coefficient of variance $= \sigma\overline{x} \times 100 = \dfrac{2.81}{5.12} \times 100 = 54.88$

Example 2.46: Standard error of mean of the sample of 20 items whose standard deviation (SD) is 2.24.

Solution: We have $N = 20$, Standard deviation $= \sigma = 2.24$

Therefore, Standard error of mean $= \dfrac{\sigma}{\sqrt{N}} = \dfrac{2.24}{\sqrt{20}} = 0.501$

Exercise 2.7

1. Find the standard deviation (SD) of the data given below:
   9, 11, 12, 14, 15, 17, 20, 22
   Ans: 4.18
2. Find the mean respiration rate per minute and its standard deviation (SD) when the rate of four cases was found to be; 16, 13, 17, and 22.
   Ans: 17 and 3.2
3. Define SD and write the merits and demerits SD.
4. Define the term standard error of mean.
5. In two series of adults aged 25 years and children 7 months old, the following values were obtained for the height. Find which series show greater variations?

	Mean height (cm)	Standard deviation (cm)
Adults	160	10
Children	60	5

   Ans: Children's heights show greater variation.
6. A Researcher collects data on the weight and length of fishes and interested in finding out which of the characters is more variable. The data are as given below:

Fish	Mean	Standard deviation
Weight (g)	350	12
Length (in.)	16	1.5

   Ans: Fishes' length show greater variability.
7. Compute standard deviation (SD) from the following data:

x	10	20	30	40	50	60
f	8	12	20	10	7	3

   Ans: 13.45
8. Steel rods are manufactured to be 3 cm in diameter but they are acceptable if they are inside the limits 2.299 cm and 3.01 cm. It is observed that 5% are rejected as undersized. Assuming that the

diameters are normally distributed. Find the standard deviation (SD) of the distribution?

Ans: 0.0061

9. A scooterist purchased petrol at the rate of Rs. 24, 29.50, and 36.85 per liter during three successive years. Calculate the average price of petrol.

a. If he purchased 150, 180, and 195 liters of petrol in the respective years, and

b. If he spent Rs. 3850, 4675, and 5825 in the 3 years

Ans: (a) Rs. 30.66 (b) 30.09

10. In a 400 m athletic competition a participant covers the distance as given below:

Distance covered (m)	Speed (m/s)
First 89	10
Next 240	7.5
Last 80	10

Ans: 8.33 m

11. A man having to drive 90 km wishes to achieve an average speed of 30 km/h. For the first half of the journey he averages only 20 km/h. What must be his speed for the second half of the journey if his overall average is to be 30 km/h?

Ans: 60 km/h

CHAPTER 3

Probability

3.1 INTRODUCTION

This chapter is devoted to the study of probability, what it is and the basic concepts of the theory surrounding it. The theory of probability is the mathematical study of random phenomena. Many experiments do not yield exactly the same results when performed repeatedly. The experiments whose outcomes are associated with uncertainties are called Random experiments or Probabilistic experiments. The outcome of Random experiment cannot be predicted with certainty. The general meaning of the term probability is chance or possibility. The term is used in two senses viz. in qualitative sense and quantitative sense. Before going to deal with the theory of probability it is necessary to deal with the following basic concepts:

Experiment

An experiment is an act or process that leads to an outcome that cannot be predicted with certainty. Experiments are usually denoted by upper case letters of English alphabet.

Tossing a coin is an experiment.

Casting a die is an experiment.

A result of an experiment is called an *outcome*. Each experiment may yield one or more outcomes. When these are not predetermined but subject to chance we have *Random experiment*. Probability is concerned with random experiments.

Tossing a coin is a Random experiment.

Event

An event is the outcome of an experiment or trial. It is also known as Random event.

Simple Event

A simple event is an outcome of an experiment that cannot be split further. It is a single outcome of an experiment.

Getting a head or a tail when a coin is tossed is a simple event.

Statistical Techniques for Transportation Engineering.
DOI: http://dx.doi.org/10.1016/B978-0-12-811555-8.00003-9

Remark:

An event is a collection of simple events.

Getting an odd number when a die is rolled is an event. It can be decomposed into simple events (1, 3, and 5).

An event is denoted by the letters A, B, C, D, etc.

There are many approaches to probability theory. The modern approach to the probability theory is based on an axiomatic approach using the fundamental set theory.

Sample Space

The set of all possible outcomes of a random experiment is called a Sample Space. It is denoted by S.

The sample space of an experiment is the collection of all its simple events where each simple event is known as an elementary event. The total number of outcomes of an experiment may be finite, countable, or uncountable. For example consider an experiment in which three fair coins are tossed simultaneously. There are eight possible outcomes. If S denotes the sample space of the experiment, then

$$S = \{\mathrm{HHH}, \mathrm{HHT}, \mathrm{HTH}, \mathrm{THH}, \mathrm{HTT}, \mathrm{THT}, \mathrm{TTH}, \mathrm{TTT}\},$$
$$(\mathrm{H} = \mathrm{Head},\ \mathrm{T} = \mathrm{Tail})$$

The first letter in S denotes the outcome in first toss.

An event of an experiment is the subset of the sample S.

Consider an experiment that involves tossing a die. The sample space of the experiment is

$$S = \{1, 2, 3, 4, 5, 6\}$$

If A is an event of even score then $A = \{2, 4, 6\}$.

Compound Event

An event that is not simple is called a compound event.

Mutually Exclusive Events

Events are said to be mutually exclusive if the occurrence of one preludes the occurrence of the others, in other words two events A and B are mutually exclusive if there is no point common in between the points belonging to A and B, i.e., $A\ B = \phi$. If A and B are mutually exclusive then they are said to be incompatible.

There are three basic ways of classifying probability:

1. Classical approach
2. Relative frequency approach
3. Subjective approach

3.2 CLASSICAL PROBABILITY

Let n be the number of all possible outcomes of a Random experiment which are likely, of which let m be favorable to the occurrence of event E. Then, the probability of E denoted by $P(E)$ is defined as follows:

$$P(E) = \frac{\text{Number of outcomes favorable to the occurence of event } E}{\text{Total number of outcomes in the experiment}}$$

or

$$P(E) = \frac{\text{Number of outcomes favorable to event } E}{\text{Total number of outcomes in } S} = \frac{n(E)}{n(S)} = \frac{m}{n}$$

where S is the sample space of the experiment.

The above definition of probability is called the classical definition of probability. It is also called the mathematical definition of probability and the probability is known as a "prior probability."

Events that have an equal probability of occurrence are said to be equally *likely events*.

When a coin is tossed the events of getting a head and getting a tail are equally likely.

Mutually likely events may or may not be equally likely. Similarly mutually exclusive events may or may not be equally likely.

A set of events is said to be *exhaustive* if one of the events must occur compulsorily, i.e., the union of the events is the entire sample space S of the experiment.

3.2.1 Properties of Classical Probability

If E is an event of a sample space S, then

1. $P(E)$ is a pure number
2. $0 \leq P(E) \leq 1$
3. If E_1, E_2, ..., E_K are k mutually exclusive and exhaustive events in sample space S, then $\sum_{i=1}^{k} E_i = 1$, i.e., $P(S) = 1$
4. The probability of an impossible event is zero, i.e., $P(\phi) = 0$
5. If $E_1 \subseteq E_2$ then $P(E_1) \leq P(E_2)$

The advantage of the classical definition of probability is that, it can be applied without experimentation. This method fails when the number of possible outcomes is infinite and the outcomes are not equally likely.

3.2.2 Probability of Failure

Suppose an event E can happen in m ways out of total n possible equally likely outcomes. Then the number of times, the event E does not occur is $n - m$. The nonoccurrence of the event is denoted by $\overline{E}$ or by E^C.

The probability of nonoccurrence of the event is denoted by $P(\overline{E})$ or by $P(E^C)$ and is given by

$$P(\overline{E}) = \frac{n-m}{n}$$

We have

$$P(\overline{E}) = \frac{n-m}{n} = \frac{n}{n} - \frac{m}{n} = 1 - \frac{m}{n} = 1 - P(E)$$

Remarks:

1. The probability of the occurrence of the event E is called the probability of its success and is usually denoted by p.
2. The probability of nonoccurrence of the event E is called the probability of its failure and is usually denoted by q.
3. We have $P(E) + P(\overline{E}) = 1$
4. The complementary events E and $\overline{E}$ cannot occur simultaneously.

3.3 RELATIVE FREQUENCY APPROACH OF PROBABILITY

Let an experiment be conducted a larger number of times under essentially homogeneous and identical conditions. If n denotes the number of times the experiment is conducted out of which m are favorable to the happening of an event E, then the probability of the event E is

$$P(E) = \lim_{n \to \infty} \frac{m}{n}$$

This definition is known as the statistical or Empirical definition of probability. By introducing relative frequency approach, some of the deficiencies of the classical probability could be removed. Mathematical probability is determined without conducting the experiment, but the empirical probability is determined after the experiment conducted and the result is obtained.

Example 3.1: Two coins are tossed simultaneously. Find the probability of getting (1) One head and (2) At least one head?

Solution: The sample space of possible outcomes, when two coins are tossed simultaneously is

$$S = \{\text{HH}, \text{HT}, \text{TH}, \text{TT}\}$$

where HH denotes the case in which both coins turn up head, HT denotes the outcome, in which the first coin turns up head and second coin turns up tail, and so on.

Number of points in the sample space S is 4, i.e., $n = 4$.

1. Let A be the event in which exactly one head turns up. Then

$$A = \{\text{HT}, \text{TH}\}$$

Number of favorable cases is $n\ (A) = m = 2$.
Therefore

$$P(A) = \frac{m}{n} = \frac{2}{4} = \frac{1}{2} = 0.5$$

2. Let B be the event in which at least one head turns up. Then

$$B = \{\text{HH}, \text{HT}, \text{TH}\}$$

Number of favorable cases is $n\ (B) = m = 2$.
Therefore

$$P(B) = \frac{m}{n} = \frac{3}{4} = \frac{1}{2} = 0.75$$

Example 3.2: A group of five men and four women agree that three should be chosen by lot to form a special committee. What is the probability that the lot consists of one man and two women?

Solution: We have 5 men + 4 women = 9

The number of possible ways in which committee consisting of 3 is selected = $C\ (9, 3)$.

$${}^9C_3 = \frac{9 \times 8 \times 7}{1 \times 2 \times 3} = 84$$

If the committee is to include one man and two women, then the number of ways it can be done is

$$C(5,1)C(4,2) = {}^5C_1\,{}^4C_2 = \frac{5}{1} \times \frac{4 \times 3}{1 \times 2} = 30$$

$n = 84$, $m = 30$

Let E be the event of forming the committee, then

$$P(E) = \frac{m}{n} = \frac{30}{94} = \frac{5}{14}$$

Exercise 3.1

1. Define the terms
   a. Event
   b. Sample space
   c. Mutually exclusive events
2. Three coins are tossed. Find the probability of getting at least one head?
   ***Ans*:** 7/8
3. Find the probability of obtaining a score of 8 with 2 dice in a single throw?
   ***Ans*:** 5/36
4. From a set of 17 cards numbered 1, 2, 3, ..., 17, one card is drawn at random. Find the probability that its number is divisible by 3 or 7?
   ***Ans*:** 6/17
5. Three bulbs are chosen at random from 15 bulbs of which 5 are defective. Find the probability that none is defective?
   ***Ans*:** 24/91
6. Dialing a telephone number an old man forgets the last two digits remembering only that these are differently dialed at random. Find the probability that the number is dialed correctly.
   ***Ans*:** 1/90
7. A group of scientific men reported 11,705 sons and 11,527 daughters. If this is for sample from general population, what is the probability that child to be born will be a boy?
   ***Ans*:** 11,705/11,527
8. There are x white and y black balls in an urn. A ball is drawn from the urn and put aside. The ball is white. Then one more ball is drawn. It is white. Find the probability that the ball is also white?
   ***Ans*:** $(x-1)/(x+y-1)$
9. A child who cannot read plays with blocks labeled with the letters of the alphabet. He takes five blocks from which the word "table" is formed. Then he scatters them and puts them side by side in arbitrary fashion. Find the probability that he will again form the same word?
   ***Ans*:** 1/5!

10. There are n particles, each of which can occupy each of m $(m > n)$ cells with the same probability $1/m$. Find the probability that there will be one particle in each of n definite cells?

Ans: $n!/m^n$

3.4 SYMBOLIC NOTATION

Let S denote the sample space of an experiment. If A and B denote any two events of the experiment, then A and B are subsets of S.

We use the following symbolic notation:

1. The nonoccurrence of the event A is denoted by A^C or $\overline{A}$
2. $A \cup B$ is an event denoting the occurrence of "either A or B"
3. $A \cap B$ or AB denotes the occurrence of both A and B
4. If $A \cap B = \phi$ (i.e., A and B are mutually exclusive) then $A \cup B$ is denoted by $A + B$.

3.5 AXIOMATIC THEORY OF PROBABILITY

The axiomatic definition of probability includes both the classical and the statistical definitions as particular cases and overcomes the deficiencies of each of them. The axioms that define probability are given below:

Let S be a sample space of ε be the class of events and let P be real valued function defined on ε. Then P is called a probability function. $P(A)$ is called the probability of event A, if the following axioms hold:

P_1: For every event A: $0 \leq P(A) \leq 1$

P_2: $P(S) = 1$

P_3: If A and B are mutually exclusive events, then $P(A \cup B) = P(A) + P(B)$

P_4: If $A_1, A_2, \ldots$ is a sequence of mutually exclusive events, then $P(A_1 + A_2 + \cdots) = P(A_1) + P(A_2) + \cdots$

We have the following theorems:

Theorem 1:

If S is a sample space then $P(\phi) = 0$, where ϕ denoted an impossible event.

Theorem 2:

If $\overline{A}$ is the complement of A, then $P(\overline{A}) = 1 - P(A)$

Theorem 3:

If $A \subseteq B$, then $P(A) \leq P(B)$

Theorem 4:
If A and B are any two events of a sample space, then

$$P(A - B) = P(A) - P(A \cap B)$$

Proof:
A can be written as the union of two mutually exclusive events $A - B$ and $A \cap B$, as follows:

$$A = (A - B) \cup (A \cap B)$$

Therefore

$$P(A) = P(A - B) + P(A \cap B)$$

Hence

$$P(A - B) = P(A) - P(A \cap B)$$

Theorem 5: (*Addition Theorem on Probability for Two Events*):
If A and B are two arbitrary events in the sample space S, then

$$P(A \text{ or } B) = P(A \cup B) = P(A) + P(B) - P(A \cap B)$$

Proof:
$A \cup B$ can be decomposed into two mutually exclusive events $A - B$ and B as follows:

$$A \cup B = (A - B) \cup B$$

$$P(A \cup B) = P(A - B) + P(B)$$

$$= P(A) - P(A \cap B) + P(B)$$

Hence

$$P(A \cup B) = P(A) + P(B) - P(A \cap B)$$

Theorem 6:
If A and B are two mutually exclusive events in the sample space S, then

$$P(A \text{ or } B) = P(A \cup B) = P(A) + P(B)$$

Corollary:
For any events A, B, and C in the sample space

$$P(A \cup B \cup C) = P(A) + P(B) + P(C) - P(A \cap B) - P(B \cap C) - P(C \cap A) + P(A \cap B \cap C)$$

Theorem 7:
If A, B, and C are mutually exclusive events, then

$$P(A \cup B \cup C) = P(A) + P(B) + P(C)$$

Remarks:
The addition rule can easily be generalized to an arbitrary number of incompatible events as follows:

$$\text{If } A_i \cdot A_j = \phi \quad \text{for} \quad i \neq j, \text{ then}$$

$$P(A_1 + A_2 + \cdots + A_n) = P(A_1)\ldots P(A_n)$$

3.6 INDEPENDENT AND DEPENDENT EVENTS

Two events A and B are said to be independent if the result of the event B is not affected by the event A, when A has already occurred first. If A and B are independent, the probability of both events, occurring is the product of the probabilities of the individual events. We can define as follows:

Definition (Independent Events): Two events A and B are said to be independent if $P(A \cap B) = P(A) \cdot P(B)$.

Note:

1. If A and B are mutually exclusive, then $P(A \cap B) = 0$. In general mutually exclusive events can never be independent.
2. If the occurrence or nonoccurrence of event A does not affect the probability of occurrence of the event B, then A and B are called independent events, we write

$$P(B/A) = P(B)$$

If A and B are independent, then $\overline{A}$ and B are independent, A and $\overline{B}$ are independent and $\overline{A}$ are also independent.

3. A is certain event if $P(A) = 1$.

Definition (Dependent events): Two events A and B are said to be dependent if

$$P(A \cap B) = P(AB) \neq P(A) \cdot P(B)$$

Remark:

If A and B are dependent events, the probability of both events occurring is the product of the probability of the first event and the probability of the second event once the first event has occurred.

Example: Consider the experiment in which two cards are drawn in succession from a pack of cards. If the first card is not replaced, the probability of the first drawn card to be a queen is $4/52$.

When the first drawn card is not replaced, the number of cards remaining in the pack for the second card to be drawn is $52-1=51$. After the first draw, the probability that the second drawn card to be a king is $4/51$. Here the probability is affected, which is a case of dependent events.

3.7 CONDITIONAL PROBABILITY

Let A and B denote two events associated with a random experiment. Then $P(B/A)$ represents the conditional probability of the occurrence of the event B, given that A has already occurred. It is also called the probaility of occurrence of B relative to A.

In particular, if S is the finite equiprobable space A and B are two events of S, then the probability of the event B occurring given that A has already occurred is the conditional probability of B given by

$$P(B/A) = \frac{P(A \cap B)}{P(A)}$$

Similarly we can write the condition probability of A, when B has already occurred as

$$P(A/B) = \frac{P(B \cap A)}{P(B)}$$

3.8 MULTIPLICATION THEOREM ON PROBABILITY

Theorem 8:
The probability of the product of two events A and B is equal to the probability of one of them say, A multiplied by the conditional probability of the other, provided the first event A has occurred

$$P(A \cap B) = P(A) \cdot P(B/A)$$

If we take B as the first event, then

$$P(A \cap B) = P(B) \cdot P(A/B)$$

Proof:
From the definition, we have

$$P(B/A) = \frac{P(A \cap B)}{P(A)}$$

Hence, we get $P(A \cap B) = P(A) \cdot P(B/A)$

Similarly we can prove that $P(A \cap B) = P(B) \cdot P(A/B)$

The multiplication rule can be generalized to an arbitrary number of events and stated as follows:

$P(A_1\, A_2 \ldots A_n) = P(A_1)\; P(A_2/A_1) \ldots P(A_n/A_1\, A_2 \ldots A_{n-1})$

Remarks:
If A and B are dependent events then

$$P(A \cap B) = P(A) \cdot P(B/A)$$

If A and B are independent events then

$$P(A \cap B) = P(A) \cdot P(B)$$

Theorem 9:
If $A_1, A_2, \ldots, A_n$ are independent events, then

$$P(A_1 \cup A_2 \cup \ldots \cup A_n) = 1 - P(\overline{A}_1)P(\overline{A}_2)\ldots P(\overline{A}_n)$$

3.8.1 Solved Examples

Example 3.3: A bag contains 6 red and 5 black balls. A ball is drawn at random from the bag and without replacing it another ball is drawn from the bag. Find the probability that both the balls drawn are red?

Solution: Total number of balls in the bag = 6 red + 5 black = 11.

Let A denote the event of drawing the first red ball and B denote the event of drawing the second red ball, after the first ball is drawn.

Number of ways of drawing 1 ball from 11 balls $= {}^{11}C_1 = 11$

Number of ways of drawing the first red ball $= {}^{6}C_1 = 6$

$$\therefore \quad P(A) = \frac{6}{11}$$

Number of balls remaining in the bag after the first red ball is drawn, i.e., after the event A is $11-1 = 10$.

Number of ways of drawing 1 ball from 10 balls $= {}^{10}C_1 = 10$

Number of red balls remaining in the bag after the event A is $6-1=5$

Number ways of drawing the second red ball = 5

The events are dependent events, therefore $P(B/A) = 5/10$

Hence

$$P(A \cap B) = P(A) \cdot P(B/A) = \frac{6}{11} \cdot \frac{5}{11} = \frac{3}{11}$$

Example 3.4: If $P(A) = 0.4$, $P(A \cup B) = 0.7$ and A and B are independent events find $P(B)$.

Solution: The events A and B are independent, therefore $P(A \cap B) = P(A) \cdot P(B)$ and

$$P(A \cup B) = P(A) + P(B) - P(A \cap B)$$

We have

$$0.7 = P(A) + P(B) - P(A) \cdot P(B)$$

or

$$0.7 = P(A) + P(B) - P(B) \cdot (1 - P(A))$$

or

$$0.7 = 0.4 + P(B)[1 - 0.4]$$

or

$$0.3 = 0.6 \times P(B)$$

$$P(B) = \frac{0.3}{0.6} = 0.5$$

3.9 BAYE'S THEOREM

Let E_1, E_2, E_3, …, E_n be n mutually and exhaustive events with nonzero probability of a Random experiment. If A is any arbitrary event of the sample space of the above experiment, then

$$P(E_i/A) = \frac{P(E_i)P(A/E_i)}{\sum_{j=1}^{n} P(E_j)P(A/E_j)}$$

Proof:
Let S be the sample space of the random experiment. Then

$$S = E_1 \cup E_2 \cup E_3 \cup \ldots \cup E_n$$

Now

$$A = S \cap A = (E_1 \cup E_2 \cup E_3 \cup \ldots \cup E_n) \cap A$$
$$= (E_1 \cap A) \cup (E_2 \cap A) \cup \ldots \cup (E_n \cap A)$$

Therefore we get $P(A) = P(E_1)P(A/E_1) + P(E_2)P(A/E_2) + \cdots + P(E_n)P(A/E_n)$

$$= \sum_{j=1}^{n} P(E_j)P(A/E_j) \tag{3.1}$$

(By multiplication theorem)
Then for any event E_i $P(E_i/A)$

$$= \frac{P(A \cap E_i)}{P(A)}$$
$$= \frac{P(E_i \cap A)}{P(A)}$$
$$= \frac{P(E_i)P(A/E_i)}{\sum_{j=1}^{n} P(E_j)P(A/E_j)} \text{ (using Eq. 3.1)}$$

Remarks:
The unconditional probability $P(E_1)$, $P(E_2)$, …, (E_n) in the Baye's theorem are called "a priori" or prior probabilities.

The unconditional probability $P(E_1/A), P(E_2/A), \ldots, P(E_n/A)$ in the theorem are called "a posterior" or inverse probabilities and $P(A) = \sum_{j=1}^{n} P(E_j)P(A/E_j)$ is known as the total probability formula.

Baye's theorem makes it possible to revise the probabilities of the hypothesis regarding the result of the experiment.

Example 3.5: In a bolt factory, machines A, B, and C manufacture respectively 25%, 35%, and 40% of the total production. Of their output 5%, 4%, and 2% are defective bolts. A bolt is drawn at random from the produce and is found to be defective. What are the probabilities that it was manufactured by machine A?

Solution: Let E_1, E_2, and E_3 denote the respective events in which the machines A, B, and C manufacture the bolts and D denote event of drawing the defective bolt.

We have

$$P(E_1) = 0.2, \quad P(E_2) = 0.35, \quad P(E_3) = 0.40$$

and

$$P(D/E_1) = 0.05, \quad P(D/E_2) = 0.04, \quad P(D/E_3) = 0.02$$

The probability in which the defective bolt is manufactured by machine A

$$P(E_1/D) = P(E_1)P(D/E_1)$$

$$P(E_1/D) = \frac{P(E_1)P(D/E_1)}{P(E_1)P(D/E_1)}$$

$$+ P(E_2)P(D/E_2)$$

$$+ P(E_3)P(D/E_{32})$$

$$= \frac{0.25 \times 0.05}{0.25 \times 0.05 + 0.35 \times 0.04 + 0.04 \times 0.02}$$

$$= 25/69$$

Exercise 3.2

1. A chord is selected at random on a fixed circle. What is the probability that its length exceeds the radius of the circle?

   Ans: 0.866
2. The probability that a communication system will have high fidelity is 0.81 and the probability that it will have high fidelity and high

selectivity is 0.18. Find the probability that a system with high fidelity will also have high selectivity?

Ans: 2/9

3. In an office 30% of employees have scooters and 25% have cars. Among those who have scooters, 90% do not have cars. What is the probability that an employee has a car given that he does not have scooter?

 Ans: 0.3142

4. In a parking lot there are 12 places arranged in a row. A man observed that there were 8 cars parked and the 4 empty spaces were adjacent to each other. Find the probability of this event?

 Ans: Probability of the event is 0.182.

5. The digits 1, 2, 7 are written in a random order such that 7! arrangements are equally likely to give 7 digited numbers. Find the probability that the number is divisible by (1) by 2 and (2) by 4?

 Ans: (1) $3 \times 6!/7!$; (2) $6!/7!$

6. A noisy communication channel is transmitting a message, which is a sequence of 0s and 1s, a1 being sent with probability p. A transmitted symbol may or may not be perturbed into the opposite symbol in the process of transmission. It is known that a1 is converted into 0 (zero) with probability p_1 and a0 is converted into 1 with probability p_2 during process of transmission. Find (1) the conditional probability that a1 was sent given that a 0 (zero) is received at the reception and (2) the conditional probability, that a 0 was sent given that a1 is received at the reception.

 Ans: (1) $P(A/B)=\dfrac{PP_1}{PP+(1-P)(1-P_2)}$ (2) $P(B/A)=\dfrac{(1-P)P_2}{(1-P)P_2)+(P_1-P_2)}$

7. Solar water heaters manufactured by a company consists of two parts, the heating panel and the insulated tank. It is found that 6% of the heaters produced by the company has defective heating panels and 8% have defective tanks. Find the percentage of nondefective heaters produced by the company?

 Ans: 86%

8. Assume that a factory has two machines. Past records show that machine I produces 20% of the items and machine II produces 80% of the items. Further 6% of the items produced by machine I are defective and only 1% of the items produced by the machine II were

defective. If a defective item is drawn at random, what is the probability that it was produced by machine I?

Ans: 0.6

9. Three machines A, B, C produce respectively 60%, 30%, 10% of the total number of items of a factory. The percentage of respective outputs of these machines are respectively 2%, 3%, and 4%. An item is selected at random and is found to be defective. Find the probability that the item was produced by machine C.

   Ans: 0.16

10. A company manufactures scooters at two plants. Plant A produces 80% and plant B manufactures 20% of the total production. 85% of the scooters produced at plant A are of standard quality, while 65% of the scooters produced at plant B are of standard quality. A scooter produced by the company is selected at random and is found to be of standard quality. What is the probability that it is manufactured at plant A?

    Ans: 68/81

11. A car manufacturing company has two plants. Plant A manufactures 70% of cars and plant B manufactures 30% of the total production. At plant A, 80% of the cars are rated to be of standard quality and at plant B, 90% of cars manufactured are of standard quality. A car manufactured by the company is picked up at random and is found to be of standard quality. What is the probability that it has come from plant A?

    Ans: 56/83

12. Police plan to enforce speed limits by using radar traps at four different locations within the city limits. The radar traps at each of the locations $L_1, L_2, L_3, \text{and} L_4$ are operated 40%, 30%, 20%, and 10% of the time and if a person is speeding on his way to work has probabilities 0.2, 0.1, 0.5, and 0.2, respectively, of passing through these locations. What is the probability that he/she will receive a speeding ticket?

    Ans: 0.23

13. In a railway reservation office, two clerks are engaged in checking reservation forms. On an average, the first clerk checks 55% of the forms while the second does the remaining. The first clerk has an error rate of 0.03 and the second clerk has an error rate of 0.02. A reservation form is selected at random from the total number of forms

checked during a day, and is found to have an error. Find the probabilities that it was checked by first and second clerk, respectively?

Ans: 11/17, 6/17

14. An oil exploration firm finds that 55 of the test cells it drills yield a deposit of natural gas. If it drills 6 wells, find the probability that at least one well yields gas?

Ans: $1-(0.95)^6$ at least

15. The following table gives the age of cars of a certain make and actual maintenance costs. Obtain the regression equation for costs related to age. Also estimate the maintenance cost for a 10-year-old car?

Age of car (years)	2	4	6	8
Maintenance cost (Rs. in hundred)	10	20	25	30

Ans: $y =$ Rs. 37.50

16. A contractor is interested in knowing whether relationship does exist between the number of building permits issued and the volume of sales of such buildings in some past years. He collects data about sales (*Y*, in thousands of rupees) and the number of building permits issued (*X*, in hundreds) in the past 10 years. The results worked out are as under:

$\sum x = 117, \sum 78, \sum xy = 981, \sum x^2 = 1491, \sum y^2 = 662$

a. What level of sales can you expect next year if it is hoped that 200 building permits would be issued?

b. What change in sales is likely to take place with an increase of 100 building permits?

Ans: (a) $Y = 0.56\ X + 1.28,\ X = 20 \Rightarrow Y = 12.448$, (b) Rs. 560

17. In an automobile factory, certain parts are to be fixed to the chassis in a section before it moves to another section. On a given day, one of the three persons A, B, or C carries out this task. A has 45%, B has 35%, and C has 20% chance of doing it. The probabilities that A, B, or C will take more than the allotted time are (1/16), (1/10), and (1/20), respectively. If it is found that one of them has taken more time, what is the probability that A has taken more time.

Ans: 5/13

18. Daily demand for transistors is having the following probability distribution:

Demand (x)	1	2	3	4	5	6
Probability (p)	0.10	0.15	0.20	0.25	0.18	0.12

Determine the expected daily demand for transistors. Also obtain the variance of the demand.

Ans: 3.52, 2.22

19. In a city seven accidents occur per week. What is the probability there will be at least one day in a week without accidents

Ans: $\frac{10_{c_5}}{10^5}$

20. What is the chance that a 4-digited number on the car license plate chosen at random in a city (1) consists of different digits, (2) includes only two pairs of identical digits, (3) contains only three identical digits, (4) includes two pairs of identical digits, and (5) consists of identical digits?

Ans: (1) $10_{c_4}a$, (2) $4_{c_2}10_{c_3}a$, (3) $90\ 4_{c_2}a$, (4) $360a$, and (5) $10a$ where $a=10^{-4}$

21. The trains R-passenger and C-express arrive at B-junction on their way to Kolkata between 8:50 a.m. and 9:15 a.m. and each of the trains wait for the other for almost 10 minutes. Supposing both the trains arrive at junction at random moments between 8:50 a.m. and 9.15 a.m. What is the probability that one person traveling by the passenger on an urgent business will be able to get express train at the junction to reach Kolkata in time?

Ans: 9/25

22. A group of four radar units scan a region in which there are five targets T_1, T_2, T_3, T_4, and T_5.The region is scanned for a time t. During this time interval, each unit independent of other units may detect each target with probability p_i. Find the expected number of targets that will be detected?

Ans: $\sum_{1}^{5}\left[1-(1-p_i)^4\right]$

23. A traffic light has a constant probability $\lambda\, dt$ of changing from red light to a green light or from a green light to a red light in an infinitesimal interval of length dt. Prove that a car arriving at a random instant has a probability 1/2 of passing through without waiting and a probability element $\frac{\lambda}{2}e^{-\lambda\omega}$ of waiting for a time ω before passing.

Ans: $\frac{\lambda}{2}e^{-\lambda\omega}\, dt$

24. Alpha particles are emitted by radioactive source at an average rate of 5 in 20 minutes interval using Poisson. Find the probability that there will be (1) two emissions, in and (2) at least emissions, in a particular 20 minute interval

Ans: 0.0842, 0.9596

25. Given that 2% of the fuses manufactured by a firm are defective. Find the probability that a box containing 200 fuses (1) has at least one of defective fuse, and (2) have three or more defective fuses?

Ans: 0.982, 0.762

26. If the number of accidents occurring on a highway each day is a Poisson variate with mean equal to 3, what is the probability that no accidents occur today?

Ans: 0.05

27. A certain screw making machine produces on an average 2 defective screws out of 100, and packs them in boxes of 500. Find the probability that a box contains 15 defective screws?

Ans: 0.035

28. The number of accidents in a year to auto drivers in a city is a Poisson variate with mean equal to 3. Out of 1000 such drivers, find approximately the number of drivers with (1) no accidents and (2) more than three accidents, in a year?

Ans: (1) 50 and (2) 353

29. If the probability that a target is destroyed on any one shot is 1/3. What is the probability that it would be destroyed in the third shot and not before?

Ans: 0.148

30. An expert shooter can hit a target 95% of the time. Find the probability that he will hit the target continuously 14 times and will miss it at the 15th attempt?

Ans: 0.2438

31. In a certain city, the probability that rain occurs on a day during June is 5/8. What is the probability that there is rain on June 5th and not earlier?

Ans: 0.0124

32. Alpha particles are emitted by a radioactive source at an average rate of 5 in a 20-minute interval. Use Poisson distribution to find the probability that there will be (1) two emissions and, (2) at least two emissions in a particular 20 minute interval.

Ans: (1) 0.082, (2) 0.9596

33. On a certain city transport route, buses ply every 30 minutes between 6:00 a.m. and 10:00 a.m. If a person reaches a bus stop on this route at a random time during this period, what is the probability that he/she will have to wait for at least 20 minutes.

***Ans*:** 1/3

34. The duration of a telephone conversation has been found to have an exponential distribution with mean 3 minutes. Find the probability that the conversation may last (1) more than 1 minute, and (2) less than 3 minutes.

***Ans*:** (1)$e^{-1/3}$ and (2) $1 - e^{-1}$

35. The number of road accidents per day in a certain city as a Gamma variate with an average of 6 and variance 18. Find the probability that there will be (1) more than 8 accidents and (2) between 5 and 8 accidents, on a particular day?

***Ans*:** (1) $\frac{1}{9} \times e^{-x/3}$ and (2) 0.199

36. The daily sales of a certain brand of bicycles in a city in excess of 1000 piece is distributed as the Gamma distribution with the parameter $\alpha = 2$ and $\beta = 500$. The city has a daily stock of 1500 pieces of the brand. Find the probability that the stock is insufficient on a particular day?

***Ans*:** 0.736

37. The time-lag between the arrivals of two successive city buses at a certain bus stop is a Gamma variate with mean 20 and variance 200. If a person has waited for 15 minutes at the bus stop, find the probability that he will have to wait at least 15 minutes more before a bus arrives.

***Ans*:** $\frac{8}{5} \times e^{-1.5}$

38. The number of accidents per day were studied for 144 days in city A and for 100 days in city B. The mean number of accidents and standard deviation were found to be 4.5 and 1.2, respectively for city A and, 5.4 and 1.5 for city B. Is city B more prone to accidents than city A?

***Ans*:** City B is more prone to accidents than city A.

39. One type of aircraft is found to develop engine trouble in 5 flights out of a total of 100 flights and in another type 7 flights out of a total of 200 flights developed engine trouble. Is there a significant difference between these two types of aircrafts so far as engine defects are concerned?

***Ans*:** No

40. Steel rods are manufactured to be 3 cm in diameter but they are acceptable if they are inside the limits 2.299 cm and 3.01 cm. It is observed that 5% are rejected as undersized. Assuming that the diameters are normally distributed. Find the standard deviation of the distribution?

***Ans*:** 0.0061

CHAPTER 4

Random Variables

4.1 INTRODUCTION

In this chapter we introduce random variables that are defined on sample spaces associated with experiments in which the outcomes are uncertain. The term random variable is one of the basic concepts of the probability theory. The values of a random variable are real numbers associated with the outcome of an experiment and the concept of the random variables is the one of the most significant in the theory of probability. Random variable is a variable whose value is subject to variations due to chance. Random variables vary from trial to trial as the experiment is repeated. It is also called stochastic variable and is defined as follows:

Definition 4.1:
A random variable is numerical valued variable defined on a sample space of an experiment.

We shall denote random variable by the capital letters X, Y, Z, etc., and the corresponding small (lower case) letters are used to denote the numerical values taken by the random variable.

Example 4.1: Consider an experiment A in which two coins are tossed simultaneously. Let S be the sample space associated with the experiment. We have,

$$S = \{\text{HH, HT, TH, TT}\}$$

If we define a random variable X, as the number of heads, then the values of X are 0, 1, and 2 corresponding to the outcomes.

$$\text{TT(0 head), HT(1 head), TH(1 head), HH(2 heads).}$$

We have the following table:

Sample point	TT	HT	TH	HH
$X = x_i$	0	1	1	2

Statistical Techniques for Transportation Engineering.
DOI: http://dx.doi.org/10.1016/B978-0-12-811555-8.00004-0

Example 4.2: The number of calls from subscribers at the telephone exchange during a definite time period is a random variable.

Example 4.3: An experiment is to fire four shots at a target. The random variable is the number of hits.

In the above experiment we have observed that each outcome is a simple event of the sample space S and the corresponding value is a real number.

Thus "A random variable is a rule that assigns one and only one numerical value to each simple event of an experiment" and we have the following definition:

Definition 4.2:

Let S be a sample space of a random experiment and R denote the set of real numbers. Then a real-valued function $X: S \rightarrow R$ is called a random variable.

The set of values which X takes is called the *Spectrum* of the random variable.

If S is a sample space of an experiment then we can define more than one random variable on S. The set $R_X = \{X(x){:}x \in S\}$ is called the range of X.

In general a real-valued function, whose domain is the sample space S of an experiment and whose range set a collection of n—tuples of real numbers, is called dimensional random variable.

If $X(x) \neq 0$ for all $X \in S$ then and $\frac{1}{X(x)}$ $|X(x)|$ are also random variables on S.

If X and Y are random variables on the sample space of an experiment then $X + Y$, $X - Y$, and XY are also random variables on S.

For all real numbers a and b, $a\,X + b\,Y$ is also a random variable on S.

The mathematical function describing the possible values of a random variable and their associated probabilities is known as a probability distribution.

Random variables can be discrete, i.e., taking any of a specified finite or countable list of values, and hence with a probability mass function (pmf) as probability distribution.

Random variables can be continuous, taking any numerical value in an interval or collection of intervals, and with a probability density function (pdf) describing the probability distribution, or a mixture of both types. The realizations of a random variable, i.e., the results of randomly choosing values according to the variable's probability distribution, are called random variables.

4.2 DISCRETE RANDOM VARIABLE

In statistics we study variables such as heights of students, number of defective bolts, number of accidents on a road, number of male children in a family, number of printing mistakes in each page of a book, and so on. Some of these quantities can vary only by finite increments, by observable jumps in value. Such values are called discrete random variables. The sample space S of a discrete random variable contains either finite number of outcomes, which are countably infinite.

Definition 4.3:
A random variable that can take only finite number of values are countably infinite number of values is called a discrete random variable.

Example 4.1: In an experiment of drawing four cards from a pack of cards, the random variable "The number of kings drawn" is a discrete random variable.

Example 4.2: Consider an experiment in which a coin is tossed four times. If X denotes the number of heads obtained then X is a discrete random variable.

Remarks:
The sample space S for discrete random variable can be discrete, continuous, or it may contain both discrete and continuous points.

Discrete random variables are the random variables whose range is finite or continuously infinite.

4.3 PROBABILITY DISTRIBUTION FOR A DISCRETE RANDOM VARIABLE

The simplest distribution for a discrete random variable X is an ordered series, which is a table whose top row contains all the values of the random variables and the bottom row contains the corresponding probabilities as shown in the following table:

$X=x_i$	x_1	x_2	...	x_i	...
$P(X=x_i)$	p_1	p_2	...	p_i	...

$x_1, x_2, \ldots, x_n$ are the values of the random variable X and $p_i = P(X = X_i)$ are the corresponding probabilities where $\sum_1^n p_i = 1$. The above distribution is an ordered series. It can also be represented graphically as a frequency polygon or a histogram.

4.3.1 Probability Mass Function

Definition 4.4:

Let X be a discrete random variable assuming the values $x = x_1$, $x = x_2, \ldots, x = x_n$ corresponding to the various outcomes of a random experiment. If the probability of occurrence of $X = x_i$, $(I = 1, 2, \ldots, n)$ is $P(X = x_i) = p_i$

Such that

1. $P(X = x_i) = p_i \geq 0$ for all i $(1 \leq i \leq n)$
2. $\sum_{i=1}^{n} P(X = x_i) = \sum_{i=1}^{n} p_i = 1$

Then the function $P(X)$ is called the probability function of the random variable X and the set

$$\{(x_1,\ P(X = x_1)),\ (x_2,\ P(X = x_2)), \ldots, (x_n,\ P(X = x_n))\} \text{ or simply}$$

$$\{(x,\ p(x_1)),\ (x_2,\ P(x_2)),\ \ldots,\ (x_n,\ P(x_n))\}$$

is called the probability distribution of the random variable X.

The probability function $P(X = x)$ is also denoted by $f(x)$ and is called the pmf of the discrete random variable X.

Remark:

Some authors refer $f(x)$ as the frequency function or a probability function.

We have

1. $f(x) = P(X = x)$
2. $f(x) \geq 0$
3. $\sum_x f(x) = 1$

Definition 4.5:

(Finite equiprobable space) A finite probability distribution where each point $X = x_i$ has the same probability for all i, is called a finite equiprobable space or uniform space.

4.3.2 Distribution Function

The probability $P(X \leq x)$ is the probability of the event$(X \leq x)$. It is a function of x. The function μ_r is denoted by $F(x)$ and is called the cumulative probability distribution function of the random variable x.

Thus $F(x) = P(X \leq x)$, $\quad (-\infty < x < \infty)$

$F(x)$ is often called the distribution function of X.

$F(x)$ possesses the following properties:

1. $F(-\infty)=0$
2. $F(\infty)=1$
3. $0 \leq F(x) \leq 1$
4. $F(x_i) \leq F(x_2)$ if $x_1 < x_2$
5. $P(x_1 < x < x_2) = F(x_2) - F(x_i)$
6. $F(x^+) = F(x)$

$F(X)$ can also be defined as follows:

Definition 4.6:

Let X be a discrete random variable, then the function

$$F(X) = P(X \leq x) = \sum_{1 \leq x} f(t), \qquad -\infty < x < \infty$$

where $f(t)$ is the value of the probability distribution is called distribution function of X.

4.3.3 Additional Properties of Distribution Function

Property 1: (Interval property) If X is a random variable and $F(x)$ is the distribution function of X, then $P(a < X \leq b) = F(b) - F(a)$

Proof: The events $a < X \leq b$ and $X \leq a$ are disjoint and we have

$$P(a < x \leq b) \cup (x \leq a) = P(x \leq b)$$

Hence

$$P(a < x \leq b) \cup (x \leq a) = P(x \leq b)$$

or

$$P(a < x \leq b) = (x \leq b) - P(x \leq a)$$
$$F(b) - F(a)$$

Property 2: If $F(x)$ is the distribution function of a random variable x, then

$$P(a \leq x \leq b) = P(x = a) + F(b) - F(a).$$

Proof: We have

$$P(a \leq x \leq b) = P(x = a) + P(a < x \leq b)$$
$$P(x = a) + F(b) - F(a) \quad \text{(by property(1))}.$$

Property 3: If $F(x)$ is the distribution function of a random variable X, then

$$P(a < x < b) = F(b) - F(a) - P(x = b).$$

Proof:

$$P(a < x < b) = P(a < x \leq b) - P(x = b)$$
$$F(b) - F(a) - P(x = b) \quad \text{(by property(1))}.$$

Property 4: $P(a \leq x < b) = F(b) - F(a) - P(x = b) + P(x = a)$
Proof: We have

$$P(a \leq x < b) = P(a < x < b) + P(x = a)$$
$$F(b) - F(a) - P(x = b) + P(x = a).$$

Property 5: (Monotone increasing property) If $F(x)$ is the distribution of a random variable X and $a < b$, then $F(a) \leq F(b)$.

Since

$$P(a < X \leq b) = 0$$

We have

$$F(b) - F(a) \geq 0$$

or

$$F(b) \geq F(a)$$

i.e.,

$$F(a) \leq F(b),$$

hence proved.

If the random variable X takes the values $x_1, x_2, \ldots, x_n$ with probabilities $p_1, p_2, \ldots, p_n$ respectively, then we have

$$P[X < x_1] = 0$$
$$P[X \leq x_1] = P[X < x_1] + P[X = x_1] = p_1$$

Therefore

$$P[X < x_2] = p_1$$
$$P[X \leq x_2] = P[X < x_2] = P[X = x_2] = p_1 + p_2,$$
$$\ldots$$
$$P[X \leq x_n] = p_1 + p_2 + \cdots + p_n$$

Example: Consider an experiment in which four coins are tossed. If we define a random variable as the number of heads obtained, then we have

$$P(X=0)=\frac{1}{16},\quad P(X=1)=\frac{4}{16}=\frac{1}{4},\quad P(X=2)=\frac{6}{16}=\frac{3}{8}$$

$$P(X=3)=\frac{4}{16}=\frac{1}{4}=\frac{1}{4},\quad P(X=4)=\frac{1}{16}$$

Therefore we get

$$F(0)=P(X=0)=\frac{1}{16}$$

$$F(1)=P(X=0)=P(X=1)=\frac{1}{16}+\frac{1}{4}=\frac{5}{16}$$

$$F(2)=P(X=0)=P(X=1)+P(X=2)=\frac{1}{16}+\frac{1}{4}=\frac{3}{8}=\frac{11}{16}$$

Hence the distribution function $F(x)$ is given by

$$F(x)=\begin{cases} 0, & \text{for} \quad X<0 \\ \frac{1}{16}, & \text{for} \quad 0\le X<1 \\ \frac{5}{16}, & \text{for} \quad 1\le X<2 \\ \frac{11}{16}, & \text{for} \quad 2\le X<3 \\ \frac{15}{16}, & \text{for} \quad 3\le X<4 \\ 1, & \text{for} \quad X\ge 4 \end{cases}$$

4.4 MEAN AND VARIANCE OF A DISCRETE DISTRIBUTION

If X is a discrete random variable, then the mean and variance of the discrete distribution can be defined as follows:

$$\text{Mean}=\mu=\sum_{i=1}^{n} x_i P(X=x_i)=\sum_{i=1}^{m} x_i p_i$$

$$\text{Variance}=Var[x]=\sum_{i=1}^{n}(x_i-\mu)^2 P(X=x_i)$$

or

$$\sigma^2=\sum_{i=1}^{n}(x_i-\mu)^2 p_i$$

where σ is the standard deviation (SD).

Since

$$\sum p_i = 1, \quad \text{we get}$$

$$\begin{aligned}
\mathrm{Var}[x] &= \sum_{i=1}^{n}(x_i-\mu)^2 p_i \\
&= \sum_{i=1}^{n}(x_i^2+\mu^2-2x_i\mu)p_i \\
&= \sum_{i=1}^{n}(x_i^2)p_i + \sum_{i=1}^{n}(\mu^2)p_i - \sum_{i=1}^{n}(2x_i\mu)p_i \\
&= \sum_{i=1}^{n}(x_i^2)p_i + \mu^2\sum_{i=1}^{n}p_i - 2\mu\sum_{i=1}^{n}x_ip_i \\
&= \sum_{i=1}^{n}(x_i^2)p_i + \mu^2 - 2\mu\mu \\
&= \sum_{i=1}^{n}(x_i^2)p_i - \mu^2
\end{aligned}$$

Thus

$$\sigma^2 = \sum_{i=1}^{n}(x_i^2)p_i - \mu^2$$

Example 4.4: Find the probability distribution of the number of blue balls drawn when 3 balls are drawn without replacement from a bag containing 4 blue and 6 red balls?

Solution: Number of balls in the bag = 4 blue + 6 red = 10

Let X be the random variable "number of blue balls"

Then X can take values 0, 1, 2, and 3.

$$P(X=0) = P(\text{no blue ball}) = P(3 \text{ red balls}) = \frac{{}^{6}C_3}{{}^{10}C_3} = \frac{1}{6}$$

$$P(X=1) = P(1 \text{ blue} + 2 \text{ red balls}) = \frac{{}^{4}C_1 \times {}^{6}C_2}{{}^{10}C_3} = \frac{1}{2}$$

$$P(X=2) = P(2 \text{ blue} + 1 \text{ red ball}) = \frac{{}^{4}C_2 \times {}^{6}C_1}{{}^{10}C_3} = \frac{3}{10}$$

$$P(X=3) = P(3 \text{ blue balls}) = \frac{{}^{4}C_3}{{}^{10}C_3} = \frac{1}{30}$$

The probability distribution is

x	0	1	2	3
$P(X=x)$	$\frac{1}{6}$	$\frac{1}{2}$	$\frac{3}{10}$	$\frac{1}{30}$

Example 4.5: Two cards are drawn successively with replacement from a well shuffled pack of cards. Find the probability distribution of the number of kings that can be drawn?

Solution: Let X denote the random variable "number of kings."

Since the number of cards drawn is 2, the random variable X can take values 0, 1, and 2

Number of kings in a pack = 4

Number of cards in a pack = 52

When one card is drawn from the pack, the probability of getting a king is

$$= \frac{{}^{4}C_1}{{}^{52}C_1} = \frac{4}{52} = \frac{1}{13}$$

Probability of failure, i.e., the probability of not getting a king is

$$1 - \frac{1}{13} = \frac{12}{13}$$

Hence we get

$$P(X=0) = P(\text{no king}) = \frac{12}{13} \cdot \frac{12}{13} = \frac{144}{169}$$

$$P(X=1) = P(\text{one king}) = \frac{1}{13} \cdot \frac{12}{13} + \frac{12}{13} \cdot \frac{1}{13} = \frac{24}{169}$$

$$P(X=2) = P(2\text{ kings}) = \frac{1}{13} \cdot \frac{1}{13} + \frac{1}{169}$$

The required probability distribution is

x	0	1	2
$P(X=x)$	$\frac{144}{169}$	$\frac{24}{169}$	$\frac{1}{169}$

Example 4.6: A random variable X has the following probability distribution:

$X=x_i$	0	1	2	3	4
$P(X=x_i)$	3k	3k	k	2k	6k

Find

1. k
2. Mean
3. $P(X > 2)$

Solution:

1. We have

$$\sum p_i = 1$$

Therefore

$3k + 3k + 3 + 2k + 6k = 1$

or

$15k = 1$

or

$$k = \frac{1}{15}$$

2. Mean $\mu = \sum_i x_i p_i$

$$= (0)(3k) + (1)(3k) + (2)(k) + (3)(2k) + (4)(6k)$$

$$= 35k = 35\left(\frac{1}{15}\right) = \frac{7}{3}$$

3. $P(X > 2) = P(X > 3) + P(X = 4)$

$$= 2k + 6k = 8k = 8\left(\frac{1}{15}\right) = \frac{8}{15}$$

Example 4.7: A random variable X has the following probability distribution:

$X = x_i$	−2	−1	0	1	2	3
$P(X = x_i)$	0.1	k	0.2	2k	0.3	k

Find

1. k
2. Mean
3. Variance of X

Solution:

1. We have

$$\sum p_i = 1$$

i.e., $0.1 + k + 0.2 + 2k + 0.3 + k = 1$
or $4k + 0.6 = 1$
or $4k = 1 - 0.6 = 0.4$
we get $k = 0.1$

2. Mean $= \mu = \sum_i x_i p_i$

$$= (-2)(0.1) + (-1)^2 k + (0)(0.2) + (1)(2k) + (2)(0.3) + (3)(k)$$
$$= 4k + 0.4 = 4(0.1) + 0.4 = 0.8$$

3. Variance $\sigma^2 = \sum_{i=1}^{n} (x_i^2) p_i = \mu^2$

$$= (0.1)(-2)^2 + (-1)^2 k + (0.2)(0)^2 + (2k)(1)^2 + (0.3)(2)^2 + (k)(3)^2 - (0.8)^2$$
$$= 0.4 + k + 0 + 2k + 1.2 + 9k - 0.64$$
$$= 12k + 1.6 - 0.64$$
$$= 12(0.1) + 1.6 - 0.64$$
$$= 2.16$$

Exercise 4.1

1. Define
  a. Random variable
  b. Discrete random variable
  c. Density function
  d. Probability distribution
  e. Spectrum

2. Obtain the probability distribution of the total number of heads in three tosses of a coin.

Ans:

x	0	1	2	3
$P(x)$	$\frac{1}{8}$	$\frac{3}{8}$	$\frac{3}{8}$	$\frac{1}{8}$

3. A fair coin is tossed 4 times. Find the probability distribution of the number of heads?

Ans:

x	0	1	2	3	4
$P(x)$	$\frac{1}{16}$	$\frac{1}{4}$	$\frac{3}{8}$	$\frac{1}{4}$	$\frac{1}{16}$

4. A random variable X has the following distribution:

X	0	1	2	3	4
$P(X)$	2λ	3λ	λ	2λ	6λ

Find

a. $P(0 < x < 2)$

b. $P(x > 2)$

***Ans*:** (a) $\frac{3}{14}$; (b) $\frac{4}{7}$

5. A bag contains 2 white, 3 red, and 4 blue balls. Two balls are drawn at random from the bag. If the random variable X denotes the "number of white balls" among the balls drawn, describe the probability distribution of X.

***Ans*:**

x	0	1	2
$P(X = x)$	$\frac{7}{12}$	$\frac{7}{18}$	$\frac{7}{36}$

6. Three balls are drawn without replacement from a bag containing 5 white and 4 red balls. Find the probability distribution of the number of red balls drawn?

***Ans*:**

x	0	1	2	3
$P(x)$	$\frac{5}{42}$	$\frac{10}{21}$	$\frac{5}{14}$	$\frac{1}{21}$

7. The probability distribution of a random variable X is given below:

X	0	1	2
$P(X)$	$3\lambda^2$	$4\lambda - 10\lambda^2$	$5\lambda - 1$

where > 0.

Find

a. λ

b. $P(X \leq 1)$

c. $P(X > 0)$

***Ans*:** (a) $\lambda = \frac{1}{3}$; (b) $P(X \leq 1) = \frac{1}{3}$; (c) $P(X > 0) = \frac{8}{9}$

8. A random variable X has the following probability distribution:

X	1	2	3	4
$P(X)$	k	$2k$	$3k$	$4k$

Find

a. k
b. $P(X<3)$
c. $P(X\geq 3)$
d. Mean

Ans: (a) $k=\frac{1}{10}$; (b) $\frac{3}{10}$; (c) $\frac{7}{10}$; (d) $\mu=3$

9. A random variable X has the following probability distribution:

$X=x_i$	-2	-1	0	1	2	3
$P(X=x_i)$	0.1	k	0.2	2k	0.3	k

Find

a. k
b. Mean
c. Variance

Ans: (a) $k=0$; (b) $\mu=0.8$; (c) $\sigma^2=2.8$

10. A box contains 6 tickets. Two of the tickets carry a prize of Rs. 5/- each, the other four tickets carry a prize of Rs. 1/- each. If one ticket is drawn what is the mean value of the prize?

Ans: $\frac{7}{3}$

11. A die is tossed twice. Getting a "number greater than 4" is considered a success. For the probability distribution of the number successes. Show that the mean is 2/3 and the variance is 4/9.

12. Obtain the probability distribution of the number of sixes in two tosses of a cubical die.

Ans:

x	0	1	2
$P(X=x)$	$\frac{25}{36}$	$\frac{10}{36}$	$\frac{1}{36}$

4.5 CONTINUOUS RANDOM VARIABLE

Let X be a random variable. If X takes noncountable infinite number of values, then X is called a continuous random variable. If X is a continuous random variable then the range of X is an interval on real line.

Example: The length of an electric bulb, the detection range of a radar, etc., are examples of continuous random variables.

4.6 PROBABILITY DENSITY FUNCTION

Definition 4.7:
Let X be a continuous random variable. If for every X in the range of X, we assign a real number $f(x)$ satisfying the conditions

1. $f(x) \geq 0$, for $-\infty < x < \infty$

2. $\int_{-\infty}^{\infty} f(x)\mathrm{d}x = 1$

The function $f(x)$ is called a pdf of X. It is also referred to as density function.

If $f(x)$ is a pdf of X then we have

$P(a \leq x \leq b) = \int_a^b f(x)\mathrm{d}x$, for any real constants a and b with $a \leq b$.

The set of values obtained from $\int_a^b f(x)\mathrm{d}x$ for various possible intervals is called a continuous probability distribution for X.

4.7 CUMULATIVE DISTRIBUTION FUNCTION

Definition 4.8:
If X is a continuous random variable having $f(x)$ as its pdf, then the function given by

$$F(X) = P(X \leq x) = \int_{-\infty}^{\infty} f(x)\mathrm{d}x, \quad \text{for} \quad -\infty < x < \infty$$

is called cumulative distribution function of X.

$F(x)$ is also referred to as the distribution function or the cumulative distribution of X.

Remark:
If the distribution function $f(x)$ of a random variable X is continuous for every x such that F′(x) exists everywhere, except may be, at individual particular points then the random variable X is said to be continuous.

The pdf of a continuous random variable X is the derivative of the distribution function, i.e., $f(x) = F'(x)$.

If a and b are real numbers with $a \leq b$, then we have $P(a \leq X|X|b) = F(b) - F(a)$.

Definition 4.9:
(Mixed random variable) Let X be a random variable and $F(x)$ be the distribution function of X. If $F(x)$ continuously increases on certain intervals but has discontinuities at particular points then the random variable is said to be mixed.

Examples of Continuous and Mixed Random Variables

Example 4.8: A random variable X has a Simpson distribution on the interval $-c$ to c. Find the expression for the pdf?

Solution: The random variable X obeys "the law of an isosceles triangle." The pdf is

$$f(x) = \frac{1}{c}\left(1 - \frac{x}{c}\right) \quad \text{for } 0 < x < c$$

$$= \frac{1}{c}\left(1 + \frac{x}{c}\right) \quad \text{for } -c < x < 0$$

$$= 0 \quad \text{otherwise}$$

i.e., $x < c$ or $x > c$

Example 4.9: A random variable X has Laplace transform defined by $f(x) = ac^{-\lambda|x|}$, where λ is a positive parameter. Find a?

Solution: Since the pdf is $f(x) = ae^{-\lambda|x|}$,

$$a\int_{-\infty}^{\infty} e^{-\lambda|x|}dx = 1$$

We have

$$8^2 \cdot \frac{1}{8} + 12^2 \cdot \frac{1}{6} + 16^2 \cdot \frac{3}{8} + 20^2 \cdot \frac{1}{4} + 24^2 \frac{1}{12} - 16^2$$

$\therefore$

$$\frac{16}{3}\left[\frac{1}{x}\right]_2^{\infty} = \frac{16}{3}\left[0 - \frac{1}{2}\right] = \frac{8}{3}$$

$$x_n = (-1)^n 2^n n^{-1} \sum x_i p_i E[x] = \sum_{i=1}^{\infty} 2^{-n}(-1)^n 2^n n^{-1} = \sum_{i=1}^{\infty} \frac{(-1)^n}{n}$$

$$\Rightarrow a = \frac{\lambda}{2}$$

4.8 MEAN AND VARIANCE OF A CONTINUOUS RANDOM VARIABLE

Let X be a continuous random variable. If $f(x)$ is the pdf of X, then the mean and variance of X are given as

$$\text{Arithmetic mean} = \mu = \int_{-\infty}^{\infty} xf(x)\mathrm{d}x$$

$$\text{Variance} = \mathrm{Var}[X] = V[x] = \sigma^2 = \int_{-\infty}^{\infty} x^2 f(x)\mathrm{d}x - \mu^2$$

4.8.1 Solved Examples

Example 4.10: If $f(x) = \begin{cases} \dfrac{x}{6} + k, & 0 \le x \le 3 \\ 0, & \text{elsewhere} \end{cases}$

is a pdf, find the value of k. Also find $P(1 \le x \le 2)$?

Solution: $f(x)$ is a pdf. Therefore we have

$$\int_{-\infty}^{\infty} f(x)\mathrm{d}x = 1$$

or

$$\int_{-\infty}^{\infty} \left(\frac{x}{6} + k\right)\mathrm{d}x = 1$$

or

$$\int_{-\infty}^{\infty} \left(\frac{x}{6} + k\right)\mathrm{d}x + \int_{0}^{3} \left(\frac{x}{6} + k\right)\mathrm{d}x + \int_{3}^{\infty} \left(\frac{x}{6} + k\right)\mathrm{d}x = 1$$

or

$$0 + \left[\frac{x^2}{12} + kx\right]_1^3 + 0 = 1$$

or

$$\frac{9}{12} + 3k = 1 \quad \text{or} \quad 3k = 1 - \frac{3}{4} = \frac{1}{4}$$

or

$$k = \frac{1}{12}$$

Now

$$P(1 \leq x \leq 2) = \int_1^2 f(x)\mathrm{d}x = \int_1^2 \left(\frac{x}{6} + \frac{1}{12}\right)\mathrm{d}x$$

$$\left[\frac{x^2}{12} + \frac{x}{12}\right]_1^2 = \left(\frac{4}{12} + \frac{2}{12}\right) - \left(\frac{1}{12} + \frac{1}{12}\right) = \frac{1}{3} + \frac{1}{6} - \frac{1}{6} = \frac{1}{3}$$

Example 4.11: Given the cumulative distribution function

$$f(x) = \begin{cases} 0, \ldots, x > 0 \\ x^2, \ldots, 0 \leq x \leq 1 \\ 1, \ldots, x > 1 \end{cases}$$

1. Find the pdf?
2. Find $P(0.5 < x \leq 0.75)$?

Solution:

1. Since $f(x) = \dfrac{\mathrm{d}}{\mathrm{d}x}(F(x))$

We get $f(x) = \begin{cases} 0, & x > 0 \\ 2x, & 0 \leq x \leq 1 \\ 1, & x > 1 \end{cases}$

or simply $f(x) = \begin{cases} 2x, & 0 \leq x \leq 1 \\ 0, & \text{otherwise} \end{cases}$

2. $$\begin{aligned} P(0.5 < x \leq 0.75) &= P(X \leq 0.75) - P(X \leq 0.5) \\ &= F(0.75) - F(0.5) \\ &= (0.75)^2 - (0.5)^2 \\ &= 0.5625 - 0.25 = 0.3125 \end{aligned}$$

Example 4.12: Find k such that $f(x)$ is a pdf of a continuous random variable X where $f(x)$ is defined as follows:

$$f(x) = \begin{cases} kxe^{-x}, & 0 < x < 1 \\ 0, & \text{otherwise} \end{cases}$$

Also find the mean.

Solution:

1. Since $f(x)$ is a pdf we have $\int_{-\infty}^{\infty} f(x)\mathrm{d}x = 1$

i.e., $\displaystyle\int_0^1 kxe^{-x}dx = 1$

or $k\left[\dfrac{xe^{-x}}{(-1)} - \dfrac{e^{-x}}{(-1)^2}\right] = 1$

or $k[(e^{-1} - 0) - (e^{-1} - e^0)] = 1$

$$\text{or } k\left(1-\frac{2}{e}\right)=1$$

$$\text{or } k=\frac{e}{e-2}$$

2. Mean $\mu=\int_{-\infty}^{\infty} xf(x)\mathrm{d}x$

$$=\int_0^1 x\left[\frac{e}{e-2}xe^{-x}\right]\mathrm{d}x$$

$$=\frac{e}{e-2}\int_0^1 x^2e^{-x}\mathrm{d}x$$

$$=\frac{e}{e-2}\left[\frac{x^2e^{-x}}{-1}-2x\frac{e^{-x}}{(-1)^2}+2\frac{e^{-x}}{(-1)^3}\right]_0^1$$

$$=\frac{e}{e-2}\left[-e^{-1}-2e^{-1}(e^{-1}-1)\right]$$

$$=\frac{e}{e-2}\left[-\frac{1}{e}-\frac{2}{e}-\frac{2}{e}+2\right]$$

$$=\frac{e}{e-2}\left[-\frac{5}{e}+2\right]=\frac{2e-5}{e-2}$$

Example 4.13: The distribution of a random variable given by

$$F(x)=\begin{cases}1-(1+x)e^{-x}, & \text{for } x\geq 0\\ 0, & \text{for } x<0\end{cases}$$

Find the corresponding density function of random variable X?

Solution: We have

$$F(x)=\frac{\mathrm{d}}{\mathrm{d}x}(F(x))$$

$$=\frac{\mathrm{d}}{\mathrm{d}x}(1-(1+x)e^{-x})$$

$$=\frac{\mathrm{d}}{\mathrm{d}x}(1-e^{-x}-xe^{-x})$$

$$=0-(-e^{-x})-[x(-e^{-x})+e^{-x}]$$

$$=e^{-x}+xe^{-x}-e^{-x}=xe^{-x}$$

Hence $f(x)=\begin{cases}xe^{-x}, & x\geq 0\\ 0, & \text{otherwise}\end{cases}$ is the required density function.

Example 4.14: Find the cumulative distribution function for the following probability function of a random variable X?

$$f(x) = \begin{cases} xe^{-x}, & x \geq 0 \\ 0, & \text{otherwise} \end{cases}$$

Solution: Since

$$f(x) = \frac{\mathrm{d}}{\mathrm{d}x}(F(x))$$

We have

$$F(x) = \int_{-\infty}^{x} f(x)\mathrm{d}x = \int_{-\infty}^{0} f(x)\mathrm{d}x + \int_{0}^{x} f(x)\mathrm{d}x$$

$$= 0 + 6\int_{-\infty}^{x} (x - x^2)\mathrm{d}x = 6\left[\frac{x^2}{2} - \frac{x^3}{3}\right]_0^x$$

$$= 6\left[\frac{3x^2 - 2x^3}{6}\right] = 3x^2 - 2x^3$$

Hence

$$F(x) = 3x^2 - 2x^3, \quad 0 \leq x \leq 1$$

Example 4.15: For the following density function:

$$F(x) = ae^{-|x|} \qquad -\infty \leq x \leq \infty$$

of a random variable X. Find 1. a, 2. Mean, 3. Variance

Solution:

1. $f(x)$ is a pdf we have,

Therefore $\int\limits_{-\infty}^{\infty} f(x)\mathrm{d}x = 1$

or $\int\limits_{-\infty}^{\infty} ae^{-|x|}\mathrm{d}x = 1$

or $2a\int\limits_{0}^{\infty} e^{-|x|}\mathrm{d}x = 1$ (since $e^{-|x|}$ is an even function.)

or $2a|-(0-1)| = 1$

Hence $a = \frac{1}{2}$

2. Mean $= \int_{-\infty}^{\infty} f(x)dx$

i.e., $\mu = \int_{-\infty}^{\infty} x\left(\frac{1}{2}e^{-|x|}\right)dx = \frac{1}{2}\int_{-\infty}^{\infty} xe^{-|x|}dx = 0$ (since $x\,e^{-|x|}$ is odd)

3. $$\text{Variance} = \int_{-\infty}^{\infty} x^2 f(x)dx - \mu^2$$

$$= \frac{1}{2}\int_{-\infty}^{\infty} x^2 e^{-|x|}dx - \theta^2$$

$$= \frac{1}{2}\cdot 2\int_{0}^{\infty} x^2 e^{-|x|}dx - \theta^2 \quad \text{(since } x^2 e^{-|x|} \text{ is odd)}$$

$$= \int_{0}^{\infty} x^2 e^{-|x|}dx - \theta^2$$

$$= [x^2(-e^{-x}) - 2xe^{-x} - 2e^{-x}]_0^{\infty}$$

$$= 0 - 0 - 2(0 - 1) = 2$$

Exercise 4.2

1. A continuous random variable X has a pdf

$$f(x) = 3x^2, \quad 0 < x \le 1$$
$$= 0, \quad \text{otherwise.}$$

Find a and b, such that $P(x \le a) = P(x > a)$ and $P(x > 4) = 0.05$?

Ans: $a = \left(\frac{1}{2}\right)^{1/3}$ $\quad b = \left(\frac{19}{20}\right)^{1/3}$

2. A random variable X has the pdf

$$f(x) = 2x, \quad 0 < x < 1$$
$$= 0, \quad \text{otherwise}$$

Find

a. $P\left(x < \frac{1}{2}\right)$

b. $P\left(\frac{1}{4} < x < \frac{1}{2}\right)$

Ans: (a) $\frac{1}{4}$; (b) $\frac{3}{16}$

3. Find the cumulative distribution function for the following probability distribution function of a random variable x?

$$f(x) = \frac{x}{4}e^{-x/2}, \quad 0 < x \le \infty$$
$$= 0, \quad \text{otherwise}$$

Ans: $f(x) = 1 - e^{-x/2} - \frac{x}{2}e^{-x/2}, \quad 0 < x < \infty$

4. If the pdf of a random variable is given by

$$f(x) = k(1 - x^2), \quad 0 < x < 1$$
$$= 0, \quad \text{otherwise}$$

Find

a. k

b. The distribution function of the random variable.

Ans: (a) $\frac{3}{2}$; (b) 1

5. If X has the pdf

$$f(x) = ke^{-3x}, \quad x > 0$$
$$= 0, \quad \text{otherwise}$$

Find k and $P(0.5 \leq x \leq 1)$

Ans: 1, 0.173

6. A continuous random variable X that can assume any value between $x = 2$ and $x = 5$ has a density function given by

$$f(x) = k(1 + x)$$

Find $P(x < 4)$?

Ans: $\frac{2}{27}, \ \frac{16}{27}$

7. Find the pdf for the random variable whose distribution function is given by

$$F(x) = 0, \quad \text{for } x \leq 0$$
$$= x, \quad \text{for } 0 < x < 1$$
$$= 1, \quad \text{for } x \geq 1$$

Ans:
$$f(x) = 1, \quad 0 < x < 1$$
$$= 0, \quad \text{otherwise}$$

8. A continuous random variable X has the following pdf:

$$f(x) = \frac{1}{2}, \quad -1 \leq x \leq 0$$

$$= \frac{1}{4}(2 - x), \quad 0 < x < 2$$

$$= 0, \quad \text{elsewhere}$$

Obtain the distribution of X.

$$f(x) = \frac{x+1}{2}, \qquad -1 \le x \le 0$$

Ans:

$$= \frac{1}{8}(4 + 4x + x^2), \quad 0 < x < 2$$

$$= 1, \qquad \text{if } x \ge 2$$

9. A random process gives measurements x between 0 and 1 with a pdf

$$f(x) = 12x^3 - 21x^2 + 10x, \quad 0 \le x \le 1$$
$$= 0, \qquad \text{elsewhere}$$

Find

a. $P\left(X \le \frac{1}{2}\right)$

b. $P\left(X > \frac{1}{2}\right)$

c. a number k such that $P(X \le k) = \frac{1}{2}$

Ans: (a) $\frac{9}{16}$; (b) $\frac{7}{16}$; (c) $\frac{1}{2}$

10. Let X be a continuous random variable with pdf

$$f(x) = ax, \qquad 0 \le x \le 1$$
$$= a, \qquad 1 \le x \le 2$$
$$= -ax + 3a, \quad 2 \le x \le 3$$
$$= 0, \qquad \text{elsewhere}$$

a. Find the value of a

b. Find $P(X < 1.5)$

Ans: $a = \frac{1}{2}$, $P(X < 1.5) = \frac{1}{2}$

Note: If X is continuous random variable and $f(x)$ is the pdf of X defined in the interval (a, b) then

a. The median M of the distribution is obtained by solving $\int_a^M f(x)dx = \frac{1}{2}$ and $= \int_M^\infty f(x)dx = \frac{1}{2}$, for M

b. The mode of the distribution is the value of x for which $f'(x) = 0$ and $f'(x) < 0$ $(a < x < b)$ hold, i.e., $f(x)$ is maximum.

c. Mean deviation about the mean is given by

$$\text{Mean deviation} = \int_a^b |x - \mu| f(x)dx$$

d. Harmonic mean $= \int_a^b \frac{1}{x} f(x)dx$

e. Geometric mean $= \log = G = \int_a^b \log x\, f(x)dx$, where G is the Geometric mean.

4.9 JOINT DISTRIBUTIONS

In the preceding sections we have discussed one-dimensional random variable as a real value function defined over a sample space of an experiment. The distributions of one-dimensional random variables are known as univariate distributions. In this section we shall be concerned with bivariate distributions. If X and Y are random variables defined over a sample space of experiment, then (X, Y) is a two-dimensional random variable or a bivariate random variable. A system of two random variables X and Y can be geometrically interpreted as a random point (X, Y) on the xy-plane (see the figure given below)

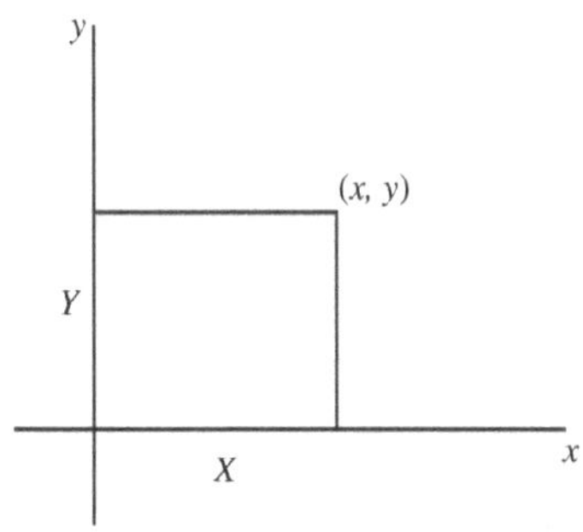

If the number of possible values (X, Y) is finite, then (X, Y) is called a two-dimensional discrete random variable. If R_X denotes the range space of X, and R_Y is the ranges space of Y, then the Cartesian product of R_X and R_Y is the range space of (X, Y). If (X, Y) takes all the values on a region R, of xy-plane the $n(X, Y)$ is called a two-dimensional continuous random variable.

If X and Y are discrete random variables the n $P(X = x, Y = y)$ is the probability of intersection of the events $X = x$ and $Y = y$. Similarly we can define the probability of continuous random variable. It is preferable to express the probability by the means of a function with the values $f(x, y) = P(X = x, Y = y)$ for any pair of values (X, Y) within the range of the random variables X and Y.

4.9.1 Joint Probability Function

Definition 4.10:

(Joint probability function) If X and Y are discrete random variables, n the function $f(x_i, y_i) = P(X = x_i, Y = y_i) = P_{ij}$ is called the joint probability function for the discrete random variables X and Y if and only if $f(x_i, y_i)$ satisfies the following conditions:

1. $f(x_i, y_i) = P_{ij} \geq 0$, for all i, j
2. $\sum_i \sum_i p_{ij} = 1$

The joint probability functions of the discrete random variables is also known as the joint pmfs of X and Y.

4.9.2 Joint Probability Distribution of Discrete Random Variables

Definition 4.11:

(Joint probability function) If X and Y are discrete random variables the n the set of triples $\{x_i, y_j, p_{ij}\}$, $i = 1, 2, 3, \ldots, n$, $j = 1, 2, \ldots, m$ is called the joint probability distribution of X and Y.

The joint probability distribution can be represented in the form of a table as shown below:

$X\backslash Y$	y_1	y_2	.	y_j	.	y_m	$P(x_i)$
x_1	p_{11}	p_{12}	...	p_{1j}	...	p_{1m}	p_1
x_2	p_{21}	p_{22}	....	p_{2j}	...	p_{2m}	p_2
$\vdots$	$\vdots$						
x_i	p_{i1}	p_{i2}	...	P_{ij}	...	P_{im}	p_i
$\vdots$	$\vdots$						
x_n	P_{n1}	p_{n2}	...	P_{nj}	...	P_{nm}	p_n
$P(y_j)$	$p_{\cdot 1}$	$p_{\cdot 2}$	...	$P_{\cdot j}$	...	$P_{\cdot m}$	1

4.9.3 Marginal Probability Function of a Discrete Random Variables

Definition 4.12:

(Marginal probability function) If $P(X = x_i, \; Y = y_i) = P_{ij}$ is the joint probability distribution of two discrete random variables X and Y the marginal probability function of X is given by

$$P(X = x_i) = p_{i\cdot} = p_{i1} + p_{i2} + \cdots + P_{ij} + \cdots$$
$$= \sum_i p_{ij} = p_i$$

The marginal probability function of X is also denoted by $f(x)$. The set $\{x_i, p_{i\cdot}\}$, is called the marginal distribution of X.

Similarly the marginal probability function of Y is given by

$$P(Y = y_i) = p_{.j} = \sum_i p_{ij} = p_{1j} + p_{2j} + \cdots$$

The marginal probability function of Y is also denoted by $f(y)$. The set $\{y_j, p_{.j}\}$, is called the marginal distribution of Y.

4.9.4 Joint Distributive Function of Discrete Random Variables

Definition 4.13:
If X and Y are discrete random variables the function given by

$$F(x, y) = P(X \leq x, Y \leq y) = \sum_{x \leq x} \sum_{1 \leq y} f(s, t) \quad \text{for} \quad -\infty < x < \infty,$$

$$-\infty < t < \infty$$

where $F(x, y)$ is the value of the joint probability distribution of X and Y at (x, y) is called the joint distribution function (Fig. 4.1).

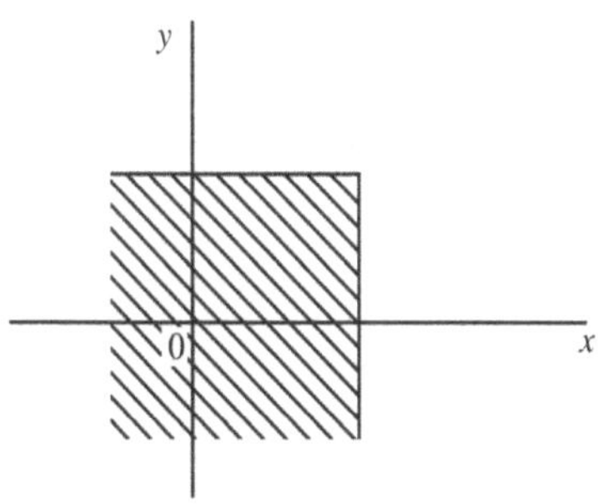

Figure 4.1 Joint distribution function.

$F(x, y)$ can be interpreted geometrically as the probability of a random point (X, Y) falling in a quadrant whose vertex is (x, y).

The joint distribution function $F(x, y)$ processes the following properties:

1. $F(\infty, y) = F(x, \infty) = 0$
2. $F(-\infty, -\infty) = 0$
3. $F(\infty, \infty) = 1$
4. $P(a_1 < X \leq b_1, a_2 < Y \leq b_2) = F(b_1, b_2) + F(a_1, b_2) - F(b_1, a_2)$

Example 4.16: For the following joint distribution of (*X*, *Y*). Find (1) F(1, 1), (2) F(1, 3)?

x\y	1	2	3	4
0	$\frac{1}{24}$	$\frac{1}{12}$	$\frac{1}{12}$	$\frac{1}{24}$
1	$\frac{1}{12}$	$\frac{1}{6}$	$\frac{1}{6}$	$\frac{1}{12}$
2	$\frac{1}{24}$	$\frac{1}{12}$	$\frac{1}{12}$	$\frac{1}{24}$

Solution:

1. We have

$$F(1,\ 1) = P(X \le 1,\ Y \le 1)$$

$$= P(0,\ 1) + P(1,\ 1)$$

$$= \frac{1}{24} + \frac{1}{12} + \frac{1+3}{24} = \frac{1}{8}$$

2. $F(1,\ 3) = P(0,\ 1) + P(0,\ 2) + P(0,\ 3) + P(1,\ 1) + P(1,\ 2) + P(1,\ 3)$

$$= \frac{1}{24} + \frac{1}{12} + \frac{1}{12} = \frac{1}{12} + \frac{1}{6} + \frac{1}{6} = \frac{15}{24} = \frac{5}{8}$$

Example 4.17: From the following table for bivariate distribution. Find

1. $P(X \le 1)$

2. $P(Y \le 3)$

X\Y	1	2	3	4	5	6
0	0	0	$\frac{1}{32}$	$\frac{2}{32}$	$\frac{2}{32}$	$\frac{3}{32}$
1	$\frac{1}{16}$	$\frac{1}{16}$	$\frac{1}{8}$	$\frac{1}{8}$	$\frac{1}{8}$	$\frac{1}{8}$
2	$\frac{1}{32}$	$\frac{1}{32}$	$\frac{1}{64}$	$\frac{1}{64}$	0	$\frac{2}{64}$

Solution:

1. $P(X \leq 1) = P(X = 0) + P(X = 1)$

$$= \left(0 + 0 + \frac{1}{32} + \frac{2}{32} + \frac{2}{32} + \frac{3}{32}\right) + \left(\frac{1}{16} + \frac{1}{16} + \frac{1}{8} + \frac{1}{8} + \frac{1}{8}\right)$$

$$= \frac{8}{32} + \frac{10}{16} = \frac{1}{4} + \frac{5}{8} = \frac{7}{8}$$

2. $P(Y \leq 3) \quad = P(Y = 1) + P(Y = 2) + P(Y = 3)$

$$= \left[0 + \frac{1}{16} + \frac{1}{32}\right] + \left[0 + \frac{1}{16} + \frac{1}{32}\right] + \left[\frac{1}{32} + \frac{1}{8} + \frac{1}{64}\right]$$

$$= \frac{3}{32} + \frac{3}{32} + \frac{1}{64} = \frac{6 + 6 + 1}{64} = \frac{13}{64}$$

4.10 CONDITIONAL PROBABILITY DISTRIBUTION

If $P(X = x_i,\ Y = y_j)$ is the probability function of discrete random variable X and Y, then the conditional probability function of X given $Y = y_j$ is defined as

$$P(X = x_i/Y = y_j) = \frac{P(X = x_1,\ Y = y_j)}{P(Y = y_i)}$$

And the conditional probability function of Y, given $X = x_i$ is denoted by

$$P(Y = y_i/X = x_j) = \frac{P(X = x_1,\ Y = y_j)}{P(Y = x_i)}$$

The conditional probability function of X given Y, and the conditional probability function of Y given X are also denoted by $f(x/y)$ and $f(y/x)$, respectively.

The conditional probability distributions are univariate probability distribution.

4.11 INDEPENDENT RANDOM VARIABLES

If $P(X = x,\ Y = y) = f(x,\ y)$ is the joint pdf of discrete random variables X and Y and marginal pdfs $f_1(x)$ and $f_2(y)$ such that $P(X = x,\ Y = y) = f_1(x)\,f_2(y)$.

Then X and Y are said to be independent random variables.

Example 4.18: The two-dimensional random variables (X, Y) has the joint density function

$$f(x, y) = \frac{x+2y}{27}, \quad x = 0, 1, 2; \; y = 0, 1, 2;$$

Find the conditional distribution of Y for $X = x$. Also find the conditional distribution of X given $Y = 1$?

Solution: The joint probability distribution of X and Y in the tabular form is:

X\Y	1	2	3
0	0	$\frac{1}{27}$	$\frac{2}{27}$
1	$\frac{2}{27}$	$\frac{3}{27}$	$\frac{4}{27}$
2	$\frac{4}{27}$	$\frac{5}{27}$	$\frac{6}{27}$

We have $P(X = x) = f(x)$.

The conditional probability distribution of Y for $X = x$ is $f(y/x) = \frac{f(x, y)}{f(x)}$ where $f(x, y)$ is the joint probability distribution of X and Y. Hence we get

$$f(y = 0)(x = 0) = \frac{f(X = x, y)}{f(x)} = \frac{f(0, 0)}{f(x = 0)} = 0$$

$$f(y = 1)(x = 0) = \frac{f(0, 1)}{f(x = 0)} = \frac{1/27}{3/27} = \frac{1}{3} \quad \left(\text{Since } f(x = 0) = \frac{3}{27}\right)$$

$$f(y = 2)(x = 0) = \frac{f(0, 2)}{f(x = 0)} = \frac{2/27}{3/27} = \frac{2}{3}$$

We have

$$f(x = 1) = \frac{2}{27} + \frac{3}{27} + \frac{4}{27} = \frac{9}{27}$$

Hence

$$f(y = 0)(x = 1) = \frac{f(0, 1)}{f(x = 1)} = \frac{2/27}{9/27} = \frac{2}{9}$$

$$f(y = 1)(x = 1) = \frac{f(1, 1)}{f(x = 1)} = \frac{3/27}{9/27} = \frac{3}{9}$$

$$f(y = 2)(x = 1) = \frac{f(2, 1)}{f(x = 1)} = \frac{4/27}{9/27} = \frac{4}{9}$$

Since

$$f(x=2)=\frac{4}{27}+\frac{5}{27}+\frac{6}{27}=\frac{15}{27}$$

We get

$$f(y=0)(x=2) = \frac{f(2,0)}{f(x=2)} = \frac{4/27}{15/27} = \frac{4}{15}$$

$$f(y=1)(x=2) = \frac{f(2,1)}{f(x=2)} = \frac{5/27}{15/27} = \frac{5}{15}$$

$$f(y=2)(x=2) = \frac{f(2,2)}{f(x=2)} = \frac{6/27}{15/27} = \frac{6}{15}$$

Therefore the conditional distribution of Y for $X=x$ is

X\Y	1	2	3
0	0	$\frac{1}{3}$	$\frac{2}{3}$
1	$\frac{2}{9}$	$\frac{3}{9}$	$\frac{4}{9}$
2	$\frac{4}{15}$	$\frac{5}{15}$	$\frac{6}{15}$

From the table, the conditional distribution of X given $Y=1$ is

x\y	$\frac{f(x,y)}{f(y=1)}$
0	$\frac{1}{9}$
1	$\frac{3}{9}$
2	$\frac{5}{29}$

4.12 JOINT PROBABILITY FUNCTION OF CONTINUOUS RANDOM VARIABLES

Definition 4.14:

If x and y are two continuous random variables, then the function $f(x, y)$ given by

$$p(a_1 \le x \le b_1,\ a_2 \le y \le b_2) = \int_{a_1}^{b_1}\int_{a_2}^{b_2} f(x,\ y)\mathrm{d}y\mathrm{d}x$$

is called the joint probability function of X and Y, if and only if it satisfies the following conditions:

1. $f(x, y) \geq 0$, for $-\infty < x < \infty,\ -\infty < y < \infty$
2. $\int_{-\infty}^{\infty} \int_{-\infty}^{\infty} f(x, y)\mathrm{d}y\mathrm{d}x = 1$

The joint probability function of X and Y is also called the joint pdf of the continuous random variables X and Y.

4.13 JOINT PROBABILITY DISTRIBUTION FUNCTION OF CONTINUOUS RANDOM VARIABLES

Definition 4.15:

If X and Y are continuous random variables, the function given by

$$\begin{aligned} f(x, y) &= P(X \leq x,\ Y \leq y) \\ &= \int_{-\infty}^{y} \int_{-\infty}^{x} f(s, t)\mathrm{d}s\mathrm{d}t \quad \text{for} \quad -\infty < x < \infty,\ -\infty < y < \infty \end{aligned}$$

where $f(s, t)$ is the value of the joint pdf of X and Y at (s, t) is called the joint probability distribution function or joint distribution function of X and Y.

We have $f(x, y) = \frac{\partial^2}{\partial x \partial y} F(x, y)$

Hence the joint distribution function is obtained by integrating the joint probability function, i.e.,

$$f(x, y) = \int_{-\infty}^{y} \int_{-\infty}^{x} f(t_1, t_2)\mathrm{d}t_1\mathrm{d}t_2$$

4.14 MARGINAL DISTRIBUTION FUNCTION

Let X and Y be continuous random variables. If $f(x, y)$ is the joint distribution function of the random variables X and Y, then the marginal distribution function of X denoted by $F(x)$ is given by

$$F(x) = \int_{-\infty}^{x} \left(\int_{-\infty}^{\infty} f(x, y)\mathrm{d}y \right) \mathrm{d}x$$

And the marginal distribution function of Y denoted by $F(Y)$ is given by

$$F(Y) = \int_{-\infty}^{y} \left(\int_{-\infty}^{\infty} f(x, y)\mathrm{d}x \right) \mathrm{d}y$$

where $f(x, y)$ is the joint probability function of X and Y.

4.14.1 Marginal Density Functions

Let X and Y be continuous random variables and $f(x, y)$ be the value of their joint probability density at (x, y). Then the function

$$f_1(x) = \int_{-\infty}^{\infty} f(x, y)\mathrm{d}x \quad \text{for} \quad -\infty < t < \infty$$

is called the marginal density function of X. Correspondingly, the function given by

$$f_2(x) = \int_{-\infty}^{\infty} f(x, y)\mathrm{d}y \quad \text{for} \quad -\infty < y < \infty$$

is called the marginal density of Y.

Example 4.19: The joint probability of continuous random variables X and Y is given by

$$\begin{aligned} f(x, y) &= 8xy, \quad 0 \le x \le 1, \ 0 \le y \le x \\ &= 0, \quad \text{otherwise} \end{aligned}$$

Find the marginal pdfs $f(x)$ and $f(y)$?

Solution: We have

$$\begin{aligned} f(x) &= \int_{-\infty}^{\infty} f(x, y)\mathrm{d}y \quad \text{(By definition)} \\ &= \int_0^x 8xy\mathrm{d}y \\ &= 8x \left| \frac{y^2}{2} \right|_x^0 \\ &= 4x^3, \quad 0 \le x \le 1 \end{aligned}$$

and

$$\begin{aligned} f(y) &= \int_{-\infty}^{\infty} f(x, y)\mathrm{d}x \\ &= \int_0^1 8xy\mathrm{d}x \\ &= 8y \left| \frac{x^2}{2} \right|_0^1 \\ &= 4y, \quad 0 \le y \le x \end{aligned}$$

4.15 CONDITIONAL PROBABILITY DENSITY FUNCTIONS

Definition 4.16:

If X and Y are continuous random variables and $f(x, y)$ is the value of the joint density function of X and Y at (x, y) then the conditional density of Y given $X = x$ is defined as

$$f(y/x) = \frac{\partial}{\partial y} F(y/x)$$

$$= \frac{f(x, y)}{f(x)}, \; f(x) \neq 0 \quad \text{for} \quad -\infty < x < \infty$$

And the conditional density of X given $Y = y$ is defined as

$$f(x/y) = \frac{f(x, y)}{f(y)}, \; f(y) \neq 0$$

Example 4.20: The joint density function of continuous random variables X and Y is given by $f(x, y) = 2$, $0 < x < y < 1$. Find the marginal and conditional pdfs?

Solution: The marginal density function of X given y is

$$f(x) = \int_{-\infty}^{\infty} f(x, y) dy = \int_{x}^{1} 2dy = [2y]_x^1 = 2(1 - x)$$

The marginal density function of Y given x is

$$f(y) = \int_{-\infty}^{\infty} f(x, y) dx = \int_{0}^{1} 2dx = [2x]_0^1 = 2$$

Conditional pdf of X given y is

$$f(y/x) = \frac{f(x, y)}{f(x)} = \frac{2}{2(1 - x)} = \frac{1}{1 - x}, \quad 0 < x < 1$$

Conditional pdf of Y given x is

$$f(x/y) = \frac{f(x, y)}{f(y)} = \frac{2}{2} = 1$$

Remarks:

The conditional distribution of a random variable entering into a system is its distribution calculated under the condition that the other random variable has assumed a definite value.

The random variables X and Y are said to be mutually independent if the conditional probability distribution of one does not depend on the value of the other, i.e.,

$$f(x/y)f(x) \quad \text{or} \quad f(y/x) = f(y)$$

The above mentioned properties of one-dimensional (univariate) and two-dimensional (bivariate) random variables can also be established for multidimensional random variables.

Exercise 4.3

1. Find the constant k so that the function

$$f(x) = \frac{1}{k}, \quad a \leq x \leq b$$
$$= 0, \quad \text{otherwise}$$

is a density function. Also find cumulative distributive function (c.d.f) of the random variable X?

Ans: $k = b - a, \quad f(x) = \dfrac{x - a}{b - a}$

2. Distribution function of a random variable X is

$$\begin{aligned} f(x) &= x, && 0 < x < 1 \\ &= 2 - x, && 1 \leq x \leq 2 \\ &= 0, && x \leq 2 \end{aligned}$$

Compute the cumulative distribution function of X.

$$f(x) = \frac{x - a}{b - a} \qquad 0 < x < 1$$

Ans: $= 2x - \dfrac{x^2}{2} - 1, \quad 1 \leq x \leq 2$

$= 1, \qquad x \geq 2$

3. A continuous random variable X that can assume any value between $x = 2$ and $x = 5$ has a density function given by $f(x) = k(1 + x)$. Find k, $P(X < 4)$?

Ans: $k = \dfrac{2}{27}; \quad P(X < 4) = \dfrac{16}{27}$

4. Suppose that the duration in minutes of a long-distance telephone conversation follows an exponential density function.

$$f(x) = \frac{1}{5}e^{-x/5}, \quad \text{for} \quad x > 0$$

Find the probability that duration of a conversation

a. Will exceed 5 minutes?

b. Will be less than 3 minutes?

Ans: (a) e^{-1} (b) $1 - e^{-3/5}$

5. If X is a continuous random variable with distribution

$$\begin{aligned} f(x) &= kx, \quad && 0 \le x \le 5 \\ &- 0, && \text{otherwise} \end{aligned}$$

Find

a. k

b. $P(1 < x < 3)$

c. $P(2 \le x \le 4)$

d. $P(x \le 3)$

Ans: (a) $k = \frac{2}{25}$; (b) $\frac{8}{25}$; (c) $\frac{12}{25}$; (d) $\frac{9}{25}$

6. Let X be a continuous random variate with pdf

$$\begin{aligned} f(x) &= ax, && 0 \le x \le 1 \\ &= a, && 1 \le x \le 2 \\ &= -ax + 3a, && 2 \le x \le 3 \\ &= 0, && \text{otherwise} \end{aligned}$$

a. Determine the constant a.

b. Compute $P(X < 1.5)$

Ans: (a) $a = \frac{1}{2}$; (b) $\frac{1}{2}$

7. The distribution function of a random variable X is given by

$$\begin{aligned} F(x) &= 1 - (1 + x)e^{-x}, \quad && x \ge 0 \\ &= 0, && \text{for } x < 0 \end{aligned}$$

Find the corresponding density function of X?

Ans: $\begin{aligned} f(x) &= xe^{-x}, \quad && x \ge 0 \\ &= 0, && x < 0 \end{aligned}$

8. Two random variables X and Y have the probability function

$$\begin{aligned} f(x, y) &= Ae^{-(2x+y)}, \quad && x, y \ge 0 \\ &= 0, && \text{otherwise} \end{aligned}$$

a. Evaluate A

b. Find the marginal pdfs.

Ans: (a) $A = 2$; (b) $\begin{aligned} f(x) &= 2e^{-2x}, && x \ge 0, \\ &= 0, && \text{otherwise} \end{aligned}$, $\begin{aligned} f(y) &= e^{-y}, && y \ge 0 \\ &= 0, && y < 0 \end{aligned}$

9. If $f(x, y) = 2 - x - y, \quad 0 \le x \le 1, \ 0 \le y \le 1$
$= 0, \quad$ elsewhere

Find

a. Marginal probability function?

b. Conditional probability function?

Ans: (a) $f(x) = \frac{3}{2} - x; \quad f(y) = \frac{3}{2} - y$

(b) $f(y/x) = \dfrac{2 - x - y}{\frac{3}{2} - x}, \quad f(x/y) = \dfrac{2 - x - y}{\frac{3}{2} - y}$

10. Let X and Y have joint density function

$$f(x, y) = e^{-(x+y)}, \quad x > 0, \ y > \infty$$
$$= 0, \quad \text{otherwise}$$

Find $P(0 < x < 1/y = 2)$?

Ans: $\dfrac{e - 1}{e}$

11. The joint density function of a bivariate distribution is given as

$$f(x, y) = x + y, \quad 0 \le x \le 1, \ 0 \le y \le 1$$
$$= 0, \quad \text{elsewhere}$$

Determine the marginal distributions of x and y.

Ans: $x + \frac{1}{2}, \quad 0 \le x \le 1, \quad y + \frac{1}{2}, \quad 0 \le y \le 1$

12. Let X and Y be the joint distributed with pdf

$$f(x, y) = \frac{1}{4}(1 + xy), \quad |x| < 1, |y| < 1$$
$$= 0, \quad \text{elsewhere}$$

Show that X and Y are independent?

13. The joint probability function of two random variables X and Y is

$$f(x, y) = c(1 + xy), \quad 0 \le x \le 6, \ 0 \le y \le 5$$
$$= 0, \quad \text{elsewhere}$$

Find

a. c

b. $F(0.1, 0.5)$

c. $f(x, 3)$

Ans: (a) $\frac{1}{255}$; (b) 0.0006102;

(c) $f(x, 3) = \frac{1}{255}(1 + 3x), \quad 0 \le x \le 6$

$= 0, \quad$ otherwise

14. Let X and Y be two continuous random variables with joint pdf

$$f(x, y) = c(x - y), \quad 0 \le x \le 2, \ -x \le y \le x$$
$$= 0, \quad \text{elsewhere}$$

a. Evaluate c
b. find $f(x)$
c. find $f(y/x)$

Ans: (a) $c = \frac{1}{8}$; (b) $\frac{x^2}{4}$, $0 < x < 2$; (c) $\frac{x - y}{2x^2}$, $0 < x < 2, \ -x < y < x$

15. Joint distribution of X and Y is given by

$$f(x, y) = 4xye^{-(x^2+y^2)}, \quad x \ge 0, \ y \ge 0$$

Show that X and Y are independent. Find the conditional density of X given $Y = y$?
Ans: $2xe^{-x^2}$

16. The joint distribution of two random variables X and Y is given by the following table:

X\Y	2	3	4
1	0.06	0.15	0.09
2	0.14	0.35	0.21

Determine the individual distributions of X and Y. Also verify that X and Y are stochastically independent.
Ans:

$X = x$	1	2
$f(x)$	0.3	0.7

$Y = y$	2	3	4
$f(y)$	0.2	0.5	0.3

17. Determine the value of k for which the function given by

$$f(x, y) = k\,xy, \quad \text{for} \quad x = 1, 2, 3, \quad y = 1, 2, 3$$

can serve as a joint probability distribution.

Ans: $k = \frac{1}{36}$

18. Given the joint pdf

$$f(x, y) = \frac{3}{5}x(y+x), \quad 0 \leq x \leq 1, \ 0 \leq y \leq 2$$
$$= 0, \qquad \text{elsewhere}$$

of two random variables X and Y. Find $P\{(X, Y) \in A\}$ where A is the region

$$\left\{(x, y): \ 0 < X < \frac{1}{2}, \quad 1 < Y < 2\right\}$$

Ans: $\frac{11}{80}$

19. Find the joint probability density of the two random variables X and Y whose joint distribution is given by

$$f(x, y) = (1 - e^{-x})(1 - e - y), \quad x > 0, \ y > 0$$
$$= 0, \qquad \text{otherwise}$$

Also use the joint probability density to determine $P(1 < X < 3, 1 < Y < 2)$.

Ans: $f(x, y) = e^{-(x+y)}, \quad x > 0, \ y > 0$
$= 0, \quad$ elsewhere

4.16 MATHEMATICAL EXPECTATION AND MOMENTS

In this section we introduce special constants, which serve to give a quantitative description of random variables. The constants of particular importance are mathematical expectation, variance, and moments of various orders. We begin our study by defining mathematical expectation, which is the central value of a variable distribution.

Mathematical Expectation: We shall first give a definition of mathematical expectation for discrete random variable.

Definition 4.17:

Let X be discrete random variable. Let $x_1, x_2, \ldots, x_n$ denote possible values of X and let $p_1, p_2, \ldots, p_n$ denote the corresponding probabilities. Then the mathematical expectation denoted by $E(X)$ is defined as

$$E(X) = \sum_{i=1}^{n} x_i p_i \quad i = 1, 2, 3, \ldots, n$$

If X is a continuous random variable having $f(x)$ is its pdf, then the mathematical expectation of X is defined as

$$E(X) = \int_{-\infty}^{\infty} xf(x)\mathrm{d}x$$

Mathematical expectation is also called the expectation or mean of the probability distribution.

Remarks:

It is possible to find mathematical expectation without computation.

Mathematical expectation does not exist for all random variables.

Expectation is also denoted by μ (or by $\overline{x}$) and is defined by $\sum xP(X = x)$ (or by $\sum(X) = \sum xf(x)$ where $f(x) = P(X = x)$).

Example: Consider an experiment in which a die is thrown we have $S = \{1, 2, 3, 4, 5, 6\}$. Let X be the number of points obtained in a single throw. Then the corresponding probabilities are $\frac{1}{6}, \frac{1}{6}, \frac{1}{6}, \frac{1}{6}, \frac{1}{6}$, and $\frac{1}{6}$.

The expected value of X is

$$E(X) = 1 \cdot \frac{1}{6} + 2 \cdot \frac{1}{6} + 3 \cdot \frac{1}{6} + 4 \cdot \frac{1}{6} + 5 \cdot \frac{1}{6} + 6 \cdot \frac{1}{6}$$

$$\frac{21}{6} = \frac{7}{2}$$

4.16.1 Properties of Mathematical Expectation

Theorem 1:

If X is a random variable and k is a real number, then

1. $E(k) = k$
2. $E(K\,x) = k\,E(x)$
3. $E(x + k) = E(x) + k$

Proof:

1. If X is a discrete random variable, then

$$E(k) = \sum_{i=1}^{n} kp_i = k\sum_{i=1}^{n} p_i = k \cdot 1 = k$$

If X is a continuous random variable, then

$$E(k) = \int_{-\infty}^{\infty} kf(x)\mathrm{d}x = k\int_{-\infty}^{\infty} f(x)\mathrm{d}x = k \cdot 1 = k$$

Thus $E(k) = k$

2. Let X be a discrete random variable. Then

$$\begin{aligned} E(kx) &= \sum_{i=1}^{n} kx_i p_i \\ &= k\sum_{i=1}^{n} x_i p_i \\ &= kE(x) \end{aligned}$$

If X is a continuous random variable, then

$$\begin{aligned} E(kx) &= \int_{-\infty}^{\infty} kxf(x)\mathrm{d}x \\ k &= \int_{-\infty}^{\infty} xf(x)\mathrm{d}x \\ &= k\,E(x) \end{aligned}$$

Remark:
In particular we have

$$E(-x) = -E(x)$$

3. If X is a discrete random variable then

$$\begin{aligned} E(X+k) &= \sum_{i=1}^{n} (k+x_i)p_i \\ &= \sum_{i=1}^{n} x_i p_i + \sum_{i=1}^{n} kp_i \\ &= \sum_{i=1}^{n} x_i p_i + k\sum_{i=1}^{n} p_i \\ &= E(X) + k\cdot 1 = E(X) + k \end{aligned}$$

If X is a continuous random variable then we have

$$\begin{aligned} E(x+k) &= \int_{-\infty}^{\infty} (x+k)f(x)\mathrm{d}x \\ &= \int_{-\infty}^{\infty} xf(x)\mathrm{d}x + \int_{-\infty}^{\infty} kf(x)\mathrm{d}x \\ &= \int_{-\infty}^{\infty} xf(x)\mathrm{d}x + k\int_{-\infty}^{\infty} f(x)\mathrm{d}x \\ &= E(X) + k\cdot 1 = E(X) + k \end{aligned}$$

Hence proved.

Remark:

The mathematical expectation of X need not be finite. For example, consider

The probability of the discontinuous random variable k defined by

$$p(k) = \frac{e^{-1}}{k!}, \quad k = 0, 1, 2, \ldots$$

The expectation $E(k!)$ is

$$E(k!) = \sum_{k=0}^{\infty} k! \frac{e^{-1}}{k!} = \sum_{k=0}^{\infty} e^{-1}$$

$$= e^{-1} \sum_{k=0}^{\infty} 1 \quad \text{which is not finite.}$$

Theorem 2:

If X and Y are random variables, then

$E(X + Y) = E(X) + E(Y)$

Proof:

Let X and Y be two discrete random variables.

Let $x_1, x_2, \ldots, x_n$ be the values assumed by the random variable X. Let $p_1, p_2, \ldots, p_n$ denote the corresponding probabilities.

Let $y_1, y_2, \ldots, y_n$ be the values assumed by the random variable Y and $p'_1, p'_2, \ldots, p'_n$ denote the corresponding probabilities. Also let $P(X = x_i, Y = y_j) = p_{ij}$

By definition, we have

$$E(X + Y) = \sum_{i=1}^{m} \sum_{j=1}^{n} (x_i + y) p_{ij}$$

$$= \sum_{i=1}^{m} \sum_{j=1}^{n} x_i p_{ij} + \sum_{i=1}^{m} \sum_{j=1}^{n} y_j p_{ij}$$

$$= \sum_{i=1}^{m} \left[\sum_{j=1}^{n} p_{ij} \right] + \sum_{j=1}^{n} p_{ij} \left[\sum_{i=1}^{m} p_{ij} \right] \tag{4.1}$$

$$= \sum_{i=1}^{m} x_i p_i + \sum_{j=1}^{n} y_j p_j$$

$$= E(X) + E(Y)$$

Let X and Y be continuous random variables and let $f(x, y)$ be the pdf of X and Y. Then

$$\begin{aligned}
E(X+Y) &= \int_{-\infty}^{\infty} (x+y)f(x, y)\mathrm{d}x\mathrm{d}y \\
&= \int_{-\infty}^{\infty}\int_{-\infty}^{\infty} xf(x, y)\mathrm{d}x\mathrm{d}y + \int_{-\infty}^{\infty}\int_{-\infty}^{\infty} yf(x, y)\mathrm{d}x\mathrm{d}y \\
&= \int_{-\infty}^{\infty} x\left[\int_{-\infty}^{\infty} (x, y)\mathrm{d}y\right]\mathrm{d}x + \int_{-\infty}^{\infty} y\left[\int_{-\infty}^{\infty} f(x, y)\mathrm{d}x\right]\mathrm{d}y \\
&= \int_{-\infty}^{\infty} xf_1(x)\mathrm{d}x + \int_{-\infty}^{\infty} yf_2(y)\mathrm{d}y
\end{aligned}$$

($f_1(x), f_2(y)$ are marginal density functions of X and Y, respectively)

$$= E(X) + E(Y) \tag{4.2}$$

From Eqs. (4.1) and (4.2), we conclude that

$$E(X+Y) = E(X) + E(Y)$$

The above theorem can be generalized and stated as follows:

Cor 1: If X, Y, Z, ..., are random variables then?

$$E(X+Y+Z+\cdots) = E(X) + E(Y) + E(Z) + \cdots$$

Definition 4.18:

Two jointly distributed random variables X and Y are statistically independent for each other if and only if the joint pdf equals the product of the two marginal pdfs,

$$\text{i.e.,}\quad f(x, y) = f_1(x)f_2(y)$$

where $f_1(x)$ and $f_2(y)$ are marginal pdfs of X and Y, respectively.

Theorem 3:

If X and Y are two independent random variables, then

$$E(XY) = E(X)E(Y)$$

Proof:

Let X and Y be two independent random variables and let X, Y be discrete.

Let X assume the m values $x_1, x_2, \ldots, x_m$. Let $p_1, p_2, \ldots, p_m$ denote the corresponding probabilities.

Let Y assume the n values $y_1, y_2, \ldots, y_n$ and corresponding probabilities be denoted as $p'_1, p'_2, \ldots, p'_n$.

Then

$$E(XY) = \sum_{i=1}^{m}\sum_{j=1}^{n} p_i p'_j \, x_i x_j$$

$$= \sum_{i=1}^{m} x_i p_i + \sum_{j=1}^{n} y_i p'_j$$

$$= E(X)\, E(Y)$$

When X and Y are continuous random variables that are independent, the theorem holds and is left as an exercise to the student.

Generalization: The above theorem can be extended to the case of several variables and stated as follows:

If $X, Y, Z, \ldots$ are independent random variables

$$E(XYZ\ldots) = E(X)\, E(Y)\, E(Z)\ldots$$

4.16.2 Variance

Definition 4.19:

Let X be a discrete random variable having probability function $P(X = x)$, then the variance of X is defined as variance of $X = E[(X - \mu)^2]$ where μ is the mean of X.

Since $\mu = E(X)$ the variance of X can be written as

Variance of $X = E[(X - E(X))^2]$

Variance of X is denoted by Var[X] [or by $V[X]$] or σ^2

If $x_1, x_2, \ldots, x_n$ are the values assumed by X, and $p_1, p_2, \ldots, p_n$ are the corresponding probabilities then

$$\sigma^2 = \mathrm{Var}[X] = \sum_{i=1}^{n} (x_i - \mu)^2 p_i$$

If X is continuous random variable then the variance of X defined as follows:

Definition 4.20:

Let X be a continuous random variable, having pdf $f(x)$ then

$$\mathrm{Var}[X] = \sigma^2 = \int_{-\infty}^{\infty} (x - \mu)^2 f(x)\mathrm{d}x$$

Remarks:
Variance gives an idea of how widely spread the values of random variables likely to be. If the value of variance is larger, then the observations are more scattered on average.

4.16.3 Properties of Variance

Theorem 4:
If X is a random variable, then $V[X] = E[x^2] - [E(X)]^2$

Proof:
Let X be discrete random variable.

By definition we have

$$\begin{aligned}\text{Var}[X] = V[X] - E[(X-\mu)^2]\\ &= E[X^2 - 2X\mu + \mu^2]\\ &= E[X^2] - 2E[X\mu] + E[\mu^2]\\ &= E[X^2 - 2\mu E[X] + \mu^2] \quad (\text{since } \mu^2 \text{ is a constant})\\ &= E[X^2] - 2E(X)E(Y) + [E(X)]^2 \quad (\text{Since } \mu = E(X))\end{aligned}$$

$$\text{Thus } V[X] = E[x^2] - [E(X)]^2$$

The above property holds good for continuous variables also.

Theorem 5:
If X is a random variable and k is a real number, then

1. $V[k\,x] = k^2\ V[X]$
2. $V[x+k] = V[x]$

Proof:

1. $$\begin{aligned}V[kx] &= E[(kx)^2] - [E(kX)]^2\\ &= E[k^2x^2] - [kE(X)]^2\\ &= k^2E[x^2] - k^2[E(X)]^2\\ &= k^2[E[x^2] - [E(X)]^2]\end{aligned}$$

$$V[kx] = k^2 V[x]$$

2. $V(X+k) = E[(X+k^2)] - [E(X+k)]^2$

$$= E[X^2 + 2kX + k^2] - [E(X)+k]^2$$
$$= E[[X^2] + 2kE(X) + E[k^2]] - E(X)^2 + 2kE(X) - k^2$$
$$= E(X^2) + k^2 - [E(X)]^2 - k^2 \quad (\text{since } E(k^2) = k^2)$$
$$= E(X^2) - [E(X)]^2$$
$$= V[x]$$

Hence proved.

Remarks:

The positive square root of variance is called the SD of X.

1. Standard deviation $= \sigma = \sqrt{V(X)}$

$$= \sqrt{E(X^2) - [E(X)]^2}$$

2. Since $[X = E(X^2)] \geq 0$, variance $X \geq 0$
3. $V[x] = 0$ if and only if X takes only one value with probability 1.
4. Variance of a constant is zero.
5. $V[x]$ is invariant to the change of origin but variance is not invariant to the change of scale.
6. Let X, Y be two random variables and $Y = \frac{X-k}{h}$ then $\sigma_y^2 = \frac{1}{h^2}\sigma_x^2$ where $\sigma_x^2 = V[x]$ and $\sigma_y^2 = V[y]$.

Example 4.21: X and Y are two random variables such that $Y \leq X$. If $E(X)$ and $E(Y)$ exist, show that $E[Y] \leq E[X]$.

Solution: We have $Y \leq X$ given,

i.e., $Y - X \leq 0$

Therefore we get $E[Y - X] \leq 0$ (applying expectation on both sides),

i.e., $E[Y] - E[X] \leq 0$

or $E[Y] \leq E[X]$.

Example 4.22: If X is a random variable and $E[X]$ exists, then show that $|E[X]| \leq E|X|$?

Solution: Since $X \leq |X|$

We have

$$E[X] \leq E|X| \tag{4.3}$$

Also we have

$$-X \leq |X|$$

Therefore

$$\begin{aligned} E[-X] &\le E|X| \\ -E[X] &\le E|X| \end{aligned} \tag{4.4}$$

From Eqs. (4.3) and (4.4), we get $|E[X]| \le E|X|$

Example 4.23: If X and Y are independent random variables with density functions

$$f(x) = \frac{8}{x^3} \quad \text{for} \quad x > 2 \text{ and}$$

$$f(y) = 2y \quad \text{for} \quad 0 < y < 1$$

Find $E[XY]$?

Solution: Since X and $Y$are independent random variables

$$f(x, y) = f(x)f(y)$$

Therefore

$$\begin{aligned} E[XY] &= \int_2^{\infty}\int_0^1 xy f(x, y)\mathrm{d}x\mathrm{d}y \\ &= \int_2^{\infty}\int_0^1 xy\left(\frac{8}{x^3}\right)(2y)\mathrm{d}x\mathrm{d}y \\ &= 16\int_2^{\infty}\left(\frac{1}{x^2}\int_0^1 y^2\mathrm{d}y\right)\mathrm{d}x \\ &= 16\int_2^{\infty}\frac{1}{x^2}\left[\frac{y^3}{3}\right]_0^1\mathrm{d}x \\ &= \frac{16}{3}\left[\frac{x^{-1}}{-1}\right]_2^{\infty} \\ &= \frac{16}{3}\left[\frac{1}{x}\right]_2^{\infty} = \frac{16}{3}\left[0 - \frac{1}{2}\right] = \frac{8}{3} \end{aligned}$$

Example 4.24: If X takes the values $x_n = (-1)^n\, 2^n\, n^{-1}$ for $n = 1, 2, \ldots$ with probabilities $p_n = 2^{-n}$, then show that $E[X] = -\log 2$

Solution: Using the definition $E[X] = \sum x_i p_i$ we get

$$\sum[X] = \sum_{i=1}^{\infty} 2^{-n}(-1)^n 2^n n^{-1}$$

$$= \sum_{i=1}^{\infty} \frac{(-1)^n}{n}$$

$$-1 + \frac{1}{2} - \frac{1}{3} + \frac{1}{4} - \cdots$$

$$= -\left[1 + \frac{1}{2} - \frac{1}{3} + \frac{1}{4} + \cdots\right]$$

$$= -\log 2$$

Example 4.25: Find variance for the following probability distribution:

x	8	12	16	20	24
$P(X = x)$	$\frac{1}{8}$	$\frac{1}{6}$	$\frac{3}{8}$	$\frac{1}{4}$	$\frac{1}{12}$

Solution: We have

$$E[X] = \sum_i x_i p_i$$

$$= 8 \cdot \frac{1}{8} + 12 \cdot \frac{1}{6} + 16\frac{3}{8} + 20 \cdot \frac{1}{4} + 24 \cdot \frac{1}{12}$$

$$= 1 + 2 + 6 + 5 + 2 = 16$$

$$\text{Mean} \quad = E[X] = 16$$

$$\text{Variance} = E(X^2) - [E(X)]^2$$

$$= \sum_i x_i^2 p_i - [E(X)]^2$$

$$= 8^2 \cdot \frac{1}{8} + 12^2 \cdot \frac{1}{6} + 16^2 \cdot \frac{3}{8} + 20^2 \cdot \frac{1}{4} + 24^2 \cdot \frac{1}{12} - 16^2$$

$$= 8 + 24 + 96 + 100 + 48 - 256$$

$$= 276 - 256 = 20$$

Therefore $V[X] = 20$.

Example 4.26: Let the variable X have the distribution $P(X=0) = P(X=2) = p$, $P(X=1) = 1-2p$, for $0 \le p \le 1/2$. For what values of p is the Var[X] maximum?

Solution: The probability distribution of X is

X:	0	1	1
$P(X=x)$:	p	$1-2p$	p

$$\text{Therefore expectation} = E(X) = 0\cdot p + 1\cdot(1-2p) + 2\cdot p$$

$$= 0 + 1 - 2p + 2p = 1$$

$$\text{Variance} = \text{Var}[X] = E(x^2) - [E(x)]^2$$

$$= 0^2\cdot p + 1^2\cdot(1-2p) + 2^2\cdot p - 1^2$$

$$= 0 + 1 - 2p + 4p - 1$$

$$= 2p$$

Clearly Var[X] is maximum when $p = 1/2$.

(since the maximum value Var[X] can take is 1 in $0 \le p \le \frac{1}{2}$ occurs is at $p = \frac{1}{2}$)

Example 4.27: A coin is tossed until a head appears. What is the expectation of the number of tosses required?

Solution: Head can appear in the first toss, or in the second toss, or in the third toss, and so on.

The favorable cases are H, TH, TTH, TTTH, . . .

The probability in which head appears in the first toss $= \dfrac{1}{2}$

The probability in which head appears in the second toss $= \left(\dfrac{1}{2}\right)^2$

The probability of getting head in the nth toss $= \left(\dfrac{1}{2}\right)^n$

Let X denote the number of tosses required to get the first head. The probability distribution of X is

$X=x$	1	2	...	n	...
$P(X=x)$	$\frac{1}{2}$	$\left(\frac{1}{2}\right)^2$	...	$\left(\frac{1}{2}\right)^n$	...

The expectation of number of tosses required is $E(X) = \sum_i x_i p_i$

i.e.,

$$E(X) = 1\cdot\frac{1}{2} + 2\cdot\left(\frac{1}{2}\right)^2 + 3\cdot\left(\frac{1}{2}\right)^3 + \cdots + n\cdot\left(\frac{1}{2}\right)^n + \cdots$$

$$= \frac{1}{2}\left[1 + 2 + 3\left(\frac{1}{2}\right)^2 + 4\left(\frac{1}{2}\right)^3 + \cdots + n\left(\frac{1}{2}\right)^{n-1} + \cdots\right]$$

$$= \frac{1}{2}\left[1 - \frac{1}{2}\right]^{-2}$$

$$= \frac{1}{2}\left(\frac{1}{2}\right)^{-2} = \frac{2^2}{2} = 2$$

Example 4.28:

1. What is the expected value of the number of points obtained in a single throw with an ordinary die. Also find the variance?
2. What is the mathematical expectation of the sum of points obtained on n dice?

Solution:

1. Let X be the random variable "the number of points" obtained X assumes the values 1, 2, 3, 4, 5, and 6, with probabilities 1/6 in each case.

   Hence the expectation of X is

$$E(X) = 1\cdot\frac{1}{6} + 2\cdot\frac{1}{6} + 3\cdot\frac{1}{6} + 4\cdot\frac{1}{6} + 5\cdot\frac{1}{6} + 6\cdot\frac{1}{6}$$

$$= \frac{1}{6}(1 + 2 + 3 + 4 + 5 + 6) = \frac{21}{6} = \frac{7}{2}$$

2. Let x_i denote the number of points ith die. Then $E(x_i) = \frac{7}{2}$, $i = 1, 2, 3, \ldots n, \ldots$

   The sum of points on n dice is $x_1 + x_2 + \cdots + x_n$

   Therefore

$$E(x_1 + x_2 + \cdots + x_n) = E(x_1) + E(x_2) + \cdots + E(x_n)$$

$$= \frac{7}{2} + \frac{7}{2} + \cdots n \text{ terms}$$

$$= n\left(\frac{7}{2}\right)$$

$$= \frac{7}{2}n$$

   Hence the expectation of sum of points on n dice $= \frac{7}{2}n$.

Example 4.29: If X and Y are random variables having joint density function

$$f(x, y) = 4xy, \quad 0 \le x \le 1,\ 0 \le y \le 1$$
$$= 0, \quad \text{otherwise}$$

Verify that

1. $E(X + Y) = E(X) + E(Y)$
2. $E(XY) = E(X)\,E(Y)$

Solution: We have

$$\begin{aligned}
E(X) &= \int_{-\infty}^{\infty}\int_{-\infty}^{\infty} xf(x, y)\mathrm{d}x\mathrm{d}y \\
&= \int_{-\infty}^{\infty}\int_{-\infty}^{\infty} x(4xy)\mathrm{d}x\mathrm{d}y \\
&= 4\int_{-\infty}^{\infty}\int_{-\infty}^{\infty} x^2y\mathrm{d}x\mathrm{d}y \\
&= 4\int_{x=0}^{\infty}\int_{y=0}^{\infty} x^2y\mathrm{d}x\mathrm{d}y \\
&= 4\int_{y=0}^{\infty} y\left(\int_{x=0}^{1} x^2\mathrm{d}x\right)\mathrm{d}y \\
&= 4\int_{y=0}^{\infty} y\left[\frac{x^3}{3}\right]_0^1 \mathrm{d}y \\
&= \frac{4}{3}\int_{y=0}^{1} y\mathrm{d}y = \frac{4}{3}\left[\frac{y^2}{2}\right]_0^1 \\
&= \frac{4}{6} = \frac{2}{3}
\end{aligned}$$

Similarly

$$\begin{aligned}
E(Y) &= \int_{-\infty}^{\infty}\int_{-\infty}^{\infty} yf(x, f)\mathrm{d}x\mathrm{d}y \\
&= \int_{-\infty}^{\infty}\int_{-\infty}^{\infty} y(4xy)\mathrm{d}x\mathrm{d}y \\
&= 4\int_{-\infty}^{\infty}\int_{-\infty}^{\infty} xy^2\mathrm{d}x\mathrm{d}y \\
&= 4\int_{x=0}^{1}\int_{y=0}^{\infty} xy^2\mathrm{d}x\mathrm{d}y = \frac{2}{3}
\end{aligned}$$

Now

1. $$E(X+Y)=\int_{-\infty}^{\infty}\int_{-\infty}^{\infty}(x+y)f(x,y)\mathrm{d}x\mathrm{d}y$$
$$=\int_{-\infty}^{\infty}\int_{-\infty}^{\infty}(x+y)(4xy)\mathrm{d}x\mathrm{d}y$$
$$=4\int_0^1\int_0^1(x^2y+xy^2)\mathrm{d}x\mathrm{d}y$$
$$=4\int_{y=0}^{1}\left(\int_{x=0}^{1}(x^2y+xy^2)\mathrm{d}x\right)\mathrm{d}y$$
$$=4\int_{y=0}^{1}y\left[\frac{x^3}{3}y+\frac{x^2}{2}y^2\right]_0^1\mathrm{d}y$$
$$=4\int_{y=0}^{1}\left[\frac{y}{3}+\frac{y^2}{2}\right]\mathrm{d}y$$
$$=4\left[\frac{y^2}{6}+\frac{y^3}{6}\right]_0^1=4\left(\frac{1}{6}+\frac{1}{6}\right)$$
$$=4\left(\frac{2}{6}\right)=\frac{4}{3}$$

Therefore $E(X+Y)=\frac{4}{3}=\frac{2}{3}+\frac{2}{3}=E(X)+E(Y)$
Also,

2. $$E(XY)=\int_{-\infty}^{\infty}\int_{-\infty}^{\infty}(xy)f(x,y)dxdy$$
$$=\int_{-\infty}^{\infty}\int_{-\infty}^{\infty}(xy)(4xy)dxdy$$
$$=4\int_0^1\int_0^1 x^2y^2dxdy$$
$$=4\int_{y=0}^{1}y^2\left[\frac{x^3}{3}\right]_0^1dy$$
$$=\frac{4}{3}\int_{y=0}^{1}y^2dy=\frac{4}{3}\left[\frac{y^3}{3}\right]_0^1$$
$$=\frac{4}{3}\cdot\frac{1}{3}=\frac{4}{9}$$

Thus $E(XY)=\frac{4}{9}=\frac{2}{3}\cdot\frac{2}{3}=E(X)E(Y)$
Hence verified.

Theorem 6:
(Cauchy–Schwartz inequality) If X, Y are random variables taking real values, then

$$[E[XY]]^2 \leq E[X^2]E[Y^2]$$

Proof:
Let t be a real variable.

Consider the expression $(X + tY)^2$
$(X + tY)^2$ is a nonnegative for all values of X and Y.
Hence $E(X + tY)^2 \geq 0$ for all i.
i.e., $E(x^2 + 2tXY + t^2Y^2) \geq 0$ for all i
i.e., $E(x^2) + 2tE(XY) + t^2\ E(Y^2) \geq 0$ for all i
$t^2\ E(Y^2) + 2tE(XY) + E(x^2)$ is of the form, $at^2 + bt + c$, a quadratic in t.
We have

$$a = E(Y^2), \quad b = 2E(XY), \quad c = E(x^2)$$
$$at^2 + bt + c \geq 0, \quad \text{implies } b^2 - 4ac \leq 0$$

Therefore

$$4[E[XY]]^2 - E[X^2]E[Y^2] \leq 0$$

or

$$[E[XY]]^2 - E[X^2]E[Y^2] \leq 0$$

or

$$[E[XY]]^2 \leq E[X^2]E[Y^2]$$

Hence proved.

Theorem 7:
(Chebyshev's inequality) If X is a random variable with mean μ and variance σ^2 then

$$P(|x - \mu| \geq k) \leq \frac{\sigma^2}{k^2}$$

or

$$P(|x - \mu| \geq k) \geq 1 - \frac{\sigma^2}{k^2}$$

where k is a real number.

Proof:

Case 1: Let X be a discrete random variable and $f(x)$ denote the probability function of X.

Then

$$\begin{aligned} \mathrm{Var}[X] = \sigma^2 &= E|(x-\mu)^2| \\ &= \sum (x-\mu)^2 f(x) \end{aligned} \tag{4.5}$$

Right hand side of the sum given by Eq. (4.5) is nonnegative.

Hence

$$\sigma^2 \geq \sum_{|x-\mu|\geq k} (x-\mu)^2 f(x) \geq \sum_{|x-\mu|\geq k} k^2 f(x)$$

i.e.,

$$\sigma^2 \geq k^2 \sum_{|x-\mu|\geq k} f(x) \tag{4.6}$$

But

$$\sum_{|x-\mu|\geq k} f(x) = P(|x-\mu| \geq k)$$

Therefore from Eq. (4.6) we get

$$\sigma^2 \geq k^2 P(|x-\mu| \geq k)$$

i.e.,

$$\frac{\sigma^2}{k^2} \geq P(|x-\mu| \geq k)$$

Thus we get

$$P(|x-\mu| \geq k) \leq \frac{\sigma^2}{k^2}$$

Case 2: Let X be a continuous random variable and $f(x)$ be the pdf of X then $\sigma^2 = \int_{-\infty}^{\infty} (x-\mu)^2 f(x)\mathrm{d}x$.

Clearly the above integrand is positive. Therefore when the range of integration is reduced, the value of the integral decreases.

Thus we have $\sigma^2 \geq \int_{|x-\mu|\geq k} (x-\mu)^2 f(x)\mathrm{d}x \geq \int_{|x-\mu|\geq k} k^2 f(x)\mathrm{d}x$

or

$$\sigma^2 \geq k^2 \int_{|x-\mu|\geq k} f(x)\mathrm{d}x$$

But

$$\int_{|x-\mu|\geq k} f(x)\mathrm{d}x = P(|x-\mu|\geq k)$$

Hence we get

$$\sigma^2 \geq k^2(|x-\mu|\geq k)$$

or

$$P(|x-\mu|\geq k)\leq \frac{\sigma^2}{k^2} \tag{4.7}$$

Also we have

$$P(|x-\mu|\geq k)+P(|x-\mu|\geq k)=1$$

Therefore we get

$$P(|x-\mu|\leq k)=1=P(|x-\mu|\geq k)$$

From Eq. (4.7), we get

$$P(|x-\mu|\leq k)\geq 1-\frac{\sigma^2}{k^2}$$

Hence proved.

Remark:

1. $P(|x-\mu|\geq k)\geq 1-\dfrac{\sigma^2}{k^2}$ can be written as $P(\mu-k<x<\mu+k)\geq 1-\dfrac{\sigma^2}{k^2}a$.
2. $\dfrac{\sigma^2}{k^2}$ is called an upper bound of $P(|x-\mu|\geq k)$.

Example 4.30: If the Chebyshev's inequality for the random variable X is given by $P(-2<X<8)\geq \frac{21}{25}$. Find $E(X)$ and $V[X]$?

Solution: Comparing $P(-2<X<8)\geq \frac{21}{25}$
with

$$P(\mu-k<x<\mu+k)\geq 1-\frac{\sigma^2}{k^2}$$

We get

$$\mu-k=-2 \tag{4.8}$$

$$\mu+k=8 \tag{4.9}$$

and

$$1 - \frac{\sigma^2}{k^2} = \frac{21}{25} \tag{4.10}$$

Adding Eqs. (4.8) and (4.9) we get

$$2\mu = 6 \text{ or } \mu = 3$$

Now $\mu + k = 8$ gives $3 + k = 8$ or $k = 5$

Substituting in Eq. (4.10) we get $1 - \dfrac{\sigma^2}{k^2} = \dfrac{21}{25}$

i.e., $\therefore\ \mu = \mu_1' + 4$

or

$$1 - \frac{21}{25} = \frac{\sigma^2}{5^2}$$

or

$$\frac{4}{25} = \frac{\sigma^2}{5^2}$$

or

$$\sigma^2 = 4$$

Hence

$E(X) = 3, \quad V[X] = 4$

Example 4.31: The number of components manufactured in a factory during one month period is a random variable with mean 600 and variance 100. What is the probability that the production will be between 500 and 700 over a month?

Solution: We have

$m = 600, \quad \sigma^2 = 100$

$P(500 < X < 700) = P(600 - 100 < X < 600 + 100)$

Comparing with

$$P(\mu - k < x < \mu + k), \quad \text{we get } k = 100$$

Using Chebyshev's inequality, we get

$$P(|x - 600| < k) \geq 1 - \frac{\sigma^2}{k^2} = 1 - \frac{100}{100^2} = 1 - \frac{1}{100}$$

or

$$P(|x - 600| < 100) \geq 1 - 0.01 = 0.99$$

i.e.,

$$P(|x - 600| < 100) \geq 0.99$$

Hence the probability of production in one month between 500 and 700 is 0.99.

Definition 4.21:

Let X and Y be two discrete random variables and $f(x, y)$ be the value of the joint probability distribution of X and Y at (x, y). If (x, y) is a real-valued function of (X, Y) then the expectation of $g(x, y)$ is given by

$$E(g(X, Y)) = \sum_x \sum_y g(x, y)F(x, y)\mathrm{d}x\mathrm{d}y.$$

If X and Y are continuous random variables and $f(x, y)$ is the value of their joint probability density of X and Y at (x, y).

Then

$$E(g(X, Y)) = \int_{-\infty}^{\infty}\int_{-\infty}^{\infty} g(x, y)f(x, y)\mathrm{d}x\mathrm{d}y$$

For example consider the joint density of X and Y is given by

$$\begin{aligned} F(x, y) &= \frac{2}{3}(x + 2y), \quad 0 < x < 1,\ 1 < y < 2 \\ &= 0, \qquad\qquad\qquad \text{elsewhere} \end{aligned}$$

The expected value of $g(x, y) = \dfrac{x}{y^3}$ is

$$\begin{aligned} E\left[\frac{x}{y^3}\right] &= \frac{2}{5}\int_1^2\int_0^1 \frac{x}{y^3}(x + 2y)\mathrm{d}x\mathrm{d}y \\ &= \frac{2}{5}\int_1^2\left[\int_0^1\left(\frac{x^2}{y^3} + \frac{2xy}{y^2}\right)\mathrm{d}x\right]\mathrm{d}y \\ &= \frac{2}{5}\int_1^2\left[\frac{x^3}{3y^3} + \frac{x^2}{y^2}\right]\mathrm{d}y \\ &= \frac{2}{5}\int_1^2\left[\frac{1}{3y^3} + \frac{1}{y^2}\right]\mathrm{d}y \\ &= \frac{2}{5}\left[\frac{1}{3}\left(\frac{y^{-2}}{-2}\right) + \frac{y^{-1}}{(-1)}\right]_2^1 \\ &= \frac{2}{5}\left[\frac{-1}{6y^2} - \frac{1}{y}\right]_1^2 \\ &= \frac{1}{4} \end{aligned}$$

4.16.4 Covariance

Definition 4.22:
Let (X, Y) be a two-dimensional (bivariate) random variable. Then covariance between X and Y denoted by Cov (X, Y) is defined as follows:

$$\text{Cov}(X, Y) = E[(X - E(X))(Y - E(Y))]$$

where $E(X)$ is the expectation of X and $E(Y)$ is the expectation of Y.

From the definition of covariance, we have

$$\begin{aligned}\text{Cov}(X, Y) &= E[(X - E(X))(Y - E(Y))] \\ &= E[XY - XE(Y) - Y\,E(X) + E(X)E(Y)] \\ &= E(XY) - E(X)E(Y) - E(Y)E(X) + E(X)E(Y) \\ &= E(XY) - E(X)E(Y)\end{aligned}$$

Remarks:

1. If X and Y are independent, then

$$\begin{aligned}\text{Cov}(X, Y) &= E(XY) - E(X)E(Y) \\ &= E(X)E(Y) - E(X)E(Y) = 0\end{aligned}$$

But the Cov $(X, Y) = 0$ does not imply that X and Y are independent.

2. If a, b are real constants then

$$\text{Cov}(aX, bY) = ab\,\text{Cov}(X, Y)$$

and

$$\text{Cov}(X + a, Y + b) = \text{Cov}(X, Y)$$

Example 4.32: For the following Joint probability distribution. Find Cov (X, Y)?

$X \backslash Y$	1	2	3	p_i
0	$\frac{2}{12}$	$\frac{1}{12}$	$\frac{1}{12}$	$\frac{4}{12}$
1	$\frac{1}{12}$	$\frac{1}{12}$	$\frac{2}{12}$	$\frac{4}{12}$
2	$\frac{3}{12}$	$\frac{1}{12}$	0	$\frac{4}{12}$
p_{ij}	$\frac{6}{12}$	$\frac{3}{12}$	$\frac{3}{12}$	1

Solution: From the given table, we get

$$E(X) = 0\cdot\frac{4}{12} + 1\cdot\frac{4}{12} + 2\cdot\frac{4}{12} = \frac{12}{12} = 1$$

$$E(Y) = 1\cdot\frac{6}{12} + 2\cdot\frac{3}{12} + 3\cdot\frac{3}{12}$$

$$= \frac{6}{12} + \frac{6}{12} + \frac{9}{12} = \frac{21}{12} = \frac{7}{4}$$

$$E(XY) = 0.1\cdot\frac{1}{12} + 0.2\frac{1}{12} + 0.3\cdot\frac{1}{12} + 1.1\frac{1}{12}$$

$$+ 1.2\cdot\frac{1}{12} + 1.3\frac{2}{12} + 2.1\cdot\frac{3}{12} + 2.2\cdot\frac{1}{12} + 2.3.0$$

$$= 0 + 0 + 0 + \frac{1}{12} + \frac{2}{12} + \frac{6}{12} + \frac{6}{12} + \frac{4}{12} + 0 = \frac{19}{12}$$

Hence

$$\text{Cov}(X, Y) = E(XY) - E(X)E(Y)$$

$$= \frac{19}{12} - 1\cdot\frac{7}{4}$$

$$= \frac{19}{12} - \frac{7}{4} = \frac{-2}{12} = \frac{-1}{6}$$

Exercise 4.4

1. Let X be random variable with the following probability distribution. Find $E(k)$?

X	3	6	9
$P(X = x)$	$\frac{1}{6}$	$\frac{1}{2}$	$\frac{1}{3}$

Ans: $\frac{13}{2}$

2. If X is a random variable and the density function of X is

$$f(x) = e^{-x}, \quad x \geq 0$$
$$= 0, \quad 0$$

Find

a. $E(X)$

b. $E(X^2)$

c. $E(X-1)^2$

d. $V(X)$

Ans: (a) 1; (b) 2; (c) 1; (d) 1

3. Define expectation of random variable.

4. A random variable X is given by

X	-2	3	1
$P(X=x)$	$\frac{1}{3}$	$\frac{1}{2}$	$\frac{1}{6}$

Find

a. $E(X)$

b. $V[X]$

Ans: (a) 1; (b) 5

5. Find the mean and variance for the distribution:

X	-1	0	1	otherwise
$P(X=x)$	$\frac{1}{8}$	$\frac{3}{8}$	$\frac{1}{2}$	0

Ans: Mean $=\frac{3}{8}$; Variance $=\frac{5}{8}$

6. The distribution of the number of raisins in a cookie is given in the following table. Find the mean and variance of raisins in a cookie?

No. of raisins (x)	0	1	2	3	4	5
Probability $P(x)$	0.05	0.1	0.2	0.4	0.15	0.1

Ans: Mean = 2.8; Variance = 1.56

7. If X is a random variable defined by the density function

$$f(x) = 3x^2, \quad 0 \le x \le 1$$
$$= 0, \quad \text{otherwise}$$

Find

a. $E(X)$

b. $E(3X-2)$

c. $E(X^2)$

d. $\text{Var}[X]$

Ans: (a) 0.75; (b) 1; (c) 0.6; (d) 0.0375

8. If X and Y are random variables each having the density functions

$$f(x) = e^{-x}, \quad x \geq 0$$
$$= 0, \quad \text{Otherwise}$$

Find

a. $E(X + Y)$
b. $E(X^2 + Y^2)$
c. $E(XY)$
d. $V[X]$ and $V[Y]$

Ans: (a) $\frac{1}{2}$; (b) 1; (c) $\frac{1}{4}$; (d) $\frac{1}{4}, \frac{1}{4}$

9. Two unbiased dice are thrown. Find the expected value of the sum of the numbers obtained on the top faces of them?

Ans: 7

10. Find the expectation of the number of failures preceding the first successes in an infinite series of independent trials with constant probability p of successes in each trial?

Ans: $\frac{q}{p}$

11. The density function of a continuous random variable X is given by

$$f(x) = \frac{1}{2}x, \quad 0 < x < 2$$
$$= 0, \quad \text{Otherwise}$$

Find $E(X)$, Var$[X]$, and $E(3x^2 - 2x)$?

Ans: $\frac{4}{3}, \frac{2}{9}, \frac{10}{3}$

12. For the density function

$$f(x) = \frac{1}{4} \quad \text{for } x = -1$$
$$= \frac{1}{4} \quad \text{for } x = 0$$
$$= \frac{1}{2} \quad \text{for } x = 2$$

Find $E(X)$ and $V[X]$?

Ans: $\frac{3}{4}, \frac{27}{16}$

13. If $a > 0$ and $n > 0$, find $E(X)$ and $V[X]$ when

$$f(x) = \left[\frac{a^{n+1} \cdot x^n}{n!}\right] e^{-ax}$$

Ans: $\frac{n+1}{a}, \frac{n+1}{a^2}$

14. From the following table, find (a) k; (b) $E(2X+3)$; (c) V(x)?

X	-3	-2	-1	0	1	2	3
$P(X=x)$	0.05	0.10	$2k$	0	0.30	k	0.10

Ans: (a) 0.15; (b) 3.5; (c) 2.887

15. The function $f(x) = \frac{1}{\pi} \cdot \frac{1}{1+x^2} \quad -\infty < x < \infty$

Defines the pdf of X. Find $E(X)$?

Ans: $E(X)$ does not exist.

16. If X is a random variable $Y = \frac{(x-\mu)}{\sigma}$, is also a random variable, show that

a. $E(Y) = 0$

b. $V(Y) = 1$

17. For a random variable with $f(x) = \frac{1}{2} e^{-|x|}$, find its SD?

Ans: $\sigma = \sqrt{2}$

18. A random variable X with unknown probability distribution has an average of 14 and variance 9. Find the probability that X will be between 7 and 21?

Ans: $\frac{40}{49}$

19. A random variable X has the density function e^{-x} for $x \geq 0$. Show that Chebyshev's inequality gives $P(|x-1| > 2) < \frac{1}{4}$.

Ans: $\frac{5}{9}$

4.17 MOMENTS

Moments about mean and an arbitrary number were discussed in Chapter 1, An Overview of Statistical Applications. In this section we introduce moments of a random variable, which serve to describe the shape of the distribution of a random variable.

Moments About Arithmetic Mean (Central Moments)

Definition 4.23:

If X is a discrete random variable, the rth moment of X about mean μ, denoted by μ_r is defined by

$$\mu_r = E(x-\mu)^r = \sum_{i=1}^{n} (x_i-\mu)^r p_i \quad r = 0,\ 1,\ 2,\ \ldots$$

If X is a continuous random variable, the rth moment about mean is defined by

$$\mu_r = E(x-\mu)^r = \int_{-\infty}^{\infty} (x-\mu)^r f(x)dx$$

Remarks:

1. Since $\mu = E(X)$, we can write

$$\mu_r = E[X - E[x]]^r$$

2. If $x_1, x_2, \ldots, x_n$ are in dual series, i.e., values of a random variable, with mean μ then the rth moment about μ is

$$\mu_r = \frac{1}{n}\sum_{i=1}^{n} (x_i-\mu)^r \quad r = 0,\ 1,\ 2,\ \ldots$$

From the definition of rth moment we have,

$$\mu_r = \sum_{i=1}^{n} (x_i-\mu)^r p_i$$

Therefore we get

$$\begin{aligned}
\mu_0 &= \sum_{i=1}^{n} (x_i-\mu)^0 p_i \\
&= \sum p_i = 1 \\
\mu_1 &= \sum_{i=1}^{n} (x_i-\mu) p_i \\
&= \sum_{i=1}^{n} x_i\, p_i - \sum_{i=1}^{n} \mu\, p_i \\
&= E[X] - \mu \sum p_i \\
&= E[X] - \mu \cdot 1 \\
&= E[X] - E[X] = 0 \quad (\text{since } \mu = E(X))
\end{aligned}$$

Therefore $\mu_0 = 1$, $\mu_1 = 0$ for any variable X.

$$\mu_2 = \sum_{i=1}^{n} (x_i - \mu)^2 p_i$$

$$= \sum_{i=1}^{n} (x_i^2 - 2\mu x_i + \mu^2)^2 p_i$$

$$= \sum_{i=1}^{n} x_i^2\ p_i - 2\mu \sum_{i=1}^{n} x_i\ p_1 + \mu^2 \sum_{i=1}^{n} p_i$$

$$= E(X^2) - \mu^2$$

or

$$\mu_2 = E(X^2) - [E(x)]^2$$

μ^2 is called the variance of the distribution of X. It is denoted by σ^2. The positive square root of σ^2, i.e., variance is called the SD.

The moments about mean are also called the central moments or simply the moments of the distribution.

4.17.1 Moments About an Arbitrary Number

Definition 4.24:
Let X be a discrete random variable and A be an arbitrary number. The rth moment about A of X denoted by μ_r', is defined as

$$\mu_r' = E[(x - A)^r] = \sum_{i=1}^{n} (x_i - \mu)^r \quad r = 0,\ 1,\ 2, \ldots$$

If X is a continuous random variable the μ_3' is defined as

$$\mu_r' = \int_{-\infty}^{\infty} (x - A)^r f(x) dx, \quad r = 0,\ 1,\ 2, \ldots$$

The moments about an arbitrary number are also called raw moments.

Relation Between μ_r and μ_r'

If μ_r denotes the rth moment of a distribution on about the mean and μ_r' denotes the rth moment about an arbitrary number, then

$$\mu_r = \mu_r' - {}^rC_1 \mu_{r+1}'\ \mu_1' + {}^rC_2\ \mu_{r-2}'\ \mu_2' + \cdots + (-1)^r (\mu_1')^r$$

Substituting $r = 2, 3, 4, \ldots$ we get

$$\mu_2 = \mu_2' - {}^2C_1\,\mu_1'\,\mu_1' + {}^2C_2(\mu_1')^2 = \mu_2' - (\mu_1')^2$$
$$\mu_3 = \mu_3' - {}^3C_1\,\mu_2'\,\mu_1' + {}^3C_2\,\mu_1'(\mu_1')^2 = {}^3C_3(\mu_1')^3$$
$$= \mu_3' - 3\mu_2'\mu_1' + 2(\mu_1')^2$$
$$\mu_4 = \mu_4'\,{}^4C_1\,\mu_3'\,\mu_1' + {}^4C_2\,\mu_2'(\mu_1')^2 - {}^4C_3\mu_1'(\mu_1')^3 + {}^4C_4(\mu_1')^4$$
$$= \mu_4' - 4\mu_3'\mu_1' + 6\mu_2'(\mu_1')^2 - 3(\mu_1')^4$$

Remarks:

1. $\mu_0' = \sum_{i=1}^{n}(x_i - \mu)^0 p_i$

   $\sum p_i = 1$

2. $$\mu_1' = \sum_{i=1}^{n}(x_i - A)p_i$$
$$= \sum_{i=1}^{n} x_i\,p_i - A\sum_{i=1}^{n} p_i$$
$$= E(X) - A$$
$$\therefore \quad \mu = \mu_1' + A$$

Example 4.33: The first four moments of a distribution about $x = 2$ are 1, 2.5, 5.5, and 26, respectively. Find the values of first four central moments?

Solution: We have $\mu' = 1$, $\mu_2' = 2.5$, $\mu_3' = 5.5$, $\mu_4' = 26$, and $A = 2$. Thus

$$\mu_1 = 0,\ \mu_2 = \mu_2' - (\mu_1')^2 = (2.5) - 1^2 = 1.5$$
$$\mu_3 = \mu_3' - 3\mu_2'\,\mu_1' + 2(\mu_1')^2$$
$$= 5.5 - 3(2.5)(1) + 2(1)^3$$
$$= 5.5 - 7.5 + 2 = 0$$

$$\mu_4 = \mu_4' - 4\mu_3'\,\mu_1' + 6\mu_2'(\mu_1')^2 - 3(\mu_1')^4$$
$$= 26 - 4(5.5)(1) + 6(2.5)(1) + 3(1)^4$$
$$= 26 - 22 + 15 - 3 = 16$$
$$\mu_1 = 0,\ \mu_2 = 1.5,\quad \mu_3 = 0,\quad \mu_4 = 16$$

4.17.2 Moments About Origin

Definition 4.25:

The rth moment about origin of a random variable X, denoted by V_r is the expected value of X^r, symbolically

$$V_r = E[X^r] = \sum_{i=1}^{n} x_i^r \, p_i$$

For $r = 0, 1, 2, \ldots$ when X is discrete, and

$$V_r = E[X^r] = \int_{-\infty}^{\infty} x^r f(x)\mathrm{d}x$$

when X is continuous.

Relation Between μ_r and V_r:

If μ is the Arithmetic mean of X, the distribution then

$$V_r = \mu_r + {}^rC_1 \, \mu_{r-1} \, \mu + {}^rC_2 \, \mu_{r-2} \, \mu^2 + \cdots + \mu^r$$

Putting $r = 1, 2, 3, \ldots$ we get

$$V_1 = \mu, \quad V_2 = \mu_2 + \mu^2, \quad V_3 = \mu_3 + 3 \, \mu_2 \, \mu + \mu^3$$

Remark:

From the definition of moment about arbitrary number A, we have

$$\mu_r' = E|(x-A)^r| = \sum_{i=1}^{n} (x_1 - A)^r \, p_i, \quad r = 0, 1, 2, \ldots$$

Taking $A = 0$, we obtain

$$\mu_r' = E[X^r] = \sum_{i=1}^{n} (x_i)^r p_i$$

i.e.,

$$\mu_r' = V_r$$

4.17.3 Skewness and Kurtosis

A measure of skewness is defined by

$$Sk_m = \frac{E[(x-\mu)^3]}{\sigma^3} = \frac{\mu_3}{\sqrt{\mu_2^3}}$$

If we define

$$\beta_1 = \frac{\mu_3^2}{\mu_2^3}$$

The moment coefficient of skewness $= \frac{\mu_3}{\sqrt{\mu_2^3}} = \pm \beta_1$

where the sign of $\sqrt{\beta_1}$ is taken as that of μ_3.

The moment of coefficient of skewness is also denoted by r_1. Thus

$$r_1 - \frac{\mu_3}{\sqrt{\mu_3^3}} = \pm \sqrt{\beta_1}$$

If measure of skewness is positive, we say that the distribution is positively skewed or right tailed. And if the measure of skewness is negative, we say that the distribution is negatively skewed or left tailed.

If Arithmetic mean of the distribution is greater than the mode, the distribution is positively skewed (i.e., right tailed), and if the Arithmetic mean is less than the value of the mode, the distribution is negatively skewed (i.e., left tailed). In a symmetrical distribution the quartiles are equidistant from the median.

Measure of kurtosis tells us the extent to which a distribution is more peaked or more flat topped than the normal curve. If the curve of a frequency distribution is more peaked than the normal curve, it is said to be "Leptokurtic." The curve of a frequency distribution is called "Platykurtic" if it is more flat topped than the normal curve. If the curve of a frequency is neither flat nor sharply peaked, then the curve of the distribution is called "Mesokurtic." The measure of kurtosis denoted by β_2 and is defined as

$$\beta_2 = \frac{\mu_4}{\mu_2^2}$$

where μ_2 and μ_4 are respectively the second and fourth central moments about mean. Kurtosis is also known as Convexity of a curve (or bulginess).

If $\beta_2 < 3$, the distribution is said to be "Platykurtic"

If $\beta_2 = 3$, the distribution is said to be "Mesokurtic"

If $\beta_2 > 3$, the distribution is said to be "Leptokurtic."

The kurtosis of distribution is also measured by the considering the value of $\beta_2 - 3$, which is denoted by r_2.

Thus

$$r_2 = \beta_2 - 3$$

If $r_2 = 0$, the distribution is "Mesokurtic"
If $r_2 < 0$, the distribution is "Platykurtic" and
If $r_2 > 0$, the distribution is "Leptokurtic".

Example 4.34: In a certain distribution the first four moments about $x = 5$ are 2, 20, 40, and 50. Calculate and state whether the distribution is Leptokurtic or Platykurtic?

Solution: We have $A = 5$

$$\mu_1' = 2, \quad \mu_2' = 20, \quad \mu_3' = 40, \quad \mu_4' = 50$$

Hence we get $\mu_1 = 0$ (always),

$$\mu_2 = \mu_2' - (\mu_1')^2 = (20) - 2^2 - 16$$

$$\mu_3 = \mu_3' - 3\mu_2' \, \mu_1' + 2(\mu_1')^2$$

$$= 40 - 3(2)(20) + 2(2)^3$$

$$= 40 - 120 + 16 = -64$$

$$\mu_4 = \mu_4' - 4\mu_3' \, \mu_1' + 6\mu_2'(\mu_1')^2 - 3(\mu_1')^4$$

$$= 50 - 4(40)(2) + 6(20)(2^2) + 3(2)^4$$

$$= 50 - 320 + 480 - 48$$

$$= 162$$

$$\beta_1 = \frac{\mu_3^2}{\mu_2^3} = \frac{(-64)^2}{16^3} = 1$$

$$\beta_2 = \frac{\mu_4}{\mu_2^2} = \frac{162}{16^2} = 0.63$$

Since $\beta_2 < 3$, the curve of the distribution is Platykurtic.

Example 4.35: The density of a random variable X is given by $f(x) = kx(2 - x)$, $0 \leq x \leq 2$. Find

a. k
b. Mean
c. Variance
d. β_1 and β_2
e. rth moment.

Solution:

a. Since $\int_{-\infty}^{\infty} f(x)\mathrm{d}x = 1$, i.e., $\int_0^2 f(x)\mathrm{d}x = 1$

We have $\int_0^2 kx(2-x)\mathrm{d}x = 1$

$$\text{or } k\left[x^2 - \frac{x^3}{3}\right]_0^2 = 1$$

$$\text{or } k\left[4 - \frac{8}{3}\right] = 1$$

$$\text{or } \frac{4k}{3} = 1$$

$$\therefore\ k = \frac{3}{4}$$

b. $$\text{Mean} = E(X) = \int_{-\infty}^{\infty} xf(x)\mathrm{d}x$$

$$= \frac{3}{4}\int_0^2 (2x^2 - x^3)\mathrm{d}x$$

$$= \frac{3}{4}\left[\frac{2}{3}x^3 - \frac{x^4}{4}\right]_0^2$$

$$= \frac{3}{4}\left[\frac{16}{3} - \frac{16}{4}\right]$$

$$= \frac{3}{4}\cdot 16\left[\frac{1}{3} - \frac{1}{4}\right] = 3.4\frac{(4-3)}{3.4}$$

$$= 1$$

c. We have

$$\mu_2' = \frac{3.8}{4.5} = \frac{6}{5}$$

$$\mu_3' = \frac{3.2^4}{5.6} = \frac{8}{5}$$

$$\mu_4' = \frac{3.2^5}{6.7} = \frac{16}{7}$$

d. Hence

$$\mu_1 = 1,$$
$$\mu_2 = \mu'_2 - (\mu'_1)^2 = \frac{6}{5} - 1^2 = \frac{1}{5}$$
$$\mu_3 = \mu'_3 - 3\mu'_2\,\mu'_1 + 2(\mu'_1)^2$$
$$= \frac{8}{5} - \frac{18}{5} + 2 = \frac{18-18}{5} = 0$$
$$\mu_4 = \mu'_4 - 4\mu'_3\,\mu'_1 + 6\mu'_2(\mu'_1)^2 - 3(\mu'_1)^4$$
$$= \frac{16}{7} - \frac{32}{5} + \frac{36}{5} - 3 = \frac{3}{35}$$

e. $$\mu'_r = \int_0^2 x^r f(x)\mathrm{d}x$$
$$= \frac{3}{4}\int_0^2 x^r(2x - x^2)\mathrm{d}x$$
$$= \frac{3}{4}\int_0^2 (2x^{r+1} - x^{r+2})\mathrm{d}x$$
$$= \frac{3}{4}\left[\frac{2}{r+2}\cdot 2^{r+2} - \frac{x^{r+3}}{(r+3)}\right]_0^2$$
$$= \frac{3}{4}\left[\frac{2}{r+2}\cdot 2^{r+2} - \frac{2^{r+3}}{r+3}\right]$$
$$= \frac{3}{4}\left[\frac{2^{r+3}}{(r+2)(r+3)}\right]$$
$$= \left[\frac{3.2^{r+1}}{(r+2)(r+3)}\right]$$

$\therefore\quad E(X) = \mu'_1 = \frac{3.2^2}{4.5} = 1$, i.e., $\mu = 1$

Therefore

$$\beta_1 = \frac{\mu_3^2}{\mu_2^3} = 0$$

and

$$\beta_2 = \frac{\mu_4}{\mu_2^2} = \frac{15}{7}.$$

Exercise 4.5

1. The first four central moments of a distribution are 0, 2.5, 0.7, and 18.75. Test the skewness and kurtosis of the distribution.
   Ans: Positively skewed: $\beta_1 = 0.031$, Mesokurtic ($\beta_2 = 3$)
2. Find the first moments about mean for the series 4, 5, 6, 1, 4.
   Ans: $\mu_1 = 0$, $\mu_2 = 2.8$, $\mu_3 = -3.6$, and $\mu_4 = 19.6$

3. The first four moments of a distribution, about the value 35 are -1.8, 240, -1020, and 14,400. Find the values for μ_1, μ_2, $\mu_{3,}$ and μ_4?
   Ans: $\mu_1 = 0$, $\mu_2 = 236.76$, $\mu_3 = 264.36$, and $\mu_4 = 141290.11$
4. The first four moments of a distribution about $x = 4$ of the variable are -1.5, 17, -30, and 108. Find the moments for μ_1, μ_2, $\mu_{3,}$ and μ_4 about mean?
   Ans: 0, 14.75, 39.75, and 142.31.
5. The first three critical moments of a distribution are 0, 2.5, and 0.7. Find the value of moment of coefficient of skewness?
   Ans: 0.177
6. The first four moments of a distribution about the value 5 of the variable are 2, 20, 40, and 50. Calculate the moment of skewness.
   Ans: -1
7. The first four central moments of a distribution are 0, 2.5, 0.7, and 18.75. Test the kurtosis of the distribution.
   Ans: Mesokurtic ($\beta_2 = 3$).
8. The first three central moments of a distribution are 0, 15, -31. Find the moment coefficient of skewness?
   Ans: -0.53
9. Calculate the first five moments about the mean for the series 4, 7, 10, 13, 16, 19, 22.
   Ans: 0, 36, 0, 22, 68
10. The first four moments of a distribution about $x = 4$ are -1.5, 17, 30, and 108. Find the moments about the mean?
    Ans: 0, 14.75, 39.75, 142.31
11. X is a random variable whose density function is

$$f(X) = Ae^x, \quad 0 < x < \infty$$
$$= 0, \quad \text{otherwise}$$

Find the value of

a. A

b. Mean of X

c. Variance of X

d. Third moment about the mean

e. Kurtosis

f. rth moment about origin.

Ans: (a) $A = 1$; (b) 1; (c) 2; (d) $\mu_3 = 2$; (e) a (Leptokurtic); (f) $r!$

12. The first three moments of a distribution about the value 2 are 1, 16, -40. Find the mean and variance?
    Ans: Mean = 3, Variance = 15

13. The first three moments of a distribution about the value 3 are 2, 10, −30. Show that the moments about $x = 0$ are 5, 31, 141. Find the mean and variance.

Ans: $\mu = 5,\ \sigma^2 = 6$

14. Given the pdf $f(x) = \begin{cases} 1 - x^2, & 0 < x < 1 \\ 0, & \text{elsewhere} \end{cases}$

Obtain

a. kth moment about the origin

b. First three moments about the mean

c. The mean and variance

Ans: (a) $\dfrac{2}{(k+1)(k+3)}$; (b) $\mu_1 = 0$; (c) $\dfrac{1}{4}, \dfrac{17}{240}$

15. Show that for the exponential distribution $P = y_0 e^{-x/a} dx$, $0 \le x < \infty$, $\sigma > 0$, y_0 being a constant, the mean and the SD are equal to σ and the SD are equal to σ and that the interquartile range is $\sigma \log_e 3$. Also find μ_r' and show that $\beta_1 = 4$ and $\beta_2 = 9$?

Hint:

$$y_0 = \int_0^{\infty} e^{-x/\sigma} dx = 1 \Rightarrow y_0 = \frac{1}{\sigma}$$

$$\mu_r' = \frac{1}{\sigma}\int_0^{\infty} x^r e^{-x/\sigma} dx = r!\sigma^r$$

We get

$$\mu_1' = \sigma,\ \ \mu_2' = 2\sigma^2,\ \ \mu_3' = 6\sigma^2,\ \ \mu_4' = 9\sigma^4$$

Hence

$$\beta_1 = \frac{\mu_3^2}{\mu_2^3} = 4,\ \ \beta_2 = \frac{\mu_4}{\mu_2^2} = 9$$

$$\frac{1}{\sigma}\int_0^{Q_1} e^{-x/\sigma} dx = \frac{1}{4} \text{ gives } e^{-Q_1/\sigma} = \frac{3}{4}$$

$$\frac{1}{\sigma}\int_0^{Q_3} e^{-x/\sigma} dx = \frac{3}{4} \text{ gives } e^{-Q_3/\sigma} = \frac{1}{4}$$

Hence $e^{\frac{Q_3 - Q_1}{\sigma}} = 3$ or $Q_3 - Q_1 = \sigma \log_e 3$

16. For the triangular distribution $dp = \frac{1}{a}\left[1 - \frac{|x-b|}{a}\right] dx$, $|x - b| < a$. Show that the mean is band variance $\frac{a^2}{6}$.

17. For the rectangular distribution $dp = dx$, $1 \le x \le 2$ show AM > GM > HM.

18. For the triangular distribution, with the density function $f(x) = 1 - |1 - x|, 0 < x < 2$. Show that the mean is 1 and variance is 1/6.

4.18 MOMENT GENERATING FUNCTION

Definition 4.26:

The moment generating function (mgf) of a random variable X, denoted by $M_X(t)$ is defined by

$$M_X(t) = E[e^{tx}] = \sum_x e^{tx} P(X = x)$$

when X is a discrete random variable, and

$$M_X(t) = E[e^{tx}] = \int_{-\infty}^{\infty} e^{ix} f(x)\mathrm{d}x \quad (t \text{ is an independent variable})$$

when X is a continuous random variable.

The mgf of X is also denoted by $M(t)$. The mgf may exist but the moments may not exist.

4.19 PROPERTIES OF MOMENT GENERATING FUNCTION

Theorem 8:

If a, c are constants and X is a random variable, then

1. $M_{cx}[t] = M_X[ct]$
2. $M_{\frac{x-a}{c}}[t] = e^{-at/c} M_X\left[\frac{t}{c}\right]$

Proof:

1. By definition, we have

$$M_{cx}[t] = E[e^{tx}]$$
$$= E[e^{x(ct)}] = M_X[ct]$$

2. $$M_{\frac{x-a}{c}}[t] = E\left[e^{t\left(\frac{x-a}{c}\right)}\right]$$
$$= E\left[e^{\frac{tx}{c}} \cdot e^{-\frac{at}{c}}\right]$$
$$= e^{\frac{-at}{c}} E\left[e^{x\left(\frac{t}{c}\right)}\right]$$
$$= e^{\frac{-at}{c}} M_X\left(\frac{t}{c}\right)$$

Theorem 9:

If $M_X[t]$ and $M_Y[t]$ are the mgf of the independent random variables X and Y, then

$$M_{X+Y}[t] = M_X[t]M_Y[t]$$

i.e., the mgf of the sum of two independent random variables is equal to the product of their respective mgf.

Proof:

$$\begin{aligned} M_{X+Y}[t] &= E[e^{t(X+Y)}] \\ &= E[e^{tX+tY}] \\ &= E[e^{tX}\, e^{tY}] \\ &= E[e^{tx}]E[e^{ty}] \\ &= M_X[t]M_Y[t] \end{aligned}$$

Hence proved.

Theorem 10:

If X is a random variable, then

$$M_X[t] = M_X[t] = 1 + \mu_1' t + \mu_2' \frac{t^2}{2!} + \mu_3' \frac{t^3}{3!} + \cdots + \mu_r' \frac{t^r}{r!} + \cdots$$

where

μ_r', $(r = 1, 2, 3, \ldots)$ are the moments about the origin.

Proof:

$$\begin{aligned} M_X[t] &= E[e^{tx}] \\ &= E\left[1 + tx + \frac{t^2x^2}{2!} + \frac{t^3x^3}{3!} + \cdots + \frac{t^rx^r}{r!} + \cdots\right] \\ &= E(1) + E(tx) + E\left(\frac{t^2x^2}{2!}\right) + E\frac{t^3x^3}{3!} + \cdots + E\left(\frac{t^rx^r}{r!}\right) + \cdots \\ &= 1 + tE(x) + \frac{t^2}{2!}E(x^2) + \frac{t^3}{3!}E(x^3) + \cdots + \frac{t^r}{r!}E(x^r) + \cdots \end{aligned} \tag{4.11}$$

From the definition, we have

$$E[(x-A)^r] = \mu_r'$$

where A is an arbitrary number.

Putting $A = 0$, we get $E[x^r] = \mu_r'$

$$E(x) = \mu_1', \quad E(x^2) = \mu_2', \quad E(x^3) = \mu_3', \ldots$$

Substituting in Eq. (4.11), we get

$$M_X[t] = 1 + \mu_1' t + \mu_2' \frac{t^2}{2!} + \mu_3' \frac{t^3}{3!} + \cdots + \mu_r' \frac{t^r}{r!} + \cdots$$

4.19.1 Solved Examples

Example 4.36: The pdf of the random variable X has the following probability law:

$$P(x) = \frac{1}{2\theta} e^{-\left|\frac{x-\theta}{\theta}\right|}, \quad -\infty < x < \infty.$$

Find the mgf of X. Hence find $E(X)$ and $\mathrm{Var}[X]$?

Solution: The moment generating function (mgf) of X is

$$M_X[t] = E\left[e^{tx}\right] = \int_{-\infty}^{\infty} e^{tx} \frac{1}{2\theta} e^{-\left|\frac{x-\theta}{\theta}\right|} dx$$

$$= \frac{1}{2\theta}\int_{-\infty}^{\theta} e^{tx} e^{-\left|\frac{x-\theta}{\theta}\right|} \mathrm{d}x + \frac{1}{2\theta}\int_{\theta}^{\infty} e^{tx} e^{-\left|\frac{x-\theta}{\theta}\right|} dx$$

$$= \frac{1}{2\theta}\int_{-\infty}^{\theta} e^{tx} e^{\left(\frac{x-\theta}{\theta}\right)} \mathrm{d}x + \frac{1}{2\theta}\int_{\theta}^{\infty} e^{tx} e^{-\left(\frac{x-\theta}{\theta}\right)} dx$$

$$= \frac{e^{-1}}{2\theta}\int_{-\infty}^{\theta} e^{\left(t+\frac{1}{\theta}\right)x} \mathrm{d}x + \frac{e}{2\theta}\int_{\theta}^{\infty} e^{\left(t-\frac{1}{\theta}\right)x} \mathrm{d}x$$

$$= \frac{e^{-1}}{2\theta}\left|\frac{e^{\left(t+\frac{1}{\theta}\right)x}}{t+\frac{1}{\theta}}\right|_{-\infty}^{\theta} - \frac{e}{2\theta}\left|\frac{e^{\left(t-\frac{1}{\theta}\right)x}}{t-\frac{1}{\theta}}\right|_{\theta}^{\infty} \tag{4.12}$$

$$= \frac{1}{2\theta}\cdot\frac{e^{t\theta}}{\left(\frac{1+t\theta}{\theta}\right)} + \frac{1}{2\theta}\cdot\frac{e^{t\theta}}{\left(\frac{1-t\theta}{\theta}\right)}$$

$$= \frac{e^{t\theta}}{2}\left[\frac{1}{1+t\theta} + \frac{1}{1-t\theta}\right] = \frac{e^{t\theta}}{1-(t\theta)^2}$$

$$= e^{t\theta}\left[1-t^2\theta^2\right]^{-1}$$

$$= e^{t\theta}\left[1-t^2\theta^2+\cdots\right]$$

$$= \left(1+t\theta+\frac{\theta^2 t^2}{2!}+\cdots\right)\left(1+\theta^2t^2+\cdots\right)$$

$$= \left(1+t\theta+\frac{3\theta^2t^2}{2!}+\cdots\right)$$

$\mu_1' = \text{Coefficient of } k \text{ in Eq. (4.12)} = \theta$
$\mu_2' = \text{Coefficient of } \frac{t^2}{2!} \text{ in Eq. (4.12)} = 3\theta^2$
Hence

$$\mu_2 = \mu_2' - (\mu_1')^2 = 3\theta^2 - \theta^2 = 2\theta^2 = 3\theta^2 - \theta^2 = 2\theta^2$$

$$E(X) = \text{mean} = \theta, \quad \text{Var}[X] = 2\theta^2$$

Example 4.37: Find the mgf of a random variable X having the density function

$$f(x) = \frac{x}{2}, \quad 0 \leq x \leq 2$$
$$= 0, \quad \text{otherwise}$$

and use it to find first four moments about its origin.

Solution: We have

$$M_X[t] = E[e^{tx}] = \int_{-\infty}^{\infty} e^{tx} f(x) dx$$

$$= \int_0^2 e^{tx} \frac{x}{2} dx = \frac{1}{2} \int_0^2 e^{tx} x dx$$

$$= \frac{1}{2} \left[x \frac{e^{tx}}{t} - 1 \cdot \frac{e^{tx}}{t^2} \right]_0^2$$

$$= \frac{1}{2} \left[\frac{2(e^{2t})}{t} - \frac{(e^{2t} - 1)}{t^2} \right]$$

$$= \frac{1}{2} \left[\frac{2(e^{2t})}{t} + 1 - \frac{(e^{2t})}{t^2} \right]$$

$$= \frac{1}{2t^2} \left[1 + 2te^{2t} - e^{2t} \right]$$

$$\therefore \quad M_X(t) = \frac{1}{2t^2} \left[1 + 2te^{2t} - e^{2t} \right]$$

Consider

$$M_X(t) = \frac{1}{2t^2}\left[1 + 2te^{2t} - e^{2t}\right]$$

$$= \frac{1}{2t^2} + \frac{1}{t}e^{2t} - \frac{1}{2t^2}e^{2t}$$

$$= \frac{1}{2t^2} + \frac{1}{t}\left[1 + \frac{2t}{1!} + \frac{2^2t^2}{2!} + \cdots\right] - \frac{1}{2t^2}\left[1 + \frac{2t}{1!} + \frac{2^2t^2}{2!} + \cdots\right]$$

$$= \frac{1}{2t^2} + \left[\frac{1}{t} + 2 + \frac{2^2t}{2!} + \frac{8t^2}{3!} + \frac{16t^3}{4!} + \cdots\right] + \left[\frac{-1}{2t^2} - \frac{1}{t} + \frac{2}{2!} - \frac{2^2t}{3!} - 2^3\frac{t^2}{4!}\cdots\right]$$

$$= 1 + \left(2 - \frac{2}{3}\right)t + \left(\frac{4}{3} - \frac{1}{3}\right)t^2 + \left(\frac{2}{3} - \frac{2}{15}\right)t^3 + \cdots$$

$$= 1 + \frac{4}{3}t + t^3 + \frac{8}{15}t^3 + \frac{2}{9}t^4 + \cdots$$

Therefore

$$M_X(t) = 1 + \frac{4}{3}t + 2\frac{t^2}{2!} + \frac{16}{5}\frac{t^3}{3!} + \frac{16t^4}{4!} + \cdots$$

is the required mgf.

Since $M_X(t) = 1 + \mu_1' t + \mu_2' \frac{t^2}{2!} + \mu_3' \frac{t^3}{3!} + \mu_4' \frac{t^4}{4!} + \cdots$

Comparing, we get $\mu_1' = \frac{4}{3}$, $\mu_2' = 2$, $\mu_3' = \frac{16}{5}$, $\mu_4' = \frac{16}{3}$.

Exercise 4.6

1. Define moment generating function (mgf).
2. Find the mgf of a random variable X having the density function

$$f(x) = \frac{x}{2}, \quad 0 \le x \le 2$$
$$= 0, \quad \text{otherwise}$$

and find the first four moments about the origin?

Ans: $\frac{4}{3}, 2, \frac{16}{5}, \frac{16}{3}$

3. A random variable X assumes values $1/2$ and $-(1/2)$ with probability $1/2$ each. Find the mgf and four moments about the origin?

Ans:

$$M_X(t) = 1 + \frac{t^2}{2!2^2} + \frac{t^4}{4!2^4} + \cdots \mu_1' = \frac{4}{3}, \mu_2' = 2, \mu_3' = \frac{16}{5}, \mu_4' = \frac{16}{3}$$

4. A random variable has the density function.

$$f(x) = e^{-x}, \quad x \geq 0$$
$$= 0, \quad \text{otherwise}$$

Determine the mgf about the origin and also about the mean.
Ans:

$$M_X(t) = 1 + t + 2\frac{t^2}{2!} + 6\frac{t^3}{3!} + \cdots$$

$$\mu_1' = 1, \ \mu_2' = 2, \ \mu_3' = 6, \ \mu_4' = 24$$

$$E(X^k) = \frac{b^{k+1} - a^{k+1}}{(k+1)(b-a)}, \ \mu_1 = 0, \ \mu_2 = 1, \ \mu_3 = 2, \ \mu_4 = 9$$

5. A random variable X has the probability distribution $f(x) = \frac{1}{8}{}^3C_x$ for $x = 0, 1, 2,$ and 3. Find the mgf of this random variable and use it to determine μ_1' and μ_2'.

Ans: $\mu_1' = \frac{b+a}{2}, \ \mu_2' = \frac{b^2+ab+a^2}{3}, \ \mu_3' = \frac{(b+a)(b^2+a^2)}{4}$

6. If a and b are constants. Prove the following:
   a. $M_{X+a}[t] = e^{at}M_x(t)$
   b. $M_{\frac{x+a}{b}}[t] = e^{at/b}M_X\left[\frac{t}{b}\right]$

   Ans: $\frac{3}{2}, 3, M_x(t) = \frac{1}{8}(1+e^t)^3$

4.20 DISCRETE PROBABILITY DISTRIBUTIONS

In this section we shall study some discrete probability distribution, which are derived using the theory of probability for the outcomes of a conceptual experiment. We discuss the following distributions:

1. Binomial distribution
2. Poisson distribution
3. Geometric distribution
4. Uniform distribution
5. Negative binomial distribution
6. Gamma distribution
7. Weibull distribution

The Binomial distribution, Poisson distribution, and Hypergeometric distribution use integers as Random variables. We begin our study by defining Bernoulli's distribution.

A random experiment with only two possible outcomes, success or failure is called a Binomial trial and random variable X which takes the values either zero or 1 is called a Bernoulli's variable. The corresponding distribution is called the Bernoulli's distribution.

It was discovered in 1713 by James Bernoulli and is defined as follows.

Let X be a Bernoulli random variable. If the probability of success is denoted by p, and the probability of failure is denoted by $q = 1 - p$, and the pmf is defined by

$$\begin{aligned} P(X = x) &= p^x q^{1-x}, \quad x = 0, 1 \\ &= 0, \qquad \text{otherwise} \end{aligned}$$

Then the probability distribution of X is called Bernoulli's distribution, thus we have (Table 4.1):

The mean of Bernoulli's distribution is

$$\begin{aligned} \mu = E(X) &= \sum_i x_i p_i \\ &= 0(1 - p) + 1 \cdot p = p \end{aligned}$$

The variance of Bernoulli's distribution is

$$\begin{aligned} \sigma^2 = \mathrm{Var}(X) &= \sum_i (x_i - \mu)^2 p_i \\ &= (0 - p)^2(1 - p) + (1 - p)^2 p \\ &= (1 - p)(p^2 + p(1 - p)) \\ &= (1 - p)(p^2 + p - p^2) \\ &= p(1 - p) \\ &= pq \end{aligned}$$

Standard distribution is $\sigma = \sqrt{pq}$.

If n is the number of Bernoulli's trials then Bernoulli's theorem states that the probability of x successes is ${}^nC_x p^x q^{n-x}$.

Table 4.1 Bernoulli's distribution

$X = x_i$	0	1
$P(X = x_i)$	$1p$	p

4.20.1 Binomial Distribution

Binomial distribution is a discrete distribution. It is a commonly used probability distribution. Then it is developed to represent various discrete phenomenons, which occur in business, social sciences, natural sciences, and medical research.

Binomial distribution is widely used due to its relation with binomial distribution. The following should be satisfied for the application of binomial distribution:

1. The experiment consists of n identical trials, where n is finite.
2. There are only two possible outcomes in each trial, i.e., each trial is a Bernoulli's trial. We denote one outcome by S (for success) and other by F (for failure).
3. The probability of S remains the same from trial to trial. The probability of S (Success) is denoted by p and the probability of failure by q (where $p + q = 1$).
4. All the trials are independent.
5. The Binomial random variable x is the number of success in n trials.

If X denotes the number of success in n trials under the conditions stated above, then x is said to follow binomial distribution with parameters n and p.

Definition:

(Binomial distribution) A discrete random variable taking the values 0, 1, 2, ..., n is said to follow binomial distribution with parameters n and p if its pmf is given by

$$P(X = x) = P(x)^nC_xp^xq^{n-x}, \quad x = 0, 1, 2, \ldots, n$$
$$0 < p < 1, \; q = 1 - p$$
$$= 0, \quad \text{otherwise}$$

If x follows Binomial distribution with parameters n and p symbolically we express $X \sim B(n, p)$ or $B(x: n, p)$.

A *binomial random variable* is the number of successes x in n repeated trials of a binomial experiment. The probability distribution of a binomial random variable is called a *binomial distribution* (It is also known as a *Bernoulli distribution*).

A *cumulative binomial probability* refers to the probability that the binomial random variable falls within a specified range.

Remark:

We have

$$\sum_{x=0}^{n} P(X=x) = \sum_{x=0}^{n} {}^nC_x p^x q^{n-x}$$
$$= (q+p)^n = 1$$

The probabilities are the terms in the binomial expansion of $(q+p)^n$ (i.e., $(p+q)^n$), hence name Binomial distribution given.

The Binomial distribution is used to analyze the error in experimental results that estimate the proportion of in a population that satisfy a condition of interest.

4.20.2 Expected Frequencies and Fitting of a Binomial Distribution

If we take large N of sets of n (Bernoulli) trials each, then the expected or theoretical frequencies of getting x success is given by

$$N.P(X=x) = N\,{}^nC_x p^x q^{n-x}$$
$$= N.{}^nC_x q^x p^{n-x} \quad x = 0, 1, 2, \ldots, n$$

The theoretical frequencies of getting 0 success, 1 success, ..., n success are respectively the 1st, 2nd,..., $(n+1)$th terms in the expansion of $N(p+q)^n$ (i.e., $N(q+p)^n$).

The expected or theoretical frequencies of Binomial distribution are as shown in the Table 4.2.

4.20.3 Recurrence Relation

Let X be a Binomial variable. Since X follows Binomial distribution, we have

$$P(x) = P(X=x) = {}^nC_x p^x q^{n-x} \quad x = 0, 1, 2, \ldots, n$$

Table 4.2 Theoretical or expected frequencies

Number of success	Expected frequency
0	$N.q^n$
1	$N.{}^nC_1 pq^{n-1}$
2	$N.{}^nC_2 p^2 q^{n-2}$
⋮	
X	$N.{}^nC_x p^x q^{n-x}$
⋮	
n	$N.p^n$

Replacing x by $x+1$ we get

$$P(x+1) = P(X = x+1) = {}^{n}C_{x+1}p^{x+1}q^{n-x-1} \quad x = 0,\ 1,\ 2,\ \ldots,\ n-1$$

Dividing we get

$$\frac{P(x+1)}{P(x)} = \frac{{}^{n}C_{x+1}q^{n-x-1}p^{x+1}}{{}^{n}C_{x}q^{n-x}p^{x}} = \frac{n!}{(x+1)(n-x-1)!} \cdot \frac{x!(n-x)!}{n!}$$

$$= \frac{(n-x)(n-x-1)!x!}{(x+1)x!(n-x-1)} \cdot \frac{p}{q}$$

or

$$P(x+1) = \frac{(n-x)}{(x+1)} \cdot \frac{p}{q} \cdot P(x) \quad x = 0,\ 1,\ 2,\ \ldots,\ n-1$$

Hence the required recurrence relation is

$$P(x+1) = \frac{(n-x)}{(x+1)} \cdot \frac{p}{q} \cdot P(x) \quad x = 0,\ 1,\ 2,\ \ldots,\ n-1$$

Using the above recurrence relation, we can write

$$P(1) = n \cdot \frac{p}{q} \cdot P(0)$$

$$P(2) = \frac{n-1}{2} \cdot \frac{p}{q} \cdot P(1)$$

$$P(3) = \frac{n-2}{3} \cdot \frac{p}{q} \cdot P(2)$$

The recurrence relation is used to find the expected or theoretical frequencies, i.e., for fitting the Binomial distribution of a given data.

4.20.4 Moments, Skewness, and Kurtosis of the Binomial Distribution

Taking an arbitrary origin at 0 successes, we get

$$\begin{aligned}
\mu_1' &= \sum_{x=0}^{n} xP(x) \\
&= \sum_{x=0}^{n} x \cdot {}^{n}C_x p^x q^{n-x} \\
&= \sum_{x=0}^{n} x \cdot \frac{n!}{x!(n-x)!} p^x q^{n-x} \\
&= np \sum_{x=1}^{n} \frac{n!}{x![n-1-(x-1)!} p^{x-1} q^{n-x} \\
&= np \sum_{x=1}^{n} {}^{n-1}C_{x-1} p^{x-1} q^{n-1-(x-1)} \\
&= np(p+q)^{n-1} \\
&= np \cdot 1 = np
\end{aligned}$$

$$\therefore \quad \text{Mean} = \mu_1' = E(X) = np$$

$$\begin{aligned}
\mu_2' &= E(X)^2 = E[X(X-1)] + E(X) \\
&= \sum_{x=0}^{n} x^{2n} C_x p^x q^{n-1} \\
&= \sum_{x=0}^{n} [x + n(x-1)]^n C_x p^x q^{n-x} \\
&= \sum_{x=0}^{n} x^n C_x p^x q^{n-x} + \sum_{x=0}^{n} x(x-1)^n C_x p^x q^{n-x} \\
&= np + n(n-1)p^2 \sum_{x=2}^{n} {}^{n-2}C_{x-2} p^{x-2} q^{n-x} \\
&= np + (n^2p^2 - np^2) \sum_{x=2}^{n} {}^{n-2}C_{x-2} p^{x-2} q^{(n-2)-(x-2)} \\
&= np + (n^2p^2 - np^2)(p+q)^{n-2} \\
&= n^2p^2 - np^2 + np
\end{aligned}$$

$$\text{Variance} = v[x] = \mu_2' - \mu_1'^2$$

i.e.,

$$\mu_2 = E(X^2) - [E(X)]^2$$
$$= n^2p^2 - np^2 + np - (np)^2$$
$$= np - np^2$$
$$= np(1 - p)$$
$$= npq$$

Again we have

$$\mu_3' = \sum_{x=0}^{n} x^3 p(x)$$

$$= \sum_{x=0}^{n} [x(x-1)(x-2) + 3x(x-1) + x]p(x)$$

$$(\text{since } x^3 = x(x-1)(x-2) + 3x(x-1) + x)$$

$$= \sum_{x=0}^{n} [x(x-1)(x-2) + 3\sum_{x=0}^{n} x(x-1) + p(x) + \sum_{x=0}^{n} xP(x)$$

$$= n(n-1)(n-2)p^3(q+p)^{n-3} + 3n(n-1)p^2(q+p)^{n-2} + np(q+p)^{n-1}$$

$$= n(n-1)(n-2)p^3 + 3n(n-1)p^2 + np$$

Since $\mu_3 = \mu_3' - 3\mu_2'\mu_1' + 2\mu_1'^3$
We get

$$\mu_3 = n(n-1)(n-2)p^3 + 3n(n-1)p^2 + np - 3(n^2p^2 - np^2 + np)(np) + 2n^3p^3$$
$$= npq(q-p)$$

Similarly we get

$$\mu_4' = n(n-1)(n-2)(n-3)p^4 + 6n(n-1)(n-2)p^3 + 7n(n-1)p^2 + np$$

Since $\mu_4 = \mu_4' - 4\mu_3'\mu_1' + 6\mu_2\mu'^2 - 3\mu_1'^4$
We get

$$\mu_4 = npq[1 + 3(n-2)pq]$$

Hence we get

$$\beta_1 = \frac{\mu_3^2}{\mu_2^3} = (npq)^2 \frac{(q-p)^2}{(npq)^3} = \frac{(1-2p)^2}{npq}$$

$$r_1 = \frac{1-2p}{\sqrt{npq}}$$

$$\beta_2 = \frac{\mu_4}{\mu_2^2} = 3 + \frac{1-6pq}{npq} \text{ and } r_2 = \frac{1-6pq}{npq}$$

When $p < (1/2)$, the distribution is positively skewed and when $p > (1/2)$, the distribution is negatively skewed. When $p = (1/2)$ the distribution is symmetric.

Since coefficient of kurtosis is given by $r_2 = \beta_2 - 3 = (1 - 6pq)/npq$
If $6pq < 1$, i.e., $pq < \frac{1}{6}$, $r_2 > 0$, the distribution is Leptokurtic.
If $6pq = 1$, i.e., $pq = \frac{1}{6}$, $r_2 = 0$, the distribution is Mesokurtic.
If $6pq > 1$, i.e., $pq > \frac{1}{6}$, $r_2 < 0$, the distribution is Platykurtic.

4.20.5 Moment Generating Function of a Binomial Distribution

From the definition we have $P(X = x) = {}^nC_x p^x q^{n-x}$

Therefore

$$\begin{aligned} M(t) &= E[e^t X] \\ &= \sum_1^n e^{tx\,n} C_x p^x q^{n-x} \\ &= \sum_1^n {}^nC_x (pe^t)^x q^{n-x} \\ &= (q+pe^t)^n \end{aligned} \tag{4.13}$$

or

$$\begin{aligned} M(t) &= q + p\left(1+t+\tfrac{t^2}{2!} + \tfrac{t^3}{3!+\cdots}\right)^n \\ &= \left(q+p+pt+\tfrac{pt^2}{2!} + \tfrac{pt^3}{3!+\cdots}\right)^n \\ &= \left(1+pt+\tfrac{pt^2}{2!} + \tfrac{pt^3}{3!+\cdots}\right)^n \end{aligned} \tag{4.14}$$

is the mgf for the Binomial distribution.

4.20.6 Characteristics of a Binomial Distribution

Binomial distribution is a discrete probability distribution. It is applied when the number of repeated trials of any experiment is finite and fixed. Each trial is a Bernoulli's trial. The trials are all independent. The probability of success of any trial is the same. It is denoted by p. If n denotes the number of trials, n and p are called the parameters of the binomial distribution. The mean of a Binomial distribution is np and the variance is npq. If $p = 0.5$ the binomial distribution is symmetric. If p is very small and less than 0.5, then the Binomial distribution is skewed to the right. If $p > 0.5$, the distribution is skewed to the left.

4.20.6.1 Solved Examples

Example 4.38: A die is thrown 4 times. Getting a number greater than 2 is a success. Find the probability of getting (1) exactly one success; (2) less than 3 success?

Solution: We have $n = 4$

$$P = \text{probability of getting a number greater than } 2 = \frac{4}{6} = \frac{2}{3}$$

$$\therefore \quad q = 1 - p = 1 - \frac{2}{3} = \frac{1}{3}$$

1. Probability of getting one success $= P(X = 1)$

$$= {}^4C_1 \left(\frac{2}{3}\right)^1 \left(\frac{1}{3}\right)^{4-1}$$

$$= 4 \cdot \frac{2}{3} \cdot \frac{1}{27} = \frac{8}{81}$$

2. Probability of getting less than three successes $= P(X < 3)$

$$= P(X = 0) + P(X = 1) + P(X = 2)$$

$$= {}^4C_0 \left(\frac{2}{3}\right)^0 \left(\frac{1}{3}\right)^4 + {}^4C_1 \left(\frac{2}{3}\right)^1 \left(\frac{1}{3}\right)^{4-1} + {}^4C_2 \left(\frac{2}{3}\right)^2 \left(\frac{1}{3}\right)^{4-2}$$

$$= \frac{1}{81} + 4 \cdot \frac{2}{3} \cdot \frac{1}{27} + 6 \cdot \frac{4}{9} \cdot \frac{1}{9}$$

$$= \frac{1 + 8 + 24}{81} = \frac{33}{81}$$

Example 4.39: If the chance that any one of five telephone lines is busy at any instant 0.01, what is probability that all the lines are busy? What is the probability that more than three lines are busy?

Solution: It is given that $n = 5$, $p = 0.01$

$$q = 1 - p = 1 - 0.01 = 0.99$$

Therefore the probability that all the lines are busy $= P(X = 5)$

$$\begin{aligned} &= {}^5C_5(0.01)^5(0.99)^0 \\ &= (0.01)^5 \end{aligned}$$

Probability that more than two lines are busy $= P(X > 3) = P(X = 4) + P(X = 5)$

$$\begin{aligned} &= {}^5C_4(0.01)^4(0.99)^1 + {}^5C_5(0.01)^5(0.99)^0 \\ &= 5(0.01)^4(0.99) + (0.01)^5 \\ &= (0.01)^4(5 \times 0.99 + 0.01) \\ &= (0.01)^4(4.95 + 0.01) \\ &= (0.01)^4(4.96) \end{aligned}$$

Example 4.40: A box contains 100 tickets each bearing one of the numbers from 1 to 100. If five tickets are drawn successively with replacement from the box, find the probability that all the tickets bear number divisible by 10?

Solution: The numbers that are divisible by 10 are 10, 20, 30, . . ., 100.

There are 10 numbers (between 1 and 100 (inclusive)), which are divisible by 10

We have, $n = 10$, $p = \frac{1}{100} = \frac{1}{10}$, $q = 1 - p = 1 - \frac{1}{10} = \frac{9}{10}$

Probability that all the tickets drawn bear the number divisible by $10 = P(X = 5)$

$$= {}^5C_5\left(\frac{1}{10}\right)^5\left(\frac{9}{10}\right)^0 = \frac{1}{10^5}$$

Example 4.41: A die is thrown three times. Getting a "3" or a "6" is considered to be success. Find the probability of getting at least two success?

Solution: Probability of getting a 3 or 6 $= p = \frac{2}{6} = \frac{1}{3}$

Probability of getting at least two successes $= P(X \geq 2) = P(X = 2) + P(X = 3)$

$$= {}^{3}C_{2}\left(\frac{1}{3}\right)^{2}\left(\frac{2}{3}\right)^{1} = {}^{3}C_{3}\left(\frac{1}{3}\right)^{3}\left(\frac{2}{3}\right)^{0}$$

$$= 3\cdot\frac{1}{9}\cdot\frac{2}{3} + \frac{1}{27} = \frac{7}{27}$$

Example 4.42: If 20% of the bolts produced by a machine are defective, determine the probability that out of 4 bolts chosen at random (a) 1 (b) 0, will be defective.

Solution:

Probability that the bolt is defective $= p = \dfrac{20}{100} = 0.2$

$$q = 1 - p = 1 - 0.2 = 0.8$$

We have $n = 4$

(a) $P(X = 1) = {}^{4}C_{1}(0.2)^{1}(0.8)^{3}$

$$= 4\,(0.2)(0.8)^{3} = 0.4096$$

(b) $P(X = 0) = {}^{4}C_{0}(0.2)^{0}(0.8)^{4} = 0.4096$

Example 4.43: Out of 1000 families of 3 children each, how many families would you expect to have two boys and one girl assuming that boys and girls are equally likely?

Solution: Probability of having a boy $= p = \dfrac{1}{2}$

We have

$$q = 1 - p = 1 - \frac{1}{2} = \frac{1}{2}, \quad n = 3, \quad N = 1000$$

Expected number of families having two boys and one girl $= 1000\cdot{}^{3}C_{2}\left(\frac{1}{2}\right)^{2}\left(\frac{1}{2}\right)^{1}$

$$= 1000\cdot 3\cdot\frac{1}{4}\cdot\frac{1}{2} = 375$$

Example 4.44: The average percentage of failures in certain examination is 40. What is the probability that out of a group of six candidates, at least four pass in the examination?

Solution: Let p be the probability that a candidate passes the examination.

$$\therefore \quad p = \frac{60}{100} \text{(given)}$$

We have $p = 0.6$, $q = 0.4$, $n = 6$

$$P(X \geq 4) = \text{Probability that at least four candidates pass the examination}$$

$$= P(X = 4) + P(X = 5) + P(X = 6)$$

$$= {}^6C_4(0.6)^4(0.4)^2 + {}^6C_5(0.6)^5(0.4) + {}^6C_6(0.6)^6(0.4)^0$$

$$= (15)(0.6)^4(0.4)^2 + (6)(0.6)^5(0.4) + (0.6)^6$$

$$= 0.5443$$

Example 4.45: X follows binomial distribution such that $4P(x = 4) = P(x = 2)$

If $n = 6$, find p the probability of success?

Solution: We have

$$4P(x = 4) = P(x = 2) \ \ (given)$$

or

$$4 \cdot {}^6C_4 \cdot p^4 q^2 = {}^6C_2 \cdot p^2 q^4$$

or

$$4p^2 = q^2$$

or

$$4p^2 = (1 - p)^2 = 1 + p^2 - 2p$$

or

$$3p^2 = 2p - 1 = 0$$

or

$$p = \frac{-2 \pm \sqrt{4 + 12}}{(2)(3)} = \frac{-2 \pm 4}{6} = \frac{-6}{6} \text{ or } \frac{2}{6}$$

$$= -1 \text{ or } \frac{1}{3}$$

p cannot be negative, therefore we consider $p = 1/3$

Thus $p = 1/3$ (since $p = -1$ is admissible)

Example 4.46: Find the maximum n such that the probability of getting no head in tossing a coin n times is greater than 0.1?

Solution: Let p denote the probability of getting a head. Then

$$p = \frac{1}{2}, \quad q = 1 - p = 1 - \frac{1}{2} = \frac{1}{2}$$

Let n be the number of trials, such that $P(X = 0) > 0.1$

i.e.,

$$^nC_0 p^0 q^{n-0} > 0.1$$

or

$$\left(\frac{1}{2}\right)^n > 0.1$$

When

$n = 1$, we get $\frac{1}{2} > 0.5$

$n = 2$, we get $\left(\frac{1}{2}\right)^2 = 0.25 > 0.1$

$n = 3$, we get $\left(\frac{1}{2}\right)^3 = 0.125 > 0.1$

$n = 4$, we get $\left(\frac{1}{2}\right)^4 = 0.0625 < 0.1$

and the maximum value for such that $P(X = 0) > .1$ holds is 3. Hence the maximum n such that $P(X = 0) > 0.1$ is $n = 4$.

Example 4.47: If the sum of the mean and variance of a binomial distribution of 5 trials is 9/5, find the binomial distribution?

Solution: Let n be the number of trials, p be the probability of success, and q be the probability of failure.

Then we have

$$np + npq = \frac{9}{5} \quad \text{(given)}, \quad n = 5$$

or

$$5p + 5p(1 - p) = \frac{9}{5}$$

or

$$25p + 25p - 25p^2 = 9$$

or

$$25p^2 - 50p + 9 = 0$$

or

$$p = \frac{50 \pm \sqrt{2500 - 900}}{(2)(25)} = \frac{50 \pm 40}{50} = \frac{9}{5} \text{ or } \frac{1}{5}$$

$p = (9/5) > 1$ is not admissible. Therefore $p = (1/5)$.

The required binomial distribution is $P(X = x) = {}^5C_x \left(\frac{1}{5}\right)^x \left(\frac{4}{5}\right)^{5-x}, 0 \leq x \leq 5.$

Example 4.48: If the probability of a defective bolt is 1/10.

(1) Find the mean; (2) Variance; (3) moment of coefficient skewness; and (4) Kurtosis for the distribution of defective bolts is a total of 400?

Solution: p = Probability of a defective bolt $= \frac{1}{10}$ (given)

We have $n = 400$, $p = \frac{1}{10}$, $q = \frac{9}{10}$

Therefore

1. Mean $= E(X) = np = 400 \cdot \frac{1}{10} = 40$
2. Variance $= npq = 400 \cdot \frac{1}{10} \cdot \frac{9}{10} = 36$
3. Moment of coefficient skewness $r_1 = \frac{Q - 6}{\sqrt{npq}} = \frac{\frac{9}{10} - \frac{1}{10}}{\sqrt{36}} = \frac{\frac{8}{10}}{6}$

$$= \frac{8}{60} = \frac{2}{15} = 0.1333$$

4. Coefficient of Kurtosis $\mathrm{r}_2 = \frac{1 - 6pq}{npq} = \frac{1 - 6 \cdot \frac{1}{10} \cdot \frac{9}{10}}{36} = \frac{1 - 0.54}{36}$

$$= \frac{0.46}{36} = 0.01277$$

Example 4.49: Using recurrence formula of the binomial distribution, compute $P(x \text{ successes})$ for $x = 1, 2, 3, 4, 5$ given $n = 5$ and $p = \frac{1}{6}$.

Solution: We have $p = \frac{1}{6}$, $q = 1 - p = 1 - \frac{1}{6} = \frac{5}{6}$, $n = 5$

Therefore

$$P(X = x) = {}^{n}C_{x}\, p^{x} q^{n-x};\ \ 0 \leq x \leq n \qquad = {}^{5}C_{x}\left(\frac{1}{6}\right)^{x}\left(\frac{5}{6}\right)^{5-x};\ \ 0 \leq x \leq 5$$

The recurrence formula of binomial distribution is $P(x+1) = \frac{n-x}{x+1} \cdot \frac{p}{q} \cdot P(x), \quad 0 \leq x \leq n-1$

We have

$$P(X = 0) = P(0) = {}^{5}C_{0}\left(\frac{1}{6}\right)^{0}\left(\frac{5}{6}\right)^{5-0} = \frac{3125}{7776}$$

Using the recurrence formula we get

$$P(1) = \frac{5-0}{0+1} \cdot \frac{\left(\frac{1}{6}\right)}{\left(\frac{5}{6}\right)} P(0) = 5 \cdot \frac{1}{5} P(0) = \frac{3125}{7776}$$

$$P(2) = \frac{5-1}{1+1} \cdot \frac{(1)}{(5)} P(1) = \frac{2}{5}\frac{3125}{7776} = \frac{1250}{7776}$$

$$P(3) = \frac{5-2}{2+1} \cdot \frac{(1)}{(5)} P(2) = \frac{3}{3} \cdot \frac{1}{5} \cdot \frac{1250}{7776} = \frac{250}{7776}$$

$$P(4) = \frac{5-3}{3+1} \cdot \frac{(1)}{(5)} P(3) = \frac{2}{4} \cdot \frac{1}{5} \cdot \frac{250}{7776} = \frac{25}{7776}$$

$$P(5) = \frac{5-4}{4+1} \cdot \frac{(1)}{(5)} P(4) = \frac{1}{5} \cdot \frac{1}{5} \cdot \frac{25}{7776} = \frac{1}{7776}$$

The required probabilities are $\frac{3125}{7776}, \frac{1250}{7776}, \frac{250}{7776}, \frac{25}{7776}$, and $\frac{1}{7776}$.

Example 4.50: In 256 sets of 12 tosses of a coin, in how many cases may one expect 8 heads and 4 tails?

Solution: Probability of getting a head in a single toss $= p = \frac{1}{2}$

We have $n = 12$, $q = \frac{1}{2}$, $x =$ number of heads.

Probability of getting 8 heads and 4 tails $= P(X = 8) = {}^{12}C_{8} p^{8} q^{4}$

$$= {}^{12}C_{8}\left(\frac{1}{2}\right)^{8}\left(\frac{1}{2}\right)^{4}$$

Expected value of getting 8 heads and 4 tails in 256 tosses is

$$N.P(X=8) = 256\left(\frac{1}{2}\right)^{12}$$
$$= 31$$

Example 4.51: Fit a binomial distribution and calculate the expected frequencies

x	0	1	2	3	4	5
f	10	20	30	15	15	10

Solution: Mean of the given distribution is

$$\mu = E(X) = \frac{(0)(10)+(1)(20)+(3)(15)+(4)(15)+(5)(10)}{10+20+30+15+15+10} = \frac{235}{100} = 2.35$$

We have $n = 5$

Since $\mu = np = 2.35$, we get $5p = 2.35$

or $p = \dfrac{2.35}{5} = 0.47$

Hence

$$q = 1 - p = 1 - 0.47 = 0.53$$
$$P(X=0) = {}^nC_0 p^0 q^{n-0}$$
$$= {}^5C_0(0.47)^0(0.53)^{5-0}$$
$$= (0.53)^5$$
$$= 0.0418$$

$$N = \sum_{i=1}^{5} f_i = 10 + 20 + 30 + 15 + 15 + 10 = 100$$

Therefore expected frequencies are

$$N.P(X=0) = 100 \times (0.0418) = 4.18 \cong 4$$
$$N.P(X=1) = 100 \times {}^5C_1(0.47)(0.53)^4$$
$$= 100 \times 5(0.47)(0.53)^4 = 18.53 \cong 19$$
$$N.P(X=2) = 100 \times {}^5C_2(0.47)^2(0.53)^3$$
$$= 32.87 \cong 33$$
$$N.P(X=3) = 100 \times {}^5C_3(0.47)^3(0.53)^2$$
$$= 29.15 \cong 29$$
$$N.P(X=4) = 100 \times {}^5C_4(0.47)^4(0.53)$$
$$= 12.93 \cong 13$$
$$N.P(X=5) = 100 \times {}^5C_5(0.47)^5(0.53)$$
$$= 2.29 \cong 2$$

The expected frequencies are 4, 19, 33, 29, 13, and 2.

The required Binomial distribution is

X	0	1	2	3	4	5
$N\cdot P(X=x)$	4	19	33	29	13	2

That is, the required Binomial distribution of fit is $N\,(p+q)^n = 100\,(0.53+0.47)^5$

Example 4.52: Five dice were thrown 96 times and the number of times an odd number actually turned one in the experiment is given below. Calculate the expected frequencies?

Number of dice showing 1 or 3 or 5	0	1	2	3	4	5
Observed frequency	1	10	24	35	18	8

Solution: Probability of getting an odd number when a dice is thrown $= \frac{3}{6} = \frac{1}{2}$

$\therefore\ q = \frac{1}{2}$

We have

$$N = 1 + 10 + 24 + 35 + 18 + 8 = 96$$
$$n = 5$$

The expected (theoretical) frequencies are

$$N\cdot P(X=0) = 96 \times {}^5C_0\left(\frac{1}{2}\right)^0\left(\frac{1}{2}\right)^{5-0} = 3$$

$$N\cdot P(X=1) = 96 \times {}^5C_1\left(\frac{1}{2}\right)^1\left(\frac{1}{2}\right)^{5-1} = 15$$

$$N\cdot P(X=2) = 96 \times {}^5C_2\left(\frac{1}{2}\right)^2\left(\frac{1}{2}\right)^{5-2} = 30$$

$$N\cdot P(X=3) = 96 \times {}^5C_3\left(\frac{1}{2}\right)^3\left(\frac{1}{2}\right)^{5-3} = 30$$

$$N\cdot P(X=4) = 96 \times {}^5C_4\left(\frac{1}{2}\right)^4\left(\frac{1}{2}\right)^{5-4} = 15$$

$$N\cdot P(X=5) = 96 \times {}^5C_5\left(\frac{1}{2}\right)^5\left(\frac{1}{2}\right)^{0} = 3$$

The binomial distribution of fit is $N(q+p)^n = 96\left(\frac{1}{2}+\frac{1}{2}\right)^5$.

Example 4.53: Show that for the binomial distribution

$$\mu_{r+1} = pq\left(n\mu_{r+1} + \frac{d\mu_r}{dp}\right)$$

where μ_r is the rth moment about the mean. Hence obtain μ_2, μ_3, and μ_4.

Solution: We know that mean of the binomial distribution is $\mu = np$. Therefore

$$\mu_r = \sum_{x=0}^{n}(x-\mu)^r p(x)$$

$$= \sum_{x=0}^{n}(x-np)^r p(x)$$

$$= \sum_{x=0}^{n}(x-np)^r \cdot {}^nC_x p^x q^{n-x}$$

Differentiating with respect to p we get

$$\frac{d\mu_r}{dp} = \sum_{x=0}^{n} -r_n(x-np)^{r-1} \cdot {}^nC_x p^x q^{n-x} + \sum_{x=0}^{n}[(x-np)^r[{}^nC_x p^x q^{n-x} + {}^nC_x p^x (n-x) q^{n-x-1}]]$$

$$= -nr\sum_{x=0}^{n}(x-np)^{r-1} p^x q^{n-x} + \frac{1}{pq}\left[\sum_{x=1}^{n} {}^nC_x p^x q^{n-x}(x-np)^r(xq-np+xp)\right]$$

$$= -nr\mu_{r-1} + \frac{1}{pq}\left[\sum_{x=0}^{n} {}^nC_x p^x q^{n-x}(x-np)^{r+1}\right]$$

$$\text{or } = \frac{d\mu_r}{dp} + -nr\mu_{r-1} = \frac{1}{pq}\left[\sum_{x=0}^{n} {}^nC_x p^x q^{n-x}(x-np)^{r+1}\right]$$

$$\text{or } pq\left[\frac{d\mu_r}{dp} + nr\mu_{r-1}\right] = \mu_{r+1}$$

(4.15)

Hence proved.
Putting $r = 1, 2, 3$ successively we get

$$\mu_2 = pq\left[\frac{d\mu_1}{dp} + n\mu_0\right] = pq(0 + n) = npq$$

$$\mu_3 = pq\left[\frac{d\mu_2}{dp} + n\mu_1\right] = p(nq - np + 0) = npq(q - p)$$

and $$\mu_4 = pq\left[\frac{d\mu_3}{dp} + 3\mu_2\right] = pq(n(6 - p^2 - 6p + 1) + 3n(np)(1 - p)$$

$$= npq[1 - 6p(1 - p) + 3npq]$$

$$= npq[1 - 6pq] + 3n^2p^2q^2]$$

Exercise 4.7

1. Eight coins are tossed simultaneously, find the probability of getting at least 6 heads?

   Ans: $\frac{37}{256}$
2. Assuming that it is true that 2 out of 10 industrial accidents are due to fatigue, find that exactly 2 of 8 industrial accidents will be due to fatigue?

   Ans: 0.29
3. The average percentage of failures in a certain examination is 40. What is the probability that out of a group of 6 candidates, at least 4 pass in the examination?

   Ans: 0.5443
4. X is a binomial random variable. If 4 $P(X = 4) = P(X = 2)$ find p?

   Ans: $\frac{1}{3}$
5. In a binomial distribution, the mean and SD are 8 and 2, respectively. Find n and p?

   Ans: $n = 16$, $p = 1$
6. An oil exploration firm finds that 55 of the test wells it drills yield a deposit of natural gas. If it drills 6 wells, find the probability that at least 1 well will yield gas?

   Ans: $1 - (0.95)^6$

7. A fair coin is tossed 6 times. Find the probability of getting at least 7 heads?

   Ans: $\frac{15}{64}$

8. Ten coins are thrown simultaneously. Find the probability of getting at least 7 heads?

   Ans: 0.1719

9. In four throws with a pair of dice, what is the probability of throwing a doublet at least twice?

   Ans: $\frac{171}{1296}$

10. In a throw of 4 dice, find the probability that at least one die shows up 4?

    Ans: $\frac{671}{1296}$

11. The probability of a man hitting a target is $1/4$. He fires 7 times. What is the probability of his hitting at least twice the target?

    Ans: $\frac{4547}{8192}$

12. The items produced by a firm are supposed to be 5% defective. What is the probability that 8 items will contain less than 2 defective items?

    Ans: $\left(\frac{19}{20}\right)^7 \left(\frac{27}{20}\right)$

13. Assuming that half of the population is vegetarian so that the chance of an individual being vegetarian is $1/2$ and assuming that 100 investigators can take samples of 10 individuals to see whether they are vegetarian, how many investigators would you expect to report that three people or less were vegetarian?

    Ans: 17

14. A box contains 100 tickets each bearing one of the numbers from 1 to 100. If 5 tickets are drawn successively with replacement from the box, find the probability that all the tickets bear number divisible by 10?

    Ans: $\frac{1}{10^5}$

15. If the probability of success is $1/20$, how many trials are necessary in order that the probability of at least one success is just greater than $1/2$?

    Ans: 14

16. If the probability that a man aged 60 will live to be 70 of 0.65. What is the probability that out of 10 men now 60, at least 7 will live to be 70?

Ans: 0.5139

17. The mean and variance of a binomial variable X with parameters n and p are 16 and 8. Find $P(X \geq 1)$ and $P(X > 2)$?

Ans: $1 - \frac{1}{2^{32}}$ and 0.999

18. In 256 sets of 12 tosses of a coin, in how many tosses one can expect 8 heads and 4 tails?

Ans: 31

19. If 3 of 20 tyres are defective and 4 of them are randomly chosen for inspection, what is the probability that only one of the defective tyre will be included?

Ans: 0.3685

20. The mean and variance of a binomial distribution are 4 and 4/3, respectively. Find $P(X \geq 1)$?

Ans: 0.998

21. Assume that 50% of all engineering students are good in mathematics, determine the probability that among 18 engineering students (1) exactly 10; (2) at least 10; (3) at most 8; (4) at least 2 and at most 9, are good in mathematics.

Ans: (1) 0.9982; (2) $\left(\frac{1}{2}\right)^{18}\left({}^{18}C_{10} + {}^{18}C_{11} + \cdots + {}^{18}C_{18}\right)$;

(3) $\left(\frac{1}{2}\right)^{18}\left({}^{18}C_0 + {}^{18}C_1 + \cdots + {}^{18}C_8\right)$;

(4) $\left(\frac{1}{2}\right)^{18}\left({}^{18}C_2 + {}^{18}C_3 + \cdots + {}^{18}C_9\right)$

22. Six dice are thrown 729 times. How many times would you expect at least three dice to show a 5 or 6?

Ans: 233

23. Out of 800 families with 4 children each, how many would you expect to have 3 boys and 1 girl? Assume equal probabilities for boys and girls.

Ans: 200

24. The mean of a binomial distribution is 3 and variance is 3/16. Find (a) the value of n; (b) $P(X \geq 7)$; (c) $P(1 \leq X \leq 6)$?

Ans: (a) $n = 12$; (b) 0.1446; (c) 0.82

25. A pair of dice is thrown. Find the probability of getting a sum of 11 (a) once; (b) twice?

Ans: (a) 250; (b) 25

26. Fit a binomial distribution for the data.

X	0	1	2	3	4	5
***f*(*x*)**	38	144	342	287	164	25

Ans: 33, 162, 316, 309, 151, 29

27. Seven coins are tossed at a time, 128 times. The number of heads observed at each throw is recorded and the results are given below:

No. of heads	0	1	2	3	4	5	6	7
Frequency	7	6	19	35	30	23	7	1

Fit a binomial distribution to the data assuming that the coins are (a) biased; (b) unbiased.

Ans: (a) $128[0.517 + 0.483]^7$; (b) 1, 7, 21, 35, 35,21, 7, 1

28. Fit a binomial distribution to the following frequency distribution:

x	0	1	2	3	4	5	6
***f*(*x*)**	13	25	52	38	32	16	4

Ans: 6, 28, 56, 60, 36, 12, 2

29. The probability of a newly generated virus attacked to the computer system will corrupt 4 files out of 20 files, which are opened in an hour. If 20 files are opened in an hour, find the probability that

a. At least 10 files are corrupted

b. Exactly 3 files are corrupted

c. All the files are corrupted

d. All the files are safe.

Ans: (a) 0.0000045; (b) 0.001073; (c) $\frac{1}{512}$; (d) 0.9999

30. Fit a binomial distribution for the following data:

x	0	1	2	3	4	5
$f(x)$	2	14	20	34	22	8

Ans: 2, 10, 26, 34, 22, 6

31. Fit a binomial distribution whose mean is 9 and SD is 3/2.

Ans: $^{12}C_x\left(\frac{3}{4}\right)^x\left(\frac{1}{4}\right)^{12-x} \quad (0 \le x \le 12)$

32. The sum of mean and variance of a binomial distribution is 15 and the sum of their squares is 117. Find the distribution?

Ans: ${}^{27}C_x\left(\frac{1}{3}\right)^x\left(\frac{2}{3}\right)^{27-x} \quad (0 \le x \le 27)$

33. If on an average 1 vessel in every 10 is wrecked. Find the probability that out of 5 vessels expected to arrive, at least 4 will arrive safely?

Ans: 0.9185

34. The probability that a bulb produced by a factory will fuse after 160 days of use is 0.06. Find the probability that out of 5 such bulbs, at most 1 bulb will fuse after 160 days in use?

Ans: 0.9681

35. Find the expectation of the number of heads in 15 tosses of a coin?

Ans: 7.5

36. The incidence of an occupational disease in an industry is such that the workmen have a 20% chance of suffering from it. What is the probability that out of 6 workmen, 4 or more will contact the disease?

Ans: 0.016

37. The probability of a defective bolt is 0.2. Find (a) Mean, (b) SD of the distribution, (c) Coefficient of skewness β_1, and (d) coefficient of kurtosis β_2?

Ans: (a) 200; (b) 12.6; (c) $\beta_1 = 0.225$; $^{(d)}$ $\beta_2 = 3.0025$

38. Show that mean of a binomial distribution is always greater than its variance.

39. Determine the binomial distribution whose mean is 5 and SD is $\sqrt{2.5}$.

Ans: ${}^{10}C_x\left(\frac{1}{2}\right)^x\left(\frac{1}{2}\right)^{10-x} \quad (0 \le x \le 10)$

40. The mean of a binomial distribution is 20 and SD is 4. Find n, p, and q?

Ans: $p = 0.2$, $q = 0.8$, $n = 100$

4.21 POISSON DISTRIBUTION

A type of probability distribution useful in describing the number of events that will occur in a specific period of time or in a specific area or volume is the Poisson distribution. It is a probability distribution discovered by French mathematician Simon Denis Poisson (1781–1840) in the year 1837. A use for this distribution was not found until 1898, when an individual named Bortkiewicz was tasked by the Prussian army to investigate accidental deaths of soldiers attributed to being kicked by horses. The initial application of the Poisson distribution to number of deaths attributed to horse kicks in the Prussian army led to its use in analyzing the accidental deaths, and service requirements.

The following are the examples of random variables for which the Poisson probability distribution provides a good model:

1. The number of car accidents in a year, on a road.
2. The number of earthquakes in a year.
3. The number of breakdowns of an electronic computer.
4. The number of printing mistakes at each page of the book.
5. The number of defective screws per box of 10 screws.

The distribution is useful in testing the randomness of a set of data and fitting of empirical data to a theoretical curve.

4.21.1 Conditions Under Which Poisson Distribution Is Used

The conditions for the applicability of the Poisson distribution are the same as those for the applicability of the Binomial distribution.

The additional requirement is that the probability of success is very small. The conditions are:

1. The random variable X is discrete.
2. The number of trials is indefinitely very large, i.e., $n \to \infty$.
3. The probability of success in a trail is very small, i.e., p is close to zero.
4. The product of n and p is constant.

A random variable that counts the number of successes in an experiment is called a Poisson variable.

4.21.2 Poisson Probability Function

In this section we show that Poisson distribution is the limiting form of a binomial distribution and define Poisson distribution.

Let an experiment satisfying the conditions for Poisson distribution be formed. Let n denote the number of trials, and p denote the probability of success. We assume that n is indefinitely very large and p is very small.

Using the definition of binomial distribution, we have,

$$\begin{aligned} P(X=x) &= \text{Probability of } x \text{ successes} \\ &= {}^{n}C_{x} p^{x} q^{n-x}, \ 0 \le x \le n, \ \text{ and } \ q = 1 - p \end{aligned}$$

Let $np = \lambda$, then $p = \frac{\lambda}{n}$ and $q = 1 - p = 1 - \frac{\lambda}{n}$

Therefore we get

$$P(X=x) = {}^{n}C_{x}p^{x}q^{n-x}$$

$$= \frac{n!}{x!(n-x)!}\left(\frac{\lambda}{n}\right)^{x}\left(\frac{1-\lambda}{n}\right)^{n-x}$$

$$= \frac{n(n-1)\ldots(n-x+1)(n-x)!}{x!(n-x)!}\left(\frac{\lambda}{n}\right)^{x}\left(1-\frac{\lambda}{n}\right)^{n-x}$$

$$= \frac{\left(\frac{n}{n}\right)\left(\frac{n-1}{n}\right)\ldots\left(\frac{n-(x-1)}{n}\right)}{x!}(\lambda)^{x}\left(1-\frac{\lambda}{n}\right)^{n}\left(1-\frac{\lambda}{n}\right)^{-x}$$

$$= \frac{\lambda^{x}}{x!}\left(1-\frac{1}{n}\right)\left(1-\frac{2}{n}\right)\ldots\left(1-\frac{x-1}{n}\right)\left(1-\frac{\lambda}{n}\right)^{n}\left(1-\frac{\lambda}{n}\right)^{-x}$$

$$\therefore \lim_{n\to\infty} P(X=x) = \lim_{n\to\infty}\left[\frac{\lambda^{x}}{x!}\left(1-\frac{1}{n}\right)\left(1-\frac{2}{n}\right)\ldots\left(1-\frac{x-1}{n}\right)\right]$$

$$\left[\left[\left(1-\frac{\lambda}{n}\right)^{\frac{-n}{\lambda}}\right]^{-\lambda}\left(1-\frac{\lambda}{n}\right)^{-x}\right]$$

$$= \lim_{n\to\infty}\frac{\lambda^{x}}{x!}\lim_{n\to\infty}\left(1-\frac{1}{n}\right)\left(1-\frac{2}{n}\right)$$

$$\ldots\left(1-\frac{x-1}{n}\right)\lim_{\frac{-n}{m}\to\infty}\left[\left(1-\frac{\lambda}{n}\right)^{\frac{-n}{\lambda}}\lim_{n\to\infty}\left(1-\frac{\lambda}{n}\right)^{-x}\right]$$

$$= \frac{\lambda^{x}}{x!}(1-0)(1-0)\ldots(1-0)\cdot e^{-x}\cdot(1-0)$$

$$= \frac{\lambda^{x}e^{-x}}{x!}$$

Therefore when n is indefinitely very large.

$$P(X=x) = \frac{\lambda^{x}e^{-x}}{x!}, \quad x = 0,\ 1,\ 2,\ 3,\ldots$$

which is called the Poisson probability function. The corresponding distribution is called Poisson distribution. λ is called the parameter of the distribution.

Definition 4.27:

(Poisson distribution) A discrete random variable X taking the values 0, 1, 2, ... is said to follow Poisson distribution with parameter λ, if its pmf is given by

$$\begin{aligned} P(X=x) &= \frac{\lambda^x e^{-x}}{x!}, \quad x=0,\ 1,\ 2,\ 3,\ldots\ldots \\ &= 0, \qquad \text{otherwise} \end{aligned}$$

In the above definition $P(X=x)$ is called Poisson probability function and the corresponding Poisson distribution is

X	0	1	2	...	x	...
$P(X=x)$	$\frac{\lambda^0 e^{-x}}{0!}$	$\frac{\lambda^1 e^{-x}}{1!}$	$\frac{\lambda^2 e^{-x}}{2!}$	....	$\frac{\lambda^x e^{-x}}{x!}$	...

4.21.3 Poisson Frequency Distribution

If an experiment satisfying the requirements of Poisson distribution is repeated N times, the expected frequency distribution of getting x successes is given by

$$F_x = N.P(X=x) = N\cdot\frac{\lambda^x e^{-x}}{x!} \quad x=0,\ 1,\ 2,\ 3,\ldots$$

Remarks:

1. Since $e^{-\lambda}>0$, $P(X=0)\geq 0$ for all $x=0,\ 1,\ 2,\ 3,\ \ldots$

2. $\sum_{x=0} P(X=x) = \sum_{x=0}^{\infty}\frac{e^{-\lambda}\lambda^x}{x!}$

$$= e^{-\lambda}\sum_{x=0}^{\infty}\frac{\lambda^x}{x!} = e^{-\lambda}\left[1+\frac{\lambda}{1!}+\frac{\lambda^2}{2!}+\ldots\right]$$

$$= e^{-\lambda}e^{\lambda} = e^0 = 1$$

From (1) and (2) $P(X=x) = \frac{\lambda^x e^{-x}}{x!}, \quad x=0,\ 1,\ 2,\ 3,\ldots$ is a pmf.

3. Poisson distribution is useful in predicting arrivals from hourly volume under the conditions of free flow one can compute the probability of

0, 1, 2, ..., x vehicles arriving per time interval of t seconds provided the hourly volume V, known. If V denotes the hourly volume, t denotes the time interval in seconds, n is the number of intervals per hour given by $n = 3600/t$ and λ denotes the average number of vehicles per second given by $\lambda = \frac{V}{3600/t} = \frac{Vt}{3600}$

Then the probability, $P(x)$, that x vehicles will arrive during any interval is

$$P(X = x) = \frac{\lambda^x e^{-\lambda}}{x!}, \quad x = 0, 1, 2, \ldots$$

The hourly frequency denoted by F_x of intervals containing r vehicles is n

$$P(x = r) = n\frac{\lambda^r e^{-\lambda}}{r!}$$

where $\lambda = \dfrac{Vt}{3600}$

If the period under consideration is different from an hour, the 3600 in the first parentheses would be replaced by the appropriate length of time in seconds. The value V, however, would still be the hourly volume.

4.21.4 Moment of a Poisson Distribution

Taking an arbitrary origin at 0 successes, we have

$$\mu_1' = \sum_{x=0}^{\infty} x \cdot \frac{e^{-\lambda}\lambda^x}{x!} = \sum_{x=0}^{\infty} \frac{e^{-\lambda}\lambda^x}{(x-1)!}$$

$$= e^{-\lambda}\left[1 + \frac{\lambda}{1!} + \frac{\lambda^2}{2!} + \ldots\right]$$

$$= \lambda e^{-\lambda}\left[1 + \frac{\lambda}{1!} + \frac{\lambda^2}{2!} + \ldots\right]$$

i.e., $E(X) = \lambda e^{-\lambda} e^x = \lambda$

Thus mean $= \mu_1' = E(X) = \lambda$

$$\mu_2' = \sum_{x=0}^{\infty} x^2 \cdot \frac{e^{-\lambda}\lambda^x}{x!} = \sum_{x=0}^{\infty}[x(x-1)+x]\frac{e^{-\lambda}\lambda^x}{x!}$$

$$= \sum_{x=0}^{\infty}[x(x-1)]\frac{e^{-\lambda}\lambda^x}{x!} + \sum_{x=0}^{\infty}[x]\frac{e^{-\lambda}\lambda^x}{x!}$$

$$= e^{-\lambda}\lambda^2\sum_{x=2}^{\infty}\frac{\lambda^{x-2}}{(x-2)!} + E(X)$$

$$= e^{-\lambda}\lambda^2 e^{\lambda} + \lambda = \lambda^2 + \lambda$$

Hence

$$\mu_2 = \mathrm{Var}[x] = \mu_2' - \mu'^2 = \lambda^2 + \lambda - \lambda^2$$
$$= \lambda$$

$$\mu_3' = \sum \lambda^3 \frac{e^{-\lambda}\lambda^x}{x!}$$

$$= e^{-\lambda}\sum_{x=2}^{\infty}\lambda^x \frac{[x(x-1)(x-2) + 3x(x-1) + x]}{x!}$$

$$= e^{-\lambda}\sum_{x=4}^{\infty}\frac{\lambda^{x-4}}{(x-4)!} + e^{-\lambda}\lambda^3\sum_{x=3}^{\infty}6\frac{\lambda^{x-3}}{(x-3)!} + e^{-\lambda}\lambda^2\sum_{x=2}^{\infty}7\frac{\lambda^{x-2}}{(x-2)!}$$

$$+ e^{-\lambda}\lambda\sum_{x=1}^{\infty}\frac{\lambda^{x-1}}{(x-1)!}$$

$$= e^{-\lambda}\lambda^4 e^{\lambda} + 6e^{-\lambda}\lambda^3 e^{\lambda} + 7e^{-\lambda}\lambda^2 e^{\lambda} + e^{-\lambda}\lambda e^{\lambda}$$

$$= \lambda^4 + 6\lambda^3 + 7\lambda^2\lambda$$

using the values of μ_1', μ_2', μ_3', μ_4' we get

$$\mu_3 = \mu_3' - 3\mu_2', \mu_1' + 2\mu_1'^3$$
$$= \lambda^3 + 3\lambda^2 + \lambda - 3(\lambda^2 + \lambda)\lambda + 2\lambda^3$$
$$= \lambda^3 + 3\lambda^2 + \lambda - 3\lambda^3 - 3\lambda^2 + 2\lambda^3$$
$$= \lambda$$

and

$$\begin{aligned}\mu_4 &= \mu_4' - 4\mu_3'\mu_1' + \mu_2'\mu_1'^{\,2} - 3\mu_1'^{\,4}\\ &= \lambda_4 + 6\lambda^3 + 7\lambda^2 + \lambda - 4(\lambda^3 + 3\lambda^2 + \lambda) + 6(\lambda^2 + \lambda)\lambda^2 - 3\lambda^4\\ &= 3\lambda^2 + \lambda\end{aligned}$$

We have

$$\beta_1 = \frac{\mu_3^2}{\mu_2^3} = \frac{\lambda^2}{\lambda^3} = \frac{1}{\lambda}, \quad r_1 = \sqrt{\beta_2} = \frac{1}{\lambda}$$

$$\beta_2 = \frac{\mu_4}{\mu_2} = \frac{3\lambda^2 + \lambda}{\lambda^2} = 3 + \frac{1}{\lambda}, \quad r_2 = \beta_2 - 3 = \frac{1}{\lambda}$$

The values of μ_1, μ_2, μ_3, μ_4 and β_1, β_2, r_1, r_2 are also called the constants of Poisson distribution. We observe that the variance and mean of a Poisson distribution are equal in magnitude.

4.21.5 Recurrence Relation

If X is Poisson variate with parameter, then we have

$$P(X = x) = \frac{\lambda^x e^{-x}}{x!}, \qquad x = 0,\ 1,\ 2,\ 3, \ldots$$

and

$$P(X = x + 1) = \frac{\lambda^{x+1} e^{-x}}{(x+1)!} = \frac{e^x e^{-x} \cdot e}{(x+1)X!}$$

Dividing, we get

$$\frac{P(X = x + 1)}{P(X = x)} = \frac{\lambda}{x + 1}$$

Therefore we obtain

$$P(X = x + 1) = \frac{\lambda}{x + 1} P(X - x)$$

or simply

$$P(x + 1) = \frac{\lambda}{x + 1} P(x)$$

which is called the relation between $P(x)$ and $P(x + 1)$. It is also known as the recurrence relation between probabilities of Poisson distribution.

4.21.6 Characteristics of Poisson Distribution

1. Poisson distribution is discrete distribution
2. If n is the number of trials, is large and p is small, the distribution gives a close approximation to the binomial distribution.
3. The distribution depends mainly on the value of the mean (i.e., λ)
4. Poisson distribution has only one parameter viz., λ
5. The experiment consists of counting the number of times a particular event occurs during a given unit of time or given area or volume.
6. The probability that an event occurs in a given unit of time, area, or volume is the same for all units.

4.21.7 Moment Generating Function of the Poisson Distribution

Mgf of a random variable X is $M(t) = \sum_{x=0}^{\infty} e^{tx} P(X = x)$

It X follows Poisson distribution then

$$\begin{aligned}
M(t) &= \sum_{x=0}^{\infty} e^{tx} \frac{\lambda^x e^{-\lambda}}{x!} \\
&= \sum_{x=0}^{\infty} \frac{e^{-\lambda}(\lambda e^t)^x}{x!} \\
&= e^{-\lambda} \sum_{x=0}^{\infty} \frac{(\lambda e^t)^x}{x!} \\
&= e^{-\lambda}\left[1 + \lambda e^t + \frac{\lambda^2 (e^t)^2}{2!} + \ldots\right] \\
&= e^{-\lambda} e^{\lambda e^t} \\
&= e^{\lambda(e^t - 1)}
\end{aligned}$$

Hence the mgf of the Poisson distribution is $M(t) = e^{\lambda(e^t - 1)}$

4.21.8 Reproductive Property of the Poisson Distribution

Theorem 11:

If two independent random variables X and Y have Poisson distributions with means λ_1 and λ_2, respectively, their sum $X + Y$ is a Poisson variable with mean $\lambda_1 + \lambda_2$.

Proof:
Let $M_X(t)$ and $M_Y(t)$ be the mgfs of X and Y and let $M(t)$ be the mgf of $X + Y$, then

$$M_X(t) = e^{\lambda_1(e^t - 1)} \quad \text{and} \quad M_Y(t) = e^{\lambda_2(e^t - 1)}$$

The mgf of the sum $X + Y$ is given by

$$\begin{aligned} M(t) &= M_X(t)M_Y(t) = e^{\lambda_1(e^t-1)}e^{x_2(e^t-1)} \\ &= e^{(\lambda_1+\lambda_2)(e^t-1)} \end{aligned} \tag{4.16}$$

Eq. (4.16) is the mgf of a Poisson distribution with mean $\lambda_1 + \lambda_2$. Hence the theorem.

4.21.8.1 Solved Examples

Example 4.54: There are 50 telephone lines in an exchange. The probability of them will be busy is 0.1. What is the probability that all the lines are busy?

Solution: Let X be the Poisson variable "number of busy line in the exchange"

By Poisson distribution $P(X = x) = \frac{\lambda^x e^{-x}}{x!}, \quad x = 0, 1, 2, 3, \ldots$

We have $n = 50$, $p = 0.1$

$$\lambda = np = (50)(0.1) = 5$$

Hence $P(\text{all lines are busy}) = P(X = 50) = \dfrac{5^{50}e^{-5}}{50!}$

Example 4.55: The probability that a bomb dropped from an envelope will strike a certain target is 1/5. If 6 bombs are dropped, find the probability that

1. Exactly 2 will strike the target.
2. At least 2 will strike the target.

Given $e^{-1.2} = 0.3012$

Solution: We have $n = 6$, $p = 1/5$

$$\lambda = np = (6)\left(\frac{1}{5}\right) = 1.2$$

By Poisson distribution $P(X = x) = \dfrac{\lambda^x e^{-\lambda}}{x!}, \quad x = 0, 1, 2, 3, \ldots$

1. $P(X = 2) = P(\text{Exactly 2 will strike the target})$

$$= \frac{(1.2)^2 e^{-1.2}}{2!} = \frac{0.3012 \times 1.414}{2}$$

$$= 0.2169$$

2. $P(\text{At least 2 will strike the target}) = P(X \geq 2) = 1 - P(X < 2)$

$$= 1 - [P(X = 0) + P(X = 1)]$$

$$= 1 - \frac{(1.2)^0 e^{-1.2}}{0!} - \frac{(1.2)^1 e^{-1.2}}{1!}$$

$$= 1 - e^{-1.2}(1 + 1.2)$$

$$= 1 - (0.3012)(2.2) = 0.3374$$

Example 4.56: If the probability that an individual suffers a bad reaction from an injection of a given serum is 0.001. Determine the probability that out of 2000 individuals exactly 3 individuals will suffer due to bad reaction.

Solution: We have

$$n = 2000,\ p = 0.001 \ \text{(given)}$$
$$\lambda = np = (2000)(0.001) = 2$$

By Poisson distribution

$$P(X = x) = \frac{\lambda^x e^{-\lambda}}{x!}, \quad x = 0, 1, 2, 3, \ldots$$

$P(X = 3) =$ Probability that 3 individuals will suffer due to bad reaction $\dfrac{2^3 e^{-2}}{3!} = 0.1805$

Example 4.57: For a Poisson distribution show that $\mu_{r+1} = r\lambda\mu_{r-1} + \lambda\frac{d\mu_r}{d\lambda}$ where λ is the mean.

Solution: The *r*th moment of the distribution is

$$\mu_r = \sum_{x=0}^{\infty}(x - \lambda)^r P(X = x)$$

$$= \sum_{x=0}^{\infty}(x - \lambda)^r \frac{\lambda^x e^{-1}}{x!}$$

Differentiating with respect to λ we get

$$\frac{d\mu_r}{d\lambda} = -r\sum_{x=0}^{\infty}(x - \lambda)^{r-1}\frac{e^{-\lambda}\lambda^x}{x!} + \sum_{x=0}^{\infty}(x - \lambda)^r\left[\frac{x\lambda^{n-1}e^{-\lambda} - \lambda^x e^{-\lambda}}{x!}\right]$$

$$= -r\mu_{r-1} + \sum_{x=0}^{\infty}(x - \lambda)^r\left[\frac{\lambda^{n-1}e^{-\lambda}(x - \lambda)}{x!}\right]$$

$$\therefore \lambda \frac{d\mu_r}{d\lambda} = -r\lambda\mu_{r-1} + \sum_{x=0}^{\infty}\left[\frac{\lambda^n e^{-\lambda}(x-\lambda)^{r+1}}{x!}\right]$$

$$= -\lambda\mu_{r-1} + \mu_{r+1}$$

Hence

$$\lambda \frac{d\mu_r}{d\lambda} + r\lambda\mu_{r-1} = \mu_{r+1}$$

i.e.,

$$\mu_{r+1} = r\lambda\mu_{r-1} + \lambda \frac{d\mu_r}{d\lambda}$$

Example 4.58: Suppose that $P(X=2) = \frac{2}{3}P(X=1)$, find $P(X=0)$?

Solution:

$$P(X=2) = \frac{2}{3}P(X=1) \quad \text{(given)}$$

$$\frac{\lambda^2 e^{-\lambda}}{2!} = \frac{2}{3}\frac{\lambda^1 e^{-\lambda}}{1!}$$

or

$$\frac{\lambda^2}{2!} = \frac{2}{3}\lambda$$

or

$$3\lambda^2 - 4\lambda = 0$$

or

$$\lambda(3\lambda - 4) = 0$$

$$\lambda = 0 \text{ or } \lambda = \frac{4}{3}$$

We consider $\lambda = 4/3$, since $\lambda = 0$ is impossible

$$P(X=0) = \frac{\lambda^0 e^{-\lambda}}{0!} = e^{-4/3}.$$

Example 4.59: A car hire firm has two cars, which it hires out day by day. The number of demands for a car on each day is distributed as

Poisson distribution with mean 1.5. Calculate the proportion of days on which neither car is used and the proportion of days on which some demand is refused ($e^{-1.5} = 0.2231$).

Solution: We have $\lambda = 1.5$ (given)

The proportion of days when no car will be required is

$$P(X=0) = \frac{\lambda^0 e^{-\lambda}}{0!} + e^{-\lambda} = e^{-1.5} = 0.2231$$

The probability that no car, one car, two cars will be required

$$\begin{aligned}
&= P(X=0) + P(X=1) + P(X=2) \\
&= e^{-\lambda} + e^{-\lambda}\frac{\lambda}{1!} + e^{-\lambda}\frac{\lambda^2}{2!} \\
&= e^{-\lambda}\left(1 + \frac{\lambda}{1!} + \frac{\lambda^2}{2!}\right) \\
&= e^{-1.5}(1 + 1.5 + 1.125) \\
&= (0.2231)(3.625) \\
&= 0.80873
\end{aligned}$$

Example 4.60: In a Poisson distribution $P(X=x)$ for $x=10$ is 10%. Find the mean, given that $\log_e 10 = 2.3026$.

Solution: Let λ be the mean

$$P(X=x) = P(X=0) = \frac{\lambda^0 e^{-\lambda}}{0!} = e^{-\lambda} = \frac{10}{100}$$

$$\therefore\ e^{-\lambda} = 0.1 = \frac{1}{10} \quad \text{or} \quad \therefore\ e^{\lambda} = 10$$

Hence

$$\lambda = \log_e 10$$

Example 4.61: At a busy traffic intersection, the probability p of an individual car having an accident is very small, say, $p = 0.0001$. However, during a certain part of the day, a large number of cars, say 1000 pass through the intersection. Under these conditions, what is the probability of two or more accidents occurring during that period?

Solution: Here we have $n = 1000$, $p = 0.0001$

$$\therefore \quad \lambda = np = 1000 \times 0.0001 = 0.1$$

$$P(X \geq 2) = 1 - P(X < 2) = 1 - [P(X = 0) + P(X = 1)]$$

$$= 1 - \left[\frac{e^{-0.1}(0.1)^0}{0!} + \frac{e^{-0.1}(0.1)^1}{1!}\right]$$

$$= 1 - e^{-0.1}(1 + 0.1)$$

$$= 1 - (0.9048)(1.1)$$

$$= 0.00472$$

Example 4.62: Using Poisson distribution, find the probability that the ace of spades will be drawn from a pack of well shuffled cards at least once in 104 consecutive trials.

Solution: Probability that the card drawn is an ace of spades = 1/52.

Since $n = 104$

Mean of the distribution $= \lambda = np = 104 \times \dfrac{1}{52} = 2$

Probability of getting ace of spades at least once $= P(X \geq 1) = 1 - P(X = 0)$

$$= 1 - \frac{\lambda^0 e^{-\lambda}}{0!} = 1 - e^{-2}$$

$$= 1 - 0.136$$

$$= 0.864$$

Example 4.63: Probability of getting no misprint in a page of a book is e^{-4}. What is the probability that a page contains more than two misprints?

Solution: It is given that $P(X = 0)$, i.e., probability of getting no misprint $= e^{-4}$

i.e.,

$$\therefore \quad \frac{\lambda^0 e^{-\lambda}}{0!} = e^{-4}$$

or

$$e^{-\lambda} = e^{-4}$$

i.e.,

$$\lambda = 4$$

Probability that a page contains more than 2 misprints $= P(X>2) = 1 - P(X\leq 2)$.

$$= 1 - [P(X=0) + P(X=1) + P(X=2)]$$

$$= 1 - \left[e^{-4} + e^{-4}0.4 + e^{-4}\frac{4^2}{2!}\right]$$

$$= 1 - e^{-4}[1+4+8]$$

$$= 0.7618$$

Example 4.64: Six coins are tossed 6400 times using Poisson distribution, what is the approximate probability of getting six heads 10 times?

Solution: Probability of getting a head when a coin is tossed $= \frac{1}{2}$

Probability of getting 6 heads when 6 coins are tossed $= \left(\frac{1}{2}\right)^6 = \frac{1}{64}$

Mean $= \lambda = np = 6400 \times \frac{1}{64} = 100$

Probability of getting 6 heads 10 times $= P(X=10) = \frac{\lambda^{10}e^{-\lambda}}{10!} = \frac{e^{-100}(100)^{10}}{100!}$.

Example 4.65: Suppose 2% of the people on the average are left handed. Find (1) the probability of finding 3 or more left handed; (2) the probability of finding none or one left handed?

Solution: Let x be the random variable the number of left handed people.

The mean is $= \lambda = 2\% = \frac{2}{100} = 0.02$

Since $P(X=x) = \frac{e^{-\lambda}\lambda^x}{x!}, \quad x = 1, 2, 3, \ldots$

1. $P(X\geq 3) = 1 - [P(X=0) + P(X=1) + P(X=2)]$

$$= 1 - e^{-0.02}\left[1 + 0.02 + \frac{0.02^2}{2}\right]$$

$$= 1 - e^{-0.02}[1.0303]$$

$$= 1.307 \times 10^{-6}$$

2. $P(X\leq 1) = P(X=0) + P(X=1)$

$$= e^{-0.02} + e^{-0.02}[0.02]$$

Example 4.66: If X is a Poisson variate such that $3P(X = 4) = \frac{1}{2}P(X = 2) + P(X = 0)$

Find

1. The mean of X
2. $P(X \leq 2)$

Solution:

1. Since X is a Poisson variate, we have

$$P(X = x) = \frac{e^{-\lambda}\lambda^x}{x!}, \quad x = 1, 2, 3, \ldots$$

Consider $3P(X = 4) = \dfrac{1}{2}P(X = 2) + P(X = 0)$ (given)

or

$$\frac{e^{-\lambda}\lambda^4}{4!} = \frac{1}{2}\frac{e^{-\lambda}\lambda^2}{2!} + \frac{e^{-\lambda}\lambda^0}{0!}$$

or

$$\frac{\lambda^4}{8} = \frac{\lambda^2}{4} + 1$$

or

$$\lambda^4 - 2\lambda^2 - 8 = 0$$

or

$$(\lambda^4 - 4)(\lambda^2 + 2) = 0$$

since $\lambda > 0$, we get $\lambda = 2$

Mean of the Poisson variate $\lambda = 2$

i.e., $\lambda = 2$

2. $P(X \leq 2) = P(X = 0) + P(X = 1) + P(X = 2)$

$$= \frac{e^{-2}2^0}{0!} + \frac{e^{-2}2^1}{1!} + \frac{e^{-2}2^2}{2!}$$

$$= 0.675$$

Example 4.67: Average number of accidents on any day on a national highway is 1.8. Determine the probability that the number of accidents (1) is at least one; (2) at most one.

Solution: Average number of accidents = 1.8
i.e., $\lambda = \text{mean} = 1.8$

1. $P(X \geq 1)$ = Probability that the number of accidents is at least one

$$\begin{aligned} &= 1 - P(X = 0) \\ &= 1 - e^{-1.8} \\ &= 1 - 0.1653 = 0.8347 \end{aligned}$$

2. Probability that the number of accidents is at most one = $P(X \leq 1)$

$$\begin{aligned} &= P(X = 0) + P(X = 1) \\ &= \frac{e^{-1.8}(1.8)^0}{0!} + \frac{e^{-1.8}(1.8)^1}{1!} \\ &= e^{-1.8}(1 + 1.8) \\ &= 0.4628 \end{aligned}$$

Example 4.68: Using the recurrence formula, find the probabilities when $x = 0, 1, 2, 3, 4,$ and 5, if the mean of the Poisson distribution is 3.

Solution: We have $\lambda = 3$ (given)

$$P(X = 0) = P(0) = e^{-\lambda} = e^{-3}$$

The recurrence formula is $P(x + 1) = \frac{\lambda}{x+1} P(x)$
Substituting $x = 0, 1, 2, 3, 4,$ and 5, we get

$$P(0 + 1) = P(1) = \frac{\lambda}{0+1} P(0) = 3(e^{-3}) = 0.147$$

$$P(1 + 1) = P(2) = \frac{\lambda}{1+1} P(1) = \frac{3}{2}(0.147) = 0.2205)$$

$$P(2 + 1) = P(3) = \frac{\lambda}{2+1} P(2) = \frac{3}{3}(0.2205) = 0.2205$$

$$P(3 + 1) = P(4) = \frac{\lambda}{3+1} P(3) = \frac{3}{4}(0.2205) = 0.1653$$

$$P(4 + 1) = P(5) = \frac{\lambda}{4+1} P(4) = \frac{3}{5}(0.1653) = 0.099$$

$$P(5 + 1) = P(6) = \frac{\lambda}{5+1} P(5) = \frac{3}{6}(0.099) = 0.0495$$

$$P(1) = 0.147, \quad P(2) = 0.2205 \quad P(3) = 0.2205,$$
$$P(4) = 0.1653, \quad P(5) = 0.099 \quad P(3) = 0.0495$$

Example 4.69: Wireless sets are manufactured with 25 soldered joints each. On an average 1 joint in 500 is defective. How many sets can be expected to be free from defective joints in a consignment of 10,000 sets?

Solution: Probability that a joint is defective $= p = \dfrac{1}{500}$.

We have $n = 25,\ p = \dfrac{1}{500}$

$\text{Mean} = \lambda = np = 25 \times \dfrac{1}{500} = 0.05$

By Poisson distribution $P(X = x) = \dfrac{e^{-\lambda}\lambda^x}{x!}, \quad x = 1,\ 2,\ 3, \ldots$

Probability that no joint is defective $= P(0) = \dfrac{e^{-\lambda}\lambda^0}{0!} = e^{-0.05}$

Thus the expected number of sets free from defects in 10,000 sets

$$= 10,000 \times P(0) = 10,000 \times e^{-0.05}$$
$$= 10,000(0.9512)$$
$$= 9512.0$$

9512 Sets are expected to be free from defects.

Example 4.70: Fit a Poisson distribution to the following data and calculate the theoretical (expected) frequencies.

x	0	1	2	3	4
$f(x)$	122	60	15	2	1

Solution: Mean of the distribution $= \dfrac{0.122 + 1.60 + 2.15 + 3.2 + 4.1}{122 + 60 + 15 + 2 + 1}$

$$\bar{x} = \frac{60 + 30 + 6 + 4}{200} = \frac{100}{200}$$

$$\therefore \quad \lambda = \bar{x} = 0.5$$

We have $N = 122 + 60 + 15 + 2 + 1 = 200$

$$N = 4, \quad \lambda = 0.5$$

$$P(0) = P(X = 0) = \frac{e^{-\lambda}\lambda^0}{0!} = e^{-0.5}$$

The theoretical frequencies are obtained as follows:

$$N.P(0) = 200 \cdot e^{-0.5} = 121$$

$$N.P(1) = 200 \cdot \frac{e^{-0.5}(0.5)^1}{1!} = 61$$

$$N.P(2) = 200 \cdot \frac{e^{-0.5}(0.5)^2}{2!} = 15$$

$$N.P(3) = 200 \cdot \frac{e^{-0.5}(0.5)^3}{3!} = 3$$

$$N.P(4) = 200 \cdot \frac{e^{-0.5}(0.5)^4}{4!} = 0$$

Hence the expected (theoretical) frequencies are 121, 61, 15, 3, 0.

Example 4.71: The variance of the Poisson distribution is 2. Find the distribution for $x = 0, 1, 2, 3, 4$, and 5 from the recurrence relation of the distribution ($e^{-2} = 0.1353$)

Solution: The variance of the Poisson distribution $\lambda = 2$ (given).

We have $P(0) = e^{-2} = 0.1353$ (given)

The recurrence relation is $P(x + 1) = \frac{\lambda}{x+1} P(x)$

Hence we get

$$P(1) = \frac{\lambda}{0+1} P(0) = \frac{2}{1} e^{-2} = 2 \times 0.1353 = 0.2706$$

$$P(2) = \frac{\lambda}{1+1} P(1) = \frac{2}{2}(0.2706) = 0.2706$$

$$P(3) = \frac{\lambda}{2+1} P(2) = \frac{2}{3}(0.2706) = 0.1804$$

$$P(4) = \frac{\lambda}{3+1} P(3) = \frac{2}{4}(0.1804) = 0.0902$$

$$P(5) = \frac{\lambda}{4+1} P(4) = \frac{2}{5}(0.0902) = 0.361$$

Example 4.72: Assuming that the typing mistakes per page committed by a typist follows a Poisson distribution, find the theoretical frequencies for the following distribution of typing mistakes?

No. of mistakes per page	0	1	2	3	4	5
No. of pages	40	30	20	15	10	5

Solution: We have $N = 40 + 30 + 20 + 15 + 10 + 5 = 120$, $n = 5$

$$\text{Mean of the distribution} = \lambda = \frac{0.40 + 1.30 + 2.20 + 3.15 + 4.10 + 5.5}{120}$$

$$= \frac{30 + 40 + 45 + 40 + 25}{120} = \frac{180}{120} = 1.5$$

Using Poisson distribution we get

$$P(0) = \frac{e^{-\lambda}\lambda^0}{0!} = e^{-\lambda} = e^{-1.5} = 0.22313$$

$$P(1) = \frac{e^{-\lambda}\lambda^1}{1!} = e.^{-\lambda}\lambda = e^{-1.5}(1.5) = 0.334695$$

$$P(2) = \frac{e^{-\lambda}\lambda^2}{2!} = \frac{e^{-1.5}(1.5)^2}{2} = 025$$

$$P(3) = \frac{e^{-\lambda}\lambda^3}{3!} = \frac{e^{-1.5}(1.5)^3}{6} = 0.13$$

$$P(4) = \frac{e^{-\lambda}\lambda^4}{4!} = \frac{e^{-1.5}(1.5)^4}{24} = 0.05$$

$$P(5) = \frac{e^{-\lambda}\lambda^5}{5!} = \frac{e^{-1.5}(1.5)^5}{120} = 0.01$$

The theoretical (expected) frequencies are

$$NP(0) = (120)(0.22313) = 27,$$
$$NP(1) = (120)(0.334695) = 40,$$
$$NP(2) = (120)(0.25) = 30,$$
$$NP(3) = (120)(0.13) = 16,$$
$$NP(4) = (120)(0.05) = 6,$$
$$NP(5) = (120)(0.01) = 1$$

i.e., 27, 40, 30, 16, 6, and 1.

Example 4.73: If X and Y are two independent random Poisson random variables such that $\text{Var}[X] + \text{Var}[Y] = 3$. Find $P(X + Y < 2)$?

Solution: Let X and Y follow Poisson distribution.

Let λ_1 be the mean of X and λ_2 be the mean of Y.

Then by additive property $X+Y$ follows Poisson distribution with mean $\lambda_1 + \lambda_2$.

We have $\text{Var}[X] + \text{Var}[Y] = \lambda_1 + \lambda_2 = 3$(given)

$\therefore \lambda = \lambda_1 + \lambda_2 = 3$ (say)

Now

$$P(X+Y<2) = P(X+Y=0) + P(X+Y=1) = e^{-\lambda} + \frac{e^{-\lambda}\lambda}{1!}$$
$$= e^{-3+}e^{-3}0.3 = e^{-3}(1+3)$$
$$= 4e^{-3} = 0.199$$

Example 4.74: In a company, a board receives 15 calls per minute. If the board is capable of handling at the most 25 calls per minute, what is the probability that the board will saturate in 1 minute?

Solution: Number of calls received = 15/minute.

$\therefore \lambda = 15$

The board reaches the saturation point if the number of calls exceeds 25.

$$\text{Hence the probability of saturation} = 1 - \sum_{x=0}^{25} \frac{e^{-15}(15)^x}{x!}$$
$$= 1 - \left[e^{-15} + \frac{e^{-15}\cdot 15}{1!} + \dots + \frac{e^{-15}(15)^{25}}{25!}\right]$$
$$= 1 - 0.994 = 0.006$$

Example 4.75: If V = Hourly volume = 37, t = length of interval = 30 seconds, then compute F_x for $x > 2$.

Solution: We have

$$\lambda = \frac{Vt}{3600} = \frac{37 \times 30}{3600} = 0.308$$

$$\text{Therefore } e^{-\lambda} = e^{-0.308} = 0.735, \quad n = \frac{3600}{t} = \frac{3600}{30} = 120$$

$$F_0 = n\frac{\lambda^0 e^{-\lambda}}{0!} = 120.e^{0.308} = 120 \times 0.735 = 88.2$$

$$F_1 = n\frac{\lambda^1 e^{-\lambda}}{1!} = \frac{\lambda}{1}F_0 = 0.308 \times 88.2 = 27.2$$

$$F_2 = n\frac{\lambda^2 e^{-\lambda}}{2!} = \frac{\lambda}{2}F_1 = \frac{0.308 \times 27.2}{2} = 4.2$$

Hence $F_{x>2} = n - (F_0 + F_1 + F_2) = 120 - (88.2 + 27.2 + 4.2) = 0.4$

Example 4.76: (One of the first published example of Poisson distribution as applied to traffic data by Adams). The rate of arrivals, i.e., the number of vehicles arriving per 10 second interval is given below. Obtain the theoretical frequencies, and hourly volume from the data.

No. of vehicles per 10 second period	0	1	2	3	>3
Observed frequency	94	63	21	2	0

Solution:

No. of vehicles per 10 second period (*x*)	Observed frequency (*f*)	Total of vehicles (*xf*)	Probability *P*(*x*)	Theoretical frequency
0	94	0	0.539	97.0
1	63	63	0.333	59.9
2	21	42	0.103	18.5
3	2	6	0.021	3.8
>3	0	0	0.004	0.8
	180	111	1.000	180

We have

$$\lambda = \frac{\text{Total of vehicles}}{\Sigma f} = \frac{111}{180} = 0.617$$

$$e^{-\lambda} = e^{-0.617} = 0.539$$

Applying the formula $P(x) = \dfrac{\lambda^x e^{-x}}{x!}$ we get

$P(0) = 0.539$, $P(1) = 0.333$, $P(2) = 0.103$, $P(3) = 0.021$, $P(>3) = 0.004$

The theoretical frequencies are $N\,P(0) = 97$, $N\,P(1) = 59.9$, $N\,P(2) = 18.5$, $N\,P(3) = 3.8$, $N\,P(X > 3) = 0.8$

where $N = 180$.

The hourly volume $= 2 \times$ no. of vehicles in 180 ten second period $= 2 \times 111 = 222$.

Exercise 4.8

1. If a random variable X follows Poisson distribution such that $P(X = 1) = P(X = 2)$. Find (1) Mean, (2) $P(X = 0)$?

   Ans: $P(X = 0) = 0.1353$

2. A hospital receives patients at the rate of 3 patients/minute on an average. What is the probability of receiving no patient in 1 minute interval?

   Ans: 0.04979

3. Suppose a book of 285 pages contains typographical errors. If these errors are randomly distributed throughout the book, what is the probability that 10 pages, selected at random will be free from errors ($e^{-0.735} = 0.4975$).

   Ans: 0.4795

4. In a Poisson distribution if $3P(X=2) = P(X=4)$. Find $P(X=3)$?

   Ans: $36e^{-6}$

5. Between the hours 2 p.m. and 4 p.m. the average number of phone calls per minute coming into the switch board of a company is 2.35. Find the probability that during one particular minute there will be at most two phone calls?

   Ans: 0.58285

6. Out of 1000 balls, 50 are red and the rest are white. If 60 balls are picked at random, what is the probability of picking up (a) 3 red balls, (b) not more than 3 red balls in the sample? Assume Poisson distribution for the number of red balls picked up in the sample where $e^{-3} = 0.0498$.

   Ans: (a) 0.2241; (b) 0.6474

7. A manufacturer, who produces medicine bottles, finds that 0.1% of the bottles are defective. The bottles are packed in boxes containing 500 boxes from how many boxes will contain (1) no defectives (2) at least 2 defectives (given $e^{-0.5} = 0.60650$).

   Ans: (a) 61; (b) 9

8. The probability that a man aged 50 years will die within a year is 0.01125. What is the probability that out of 12 such men at least 11 will reach their 51st birthday ($e^{-0.135} = 0.87371$)?

   Ans: 0.9917

9. A pair of dice is thrown 6 times. If getting a total of 7 is considered to be a success, what is the probability of at least 4 successes?

   Ans: $\frac{406}{46650}$

10. There are 20% chances for a worker of an industry to suffer from an occupational disease. Fifty workers were selected at random and examined for the occupational disease. Find the probability that (a) only one worker is found suffering from the disease, (b) more than 3 are suffering from the disease, (c) None is suffering from the disease?

    Ans: (a) $10\left(\frac{4}{5}\right)^{49}$; (b) $1 - \left(\frac{4^{47}}{550}\right)(25364)$; (c) $\left(\frac{4}{5}\right)^{50}$

11. If a random variable X follows a Poisson distribution such that $P(X=2)=9\ P(X=4)+90\ P(X=6)$. Find the mean and variance of X?

Ans: Mean = 21, Variance = 1

12. A random variable has a Poisson distribution such that $P(1)=P(2)$. Find the (a) Mean of the distribution, (b) $P(4)$, (c) $P(X \geq 1)$, (d) $P(1 < X < 4)$?

Ans: (a) 1.2; (b) 0.0902; (c) 0.8647; (d) 0.4511

13. Fit a Poisson distribution for the following data and calculate the expected frequencies.

x	0	1	2	3	4
$f(x)$	109	65	22	3	1

Ans: 108, 67, 66, 20, 4, 11, 1

14. A distributor of bean seeds determines from extensive tests that 5% of large batch of seeds will not germinate. He sells the seeds in packets of 200 and generates 90% germination. Determine the probability that a particular packet will violate the guarantee.

Ans: 0.0016

15. The number of accidents in a year to the taxi drivers in a city follows a Poisson distribution with mean 3. Out of 1000 taxi drivers, find approximately the number of drivers with (a) no accidents in a year, (b) more than 3 accidents in a year.

Ans: (a) 50; (b) 350

16. If the mean of a Poisson distribution is 4, find (a) SD, (b) β_1, (c) β_2, (d) μ_3, (e) μ_4.

Ans: (a) 2; (b) 0.25; (c) 3.25; (d) 4; (e) 52

17. One-hundred car radios are inspected as they come off the production line and the number of defects per set is recorded below:

No. of defects	0	1	2	3	4
No. of sets	79	18	2	1	0

Find the Poisson distribution to the above data and calculate the expected frequencies of 0, 1, 2, 3, and 4 defects?

Ans: 78, 20, 2, 0, 0

18. A book contains 100 misprints distributed randomly throughout its 100 pages. What is the probability that a page observed at random contains at least two misprints? Assume Poisson distribution.

Ans: 0.264

19. If X and Y are independent Poisson random variables with means 2 and 4, respectively, find (a) $P\left[\frac{X+Y}{2} < 1\right]$, (b) $P[3(X+Y) \geq 9]$?

Ans: (a) 0.017352; (b) 0.93803

20. For a Poisson distribution $P(X=1) = 0.03$ and $P(X=2) = 0.2$, find (a) $P(X=0)$ and (b) $P(X=3)$?

Ans: (a) 0.00225; (b) 0.888899

21. Fit a Poisson distribution to the following data which gives the number of yeast cells per square for 400 squares.

No. of cells per square	0	1	2	3	4	5	6	7	8	9	10
No. of squares	103	143	98	42	8	4	2	0	0	0	0

Ans: 107, 14, 93, 41, 14, 4, 0, 0, 0, 0

22. The number of items collected at a center is Poisson distribution, where the probability of no items collected is 0.4. Find the probability that 3 or less items are collected?

Ans: 0.9856

23. A shopkeeper receives 5 bad currencies every day. What are the probabilities that he will receive (a) 3 bad currencies on a particular day, (b) 8 bad currencies for 2 consecutive days?

Ans: 0.1113

24. The following table compares the predicted frequencies with frequencies observed in a field study:

No. of vehicles arriving per interval	0	1	2	3
Observed frequency	88	27	5	0

Calculate the predicted frequencies.

Ans: 88.2, 27.2, 4.2, 0.4

25. Fit Poisson distribution by calculating predicted (theoretical) frequencies from the table given below:

No. of vehicles per 10 second period	0	1	2	>2
Observed frequency	88	27	5	0

Ans: 88.2, 27.2, 4.2, 0.4

26. Suppose that the chance of an individual coal miner is killed in a mining accident during a year is 1/1400. Use the Poisson distribution to calculate the probability that in the mine employing 350 minors, there will be at least one fatal accident.

Ans: 0.22

27. A car hire firm has two cars, which it hires day by day. The number of demands for a car on each day is distributed as a Poisson distribution with mean 1.5. Calculate the proportion of days on which no car is used and proportion of days on which some demand is refused.
  Ans: 0.1913

28. Between the hours 2 p.m. and 4 p.m., the average number of phone calls per minute coming into the switch board is 2.35. Find the probability that during one particular unit there will be at most 2 phone calls?
  Ans: 0.582854

29. The quality control manager of a tyre company has a sample of 100 tyres and has found the lifetime to be 30.214 km. The population SD is 860. Construct a 95% confidence interval for the mean life time for this particular brand of tyres.
  Ans: (30045. 44, 303 82 .56)

30. In a random selection of 64 of 600 road crossings in a town, the mean number of automobile accidents per year was found to be 4.2 and the sample SD was 0.8. Construct a 95% confidence interval for the mean number of automobile accidents per crossing per year.
  Ans: 4.023, 4.377

31. On a certain day 74 trains were arriving on time at Delhi station during the rush hours and 83 were late. At New Delhi there were 65 on time and 107 were late. Is there any difference in the proportions arriving on time at the two stations?
  Ans: Null hypothesis (H_0): There is no difference of trains arriving "on time" at two stations is accepted.

32. The number of accidents in a year attributed to the state transport bus driver follows Poisson distribution with mean 3. Out of 1000 bus drivers, how many drivers had more than 1 accident in a year?
  Ans: 801

33. For 10,000 insured cars of an insurance company, the average number of claims per year is said to be 2000. Using Poisson distribution, find the probability that there is 1 claim in a particular year?
  Ans: 0.1638

34. A train runs 25 km at a speed of 30 km/h, another 50 km at a speed of 40 km/h, then due to repairs of the track it travels for 6 minutes at a speed of 10 km/h and finally covers the remaining distance of 24 km at a speed of 24 km/h. What is the average speed in kilometers per hour?
  Ans: 31.41 km/h

4.22 DISCRETE UNIFORM DISTRIBUTION

If a random variable X can take on k different values with equal probability, we say that X has a discrete uniform distribution, which is defined as follows:

Definition 4.28:
Let X be a discrete random variable taking the values 1, 2, 3, ….., n.
X is said to follow a uniform distribution if its pmf is given by

$$P(X = x) = \frac{1}{k}, \quad x = 1, 2, 3, \ldots k$$
$$= 0, \quad \text{otherwise}$$

k is called the parameter of the distribution. Discrete uniform distribution is applied whenever all the values of the random variable X are equally likely. The mean of the discrete uniform distribution is $\frac{k+1}{2}$ and the variance is $\frac{k^2-1}{12}$. Coefficient of skewness of the uniform distribution is zero. Hence uniform distribution is symmetric.

Example 4.77: Let X denote the number on the face of a n unbiased die, when it is rolled.

We have

$$P(X = x) = \frac{1}{6}, \quad x = 1, 2, 3, \ldots 6$$
$$= 0, \quad \text{otherwise}$$

Clearly X follows uniform distribution.

4.23 THE NEGATIVE BINOMIAL AND GEOMETRIC DISTRIBUTION

Consider an experiment consisting of repeated Bernoulli (independent) trials. The number of the trial on which kth success occurs is a random variable having a negative binomial distribution which defined as follows:

Definition 4.29:
A random variable X has a negative binomial distribution if and only if

$$P(X = x) = {}^{x-1}C_{k-1} p^k q^{x-k}, \quad \text{for } x = k, k+1, k+2, \ldots$$

where p is the probability of successes, q in the probability of failure given by $q = 1 - p$.

In this case the random variable X is called the negative binomial random variable.

Negative binomial distribution is also called as Pascal or waiting time distribution.

The mean of negative binomial distribution is kq/p and the variance of negative binomial distributions $k(1-p)/p^2$.

The mgf of a negative binomial distribution is $p^k(1-qe^t)^{-k}$ for a negative binomial distribution Var $[X] > E(X)$. If $k = 1$, the distribution is called geometrical distribution.

4.24 GEOMETRIC DISTRIBUTION

Definition 4.30:

A random variable X is said to follow geometric distribution, if it assumes only nonnegative vales and its pmf is

$$P(X = x) = pq^{x-1}, x = 1, 2, 3, \ldots \text{ where } q = 1 - p$$

The mean of geometric distribution is $1/p$ and the variance is $2q/p^2$.
The mgf of geometric distribution is $pe^t/(1-(1-p)e^t)$.

Example 4.78: A typist types 3 letters erroneously for every 100 letters. What is the probability that 10th letter typed is the first erroneous letter?

Solution: We have

$$P = \frac{3}{100} = 0.03$$

$$q = 1 - p = 0.97$$

$$x = 10$$

$$\begin{aligned} P(X = x) &= pq^{x-1} \\ &= (0.03)(0.97)^{10-1} \\ &= (0.03)(0.97)^9 \end{aligned}$$

Exercise 4.9

1. If the probability is 0.75 that an applicant for a driver's license will pass the road test an any given try, what is the probability that an applicant will finally pass the test on fifth try?
   Ans: $(0.75)(0.25)^4$

2. A typist types 2 letters erroneously for every 100 letters, what is the probability that the truth letter typed is the first erroneously letter?
   Ans: $(0.02)(0.98)^9$
3. If the probability is 0.40, that a child exposed to a certain contagious disease will catch it, what is the probability that the 10th child exposed to the disease will be the third to catch it?
   Ans: ${}^9C_2(0.40)^3(0.60)^7$
4. Find the probability that in tossing four coins one will get either all heads or tails for the third time on seventh toss?
   Ans: ${}^6C_2\left(\frac{1}{8}\right)^3\left(\frac{7}{8}\right)^3$
5. The probability of a student passing a subject is 0.6. What is the probability that he will pass in the subject in his third attempt?
   Ans: $(0.6)(0.4)^2$
6. In a company, a group of particular items is accepted if more items are tested before the first defective is found. If $m = 5$, and the group consists of 15% defective items, what is the probability that it will be accepted?
   Ans: 0.04437

4.25 CONTINUOUS PROBABILITY DISTRIBUTIONS

In this section we discuss the method of finding the probability distributions associated with continuous random variables and also how to use the mean and SD to describe these distributions.

4.25.1 Uniform Distribution

Let X be a continuous random variable. X is said to have uniform distribution, if its pdf is given by

$$\begin{aligned} f(x) &= k, \quad a < x < b \\ &= 0, \quad \text{otherwise} \end{aligned}$$

where k is a constant given by $\int_a^b f(x)\mathrm{d}x = 1$

i.e.,

$$\int_a^b k\mathrm{d}x = 1$$

or

$$k[x]_a^b = 1$$

or

$$k(b - a) = 1$$

or

$$k = \frac{1}{b - a}$$

Hence

$$f(x) = \frac{1}{b - a}, \quad a < x < b$$
$$= 0, \quad \text{otherwise}$$

Uniform distribution is also known as rectangular distribution.

The distribution function $f(x)$ of X is given by

$$F(x) = \begin{cases} 0, & -\infty < x < a \\ \dfrac{x - a}{b - a}, & a \le x \le b \\ 1, & b < x < \infty \end{cases}$$

Clearly $F(x)$ is not continuous at the two end points, $x = a$ and $x = b$. Hence $F(x)$ is not differentiable at $x = a$ and $x = b$.

The pdf of a uniform variate X is $(-a, a)$ is given by

$$f(x) = \begin{cases} \dfrac{1}{2a}, & -a < x < a \\ 0, & \text{otherwise} \end{cases}$$

4.25.1.1 Moments of the Uniform Distribution

We have $\mu'_r = r$th moment about the origin

$$= \int_a^b x^r f(x) dx$$
$$= \int_a^b x^r \frac{1}{b - a} dx$$
$$= \frac{1}{b - a} \left[\frac{x^{r+1}}{r+1} \right]_a^b$$
$$= \frac{b^{r+1} - a^{r+1}}{(r + 1)(b - a)}$$

Putting $r = 1, 2, 3, 4$ we obtain,

$$\mu_1' = \frac{b^2 - a^2}{2(b-a)} = \frac{b+a}{2}$$

$$\mu_2' = \frac{b^3 - a^3}{3(b-a)} = \frac{b^2 + ba + a^2}{2}$$

$$\mu_3' = \frac{b^4 - a^4}{4(b-a)} = \frac{(b^2 + a^2)(b+a)}{4}$$

$$\mu_4' = \frac{b^5 - a^5}{5(b-a)} = \frac{b^4 + b^3 a + b^2 a^2 + ba^3 + a^4}{5}$$

4.25.1.2 Mean of Uniform Distribution

$$\mu_1' = E(X) = \frac{b+a}{2}$$

4.25.1.3 Variance of Uniform Distribution

$$\mu_2 = \mu_2' - \mu_1'^2 \frac{b^2 + ab + a^2}{3} - \frac{(b+a)^2}{4}$$

$$= \frac{1}{12}\left[4b^2 + 4ab + 4a^2 - 3b^2 - 3a^2 - 6ab\right]$$

$$= \frac{1}{12}\left[b^2 + a^2 - 2ab\right]$$

$$= \frac{(b-a)^2}{12}$$

4.25.1.4 Moment Generating Function of the Uniform Distribution

We have

$$M(t) = E[e^{tx}]$$

$$= \int_a^b e^{tx} \frac{1}{b-a} \mathrm{d}x = \frac{e^{bt} e^{at}}{(b-a)t}$$

$$= \frac{\left(1 + \frac{bt}{1!} + \frac{b^2 t^2}{2!} + \cdots\right) - \left(1 + \frac{at}{1!} + \frac{a^2 t^2}{2!} + \cdots\right)}{(b-a)t}$$

$$= 1 + \frac{b+a}{2} \cdot t + \frac{b^3 - a^3}{3(b-a)} \cdot \frac{t^2}{2} + \frac{b^4 - a^4}{4(b-a)} \cdot \frac{t^3}{3} + \cdots$$

Example 4.79: If X is uniformly distributed over $(-a, a)$. Find a so that $P(X > 1) = 1/3$.

Solution: Therefore the pdf of X is

$$f(x) = \begin{cases} \frac{1}{30}, & 0 < x < 30 \\ 0, & \text{otherwise} \end{cases}$$

We have

$$P(X > 1) = \frac{1}{3}(\text{given})$$

i.e.,

$$\int_1^a f(x)\mathrm{d}x = \frac{1}{3}$$

or

$$\int_1^a \frac{1}{2a}\mathrm{d}x = \frac{1}{3}$$

or

$$\frac{1}{2a}[x]_1^a = \frac{1}{3}$$

or

$$\frac{a-1}{2a} = \frac{1}{3}$$

i.e., $3a-3 = 2a$

Hence $a = 3$

Example 4.80: Subway trains on a certain line run after every half an hour between midnight and 6:00 in the morning. What is the probability that a man entering the station at a random time during this period will have to wait at least 20 minutes?

Solution: Let the waiting time for the next train by the man be denoted by X.

Clearly X is a random variable which in uniformly distributed in (0, 30) when the man arrived at station. The pdf of X is given by

$$f(x) = \begin{cases} \frac{1}{30}, & 0 < x < 30 \\ 0, & \text{otherwise} \end{cases}$$

Probability that a man entering the station at random waits at least 20 minutes

$$= P(X \geq 20)$$

$$= \int_{20}^{30} f(x)\mathrm{d}x = \int_{20}^{30} \frac{1}{30}\mathrm{d}x = \frac{1}{30}[x]_{20}^{30} = \frac{30-20}{30} = \frac{1}{3}$$

4.25.2 Exponential and Negative Exponential Distribution

Let X be a continuous random variable, X is said to have an exponential distribution if the pdf of x is given by

$$f(x) = \begin{cases} e^x, & x \geq 0 \\ 0, & \text{otherwise} \end{cases}$$

and X is said to have a negative exponential distribution with parameter $\lambda > 0$. If the pdf is given by

$$f(x) = \begin{cases} \lambda e^{-\lambda x}, & x > 0 \\ 0, & x \leq 0 \end{cases}$$

Mean of negative exponential distribution is $1/\lambda$ and the variance is $1/\lambda^2$, the mgf of the negative exponential distribution is $\frac{\lambda}{\lambda - t}, (\lambda > t)$ rth moment about origin is $r!/\lambda^r$, $r = 1, 2, 3, \ldots$.

4.26 NORMAL DISTRIBUTION

The Normal distribution is a continuous distribution. It was first discovered by Abraham De moivre (1667–1754). He derived normal probability function as the limiting form of the binomial distribution, simultaneously Karl Friedrich Gauss (1777–1853) has also derived normal distribution as law of errors. Hence the normal probability distribution is also called as the Gaussian distribution.

Definition 4.31:

A continuous random variable X is said to have a Normal distribution if its probability function is given by

$$f(x) = \frac{1}{\sigma\sqrt{2\pi}} e^{-\frac{1}{2}\left[\frac{x-\mu}{\sigma}\right]^2}, (-\infty < x < \infty)$$

where μ = mean of the normal random variable; σ = Standard deviation of the normal variable X; σ^2 = Variance of the normal variable X and are the parameters of the distribution. If X follows Normal distribution then it is denoted by $X \sim N(\mu, \sigma)$.

4.26.1 Standard Normal Variable

Computing the area over intervals under the normal probability distribution is difficult task. Consequently we will use a table of areas as a function of Z-score. The Z-score give the distance between the measurement and the mean in units equal to the SD. Z is always a normal random variable with mean and SD 1. For this reason Z is often referred to as the standard normal random variable. Z is a continuous random variable.

Definition 4.32:
The standard normal random variable Z is defined by

$$Z = \frac{x - \mu}{\sigma}$$

where μ = mean of the normal random variable X; σ = Standard deviation of the normal variable X; X = normal random variable.

Definition 4.33:
A random variable $Z = (x - \mu)/\sigma$ is said to have a standard normal distribution if its probability function is defined by

$$f(z) = \frac{1}{\sqrt{2\pi}} e^{-\frac{1}{2}Z^2}, (-\infty < Z < \infty)$$

A standard normal variate is denoted by $N(0, 1)$. Standard normal distribution is also known as Z-distribution or unit Normal distribution. Standardization of Normal distribution helps us to make use of the tables of area of standard curve.

$$\frac{1}{\sqrt{2\pi}} e^{-\frac{1}{2}Z^2}$$

For various points along the x-axis

The area between the points say Z_1 and Z_2 under the standard normal curve will represent the probability, the z will lie between Z_1 and Z_2. It is denoted by

$$P(Z_1 \leq Z \leq Z_2)$$

The general equation of the normal curve is

$$y = f(x) = \frac{1}{\sigma\sqrt{2\pi}} e^{-\frac{1}{2}\left[\frac{x-\mu}{\sigma}\right]^2}, (-\infty < x < \infty)$$

If the corresponding total frequency is N then

$$y = f(x) = \frac{N}{\sigma\sqrt{2\pi}} e^{-\frac{1}{2}\left[\frac{x-\mu}{\sigma}\right]^2}$$

is the equation if best fitting normal probability curve.

4.26.2 Distribution Function $\varphi(Z)$ of Standard Normal Variate

The distribution function $\varphi(Z)$ of standard normal variate, Z is defined by properties of

$$\varphi(Z) = P(Z \leq z) = \int_{-\infty}^{Z} f(t)\mathrm{d}t$$

$$= \frac{1}{\sqrt{2\pi}} \int_{-\infty}^{Z} \cdot e^{-\frac{1}{2}Z^2}$$

where

$$Z = \frac{x-\mu}{\sigma}$$

Properties of $\varphi(Z)$:

1. $\varphi(-Z) = 1 - \varphi(Z)$
2. $P(a \leq X \leq b) = \varphi\left(\frac{b-\mu}{\sigma}\right) - \varphi\left(\frac{a-\mu}{\sigma}\right)$
3. $P(Z \leq a) = P(Z \geq -a)$

4.26.3 Area Under Normal Curve

The curve of normal distribution is unimodal and is bell shaped with the highest point over the mean μ. It is symmetrical about a vertical line through μ.

The normal curve has the following features (Fig. 4.2):

1. The curve is symmetrical about the coordinate at its mean, which locates the peak of the bell.
2. The values of mean, median, and mode are equal.
3. The curve extends from $-\infty$ to ∞.

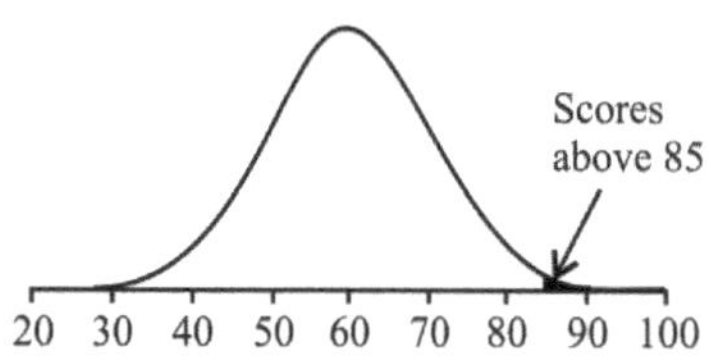

Figure 4.2 Normal curve.

4. The area covered between $\mu - \sigma$ and $\mu + \sigma$ is 0.6826 (i.e., 68.26% of the total area).
5. The area covered between the limits $\mu - 2\sigma$ and $\mu + 2\sigma$ is 0.9544 (i.e., 95.44% of area).
6. The area covered between $\mu - 3\sigma$ and $\mu + 3\sigma$ is 0.9974 (i.e., 99.74% of area).

4.26.4 Area Under Standard Normal Curve

Let X be a normal random variable and if Z be the corresponding standard normal variable then Z and X have identical curves. The curve of Z is called standard normal curve. The area bounded by a standard normal variable Z, the Z axis and the ordinates at $Z = \infty$ and any positive value of Z is provided by a standard table.

The standard normal curve covers 68.26% area between $Z = -1$ and 1. Therefore

$$P(-1 \leq Z \leq 1) = 0.6826$$

The curve covers 95.44% of the total area between $Z = -2$ and 2. Therefore

$$P(-2 \leq Z \leq 2) = 0.9544$$

And the area covered between $Z = -3$ and 3 is 99.74% of the total area.

Hence

$$P(-3 \leq Z \leq 3) = 0.9474$$

4.26.5 Properties of Normal Curve

1. The mean μ and variance σ^2 of a normal distribution are called the parameters of the distribution.
2. The mean, median, and mode of a normal distribution coincide with each other.
3. In a normal distribution, mean deviation about mean is approximately equal to 4/3 times its SD.
4. In a normal distribution, the quartiles Q_1 and Q_3 are equidistant from the median.
5. The normal curve is bell shaped and is symmetrical about $X = \mu$.
6. The area bound by the normal curve and x-axis equal to 1.

7. The tails of the curve of a normal distribution extend identically in both sides of $x = \mu$ and never touch the x-axis.
8. The probability that x lies between a and b is equal to the area bounded by the curve X-axis and ordinates $X = a$ and $X = b$.

4.26.6 Mean of Normal Distribution

Consider the normal distribution with as the parameter then,

$$f(x) = \frac{1}{\sigma\sqrt{2\pi}} e^{-\frac{1}{2}\left[\frac{x-\mu}{\sigma}\right]^2}, (-\infty < x < \infty)$$

Mean of $X = E[x]$
Therefore

$$\mu = E[X] = \int_{-\infty}^{\infty} xf(x)dx$$

$$= \frac{1}{\sigma\sqrt{2\pi}} \int_{-\infty}^{\infty} e^{-\frac{1}{2}\left[\frac{x-\mu}{\sigma}\right]^2} dx$$

Putting

$Z = \frac{x-\mu}{\sigma}$, we have

$dZ = \frac{dx}{\sigma}$ i.e., $dx = \sigma dZ$

and

$$x = \mu + \sigma Z$$

Hence we get

$$\mu = \frac{1}{\sqrt{2\pi}} \int_{-\infty}^{\infty} (\sigma z + \mu) e^{-\frac{1}{2}Z^2} dz$$

$$= \frac{\sigma}{\sqrt{2\pi}} \int_{\infty}^{\infty} Ze^{-\frac{1}{2}z^2} dz + \frac{\mu}{\sqrt{2\pi}} \int_{-\infty}^{\infty} e^{-\frac{1}{2}z^2} dz$$

$$= \frac{\sigma}{\sqrt{2\pi}} \int_{-\infty}^{\infty} Ze^{-\frac{1}{2}z^2} dz + \frac{2\mu}{\sqrt{2\pi}} \int_{0}^{\infty} e^{-\frac{1}{2}z^2} dz$$

$$= \frac{\sqrt{2}\mu}{\sqrt{\pi}} \cdot \frac{\sqrt{\pi}}{\sqrt{2}} = \mu$$

4.26.7 Variance of Normal Distribution

$$\text{Variance} = E(X-\mu)^2$$

i.e.,

$$V[X] = \int_{-\infty}^{\infty} (x-\mu)^2 f(x)\mathrm{d}x$$

$$= \frac{1}{\sigma\sqrt{2\pi}}\int_{-\infty}^{\infty} (x-\mu)^2 e^{-\frac{1}{2}\left[\frac{x-\mu}{\sigma}\right]^2}\mathrm{d}x$$

Putting $Z = \frac{x-\mu}{\sigma}$, we get

$$V[X] = \frac{1}{\sigma\sqrt{2\pi}}\int_{-\infty}^{\infty} \sigma^2 z^2 e^{-\frac{1}{2}z^2}(\sigma \mathrm{d}z)$$

Putting $Z^2/2 = t$, we obtain

$$V[X] = \frac{2\sigma^2}{\sqrt{2\pi}}\int_{0}^{\infty} 2te^{-t}\frac{\mathrm{d}t}{\sqrt{2t}} = \frac{2\sigma^2}{\sqrt{\pi}}\int_{0}^{\infty} e^{-t}\sqrt{t}\mathrm{d}t$$

$$= \frac{2\sigma^2}{\sqrt{\pi}}\int_{0}^{\infty} e^{-t} t^{\frac{3}{2}-1}\mathrm{d}t$$

$$= \frac{2\sigma^2}{\sqrt{\pi}}\Gamma\left(\frac{3}{2}\right) = \frac{2\sigma^2}{\sqrt{\pi}}\frac{3}{2}\Gamma\left(\frac{1}{2}\right)$$

$$= \frac{\sigma^2}{\sqrt{\pi}}.\sqrt{\pi}$$

Hence variance $= \sigma^2$

4.26.8 Mode of Normal Distribution

We have

$$f(x) = \frac{1}{\sigma\sqrt{2\pi}}e^{-\frac{1}{2}\left[\frac{x-\mu}{\sigma}\right]^2}$$

Applying logarithms on both sides, we get

$$\operatorname{Log} f(x) = -\log\sigma\sqrt{2\pi} - \frac{1}{2\sigma^2}\cdot(x-\mu)^2$$

Differentiating with respect to x we get

$$\frac{f'(x)}{f(x)} = -\frac{1}{\sigma^2}(x-\mu)$$

or

$$f'(x) = -\frac{1}{\sigma^2}(x-\mu)f(x) \tag{4.17}$$

Again differentiating, we get

$$f''(x) = -\frac{1}{\sigma^2}\left[1.f(x) + (x-\mu)f'(x)\right]$$

$$= \frac{f(x)}{\sigma^2}\left[1 - \left(\frac{(x-\mu)}{\sigma}\right)^2\right] \text{ using Eq. 4.17}$$

$f'(x) = 0$ gives $x = \mu$ (since $f(x) \neq 0$)

At $x = \mu$, we have

$$f''(x) = \frac{1}{\sigma^2} f(u)$$

$$= -\frac{1}{\sigma^2}\frac{1}{\sigma\sqrt{2\pi}}$$

$$= -\frac{1}{\sigma^3\sqrt{2\pi}} < 0$$

Hence $x = \mu$ is the mode of the distribution.

4.26.9 Median of the Normal Distribution

Let M denote the median of normal distribution

$$\int_{-\infty}^{M} f(x)\mathrm{d}x = \frac{1}{2}$$

i.e.,

$$\frac{1}{\sigma\sqrt{2\pi}}\int_{-\infty}^{\mu} e^{-\frac{1}{2}\left[\frac{x-\mu}{\sigma}\right]^2}\mathrm{d}x + \frac{1}{\sigma\sqrt{2\pi}}\int_{\mu}^{M} e^{-\frac{1}{2}\left[\frac{x-\mu}{\sigma}\right]^2}\mathrm{d}x = \frac{1}{2} \tag{4.18}$$

Consider

$$\frac{1}{\sigma\sqrt{2\pi}}\int_{-\infty}^{\mu} e^{-\frac{1}{2}\left[\frac{x-\mu}{\sigma}\right]^2}\mathrm{d}x$$

Putting
$\frac{x-\mu}{\sigma} = Z$, we get $dx = \sigma dz$
Therefore

$$\frac{1}{\sigma\sqrt{2\pi}}\int_{-\infty}^{\mu} e^{-\frac{1}{2}\left[\frac{x-\mu}{\sigma}\right]^2} dx = \frac{1}{\sqrt{2\pi}}\int_{-\infty}^{0} e^{-\frac{1}{2}z^2} dz$$

$$= \frac{1}{\sqrt{2\pi}}\int_{0}^{\infty} e^{-\frac{1}{2}z^2} dz \quad \text{(by symmetry)} \tag{4.19}$$

$$= \frac{1}{\sqrt{2\pi}}\frac{\sqrt{\pi}}{\sqrt{2}} = \frac{1}{2}$$

from Eqs. (4.18) and (4.19), we obtain

$$\frac{1}{2} + \frac{1}{\sigma\sqrt{2\pi}}\int_{\mu}^{M} e^{-\frac{1}{2}\left[\frac{x-\mu}{\sigma}\right]^2} dx = \frac{1}{2}$$

thus

$$\frac{1}{\sigma\sqrt{2\pi}}\int_{\mu}^{M} e^{-\frac{1}{2}\left[\frac{x-\mu}{\sigma}\right]^2} dx = 0$$

or

$$\int_{\mu}^{M} e^{-\frac{1}{2}\left[\frac{x-\mu}{\sigma}\right]^2} dx = 0$$

which is possible when $\mu = M$

Hence

$$\mu = M$$

i.e., Median of normal distribution $= \mu$

4.26.10 Moment Generating Function of Normal Distribution With Respect to Origin

We have

$$M_x(t) = E(e^{tx}) = \int_{-\infty}^{\infty} e^{tx} f(x) dx$$

$$= \frac{1}{\sigma\sqrt{2\pi}}\int_{-\infty}^{\infty} e^{tx} e^{-\frac{1}{2}\left[\frac{x-\mu}{\sigma}\right]^2} dx$$

Putting
$\frac{x-\mu}{\sigma} = Z$, we get

$$M_x(t) = \frac{e^{\mu t}}{\sqrt{2\pi}} \int_{-\infty}^{\infty} e^{-\frac{1}{2}z^2 + (t\sigma)^2} dz$$

$$= \frac{e^{\mu t + \frac{1}{2}t^2\sigma^2}}{\sqrt{2\pi}} \int_{-\infty}^{\infty} e^{-\frac{1}{2}(z^2 - t\sigma)^2} dz$$

$$= \frac{e^{\mu t + \frac{1}{2}t^2\sigma^2}}{\sqrt{2\pi}} \cdot 2\frac{\sqrt{\pi}}{\sqrt{2}}$$

$$= e^{\mu t + \frac{1}{2}t^2\sigma^2}$$

4.26.11 Mean Deviation of Normal Distribution

We have mean deviation $= \int_{-\infty}^{\infty} |x-\mu| f(x) dx$

$$= \frac{1}{\sigma\sqrt{2\pi}} \int_{-\infty}^{\infty} e^{-\frac{1}{2}\left[\frac{x-\mu}{\sigma}\right]^2} |x-\mu| dx$$

$$= \frac{\sigma}{\sqrt{2\pi}} \int_{-\infty}^{\infty} e^{-\frac{1}{2}z^2} |z| dz \text{ where } Z = \frac{x-\mu}{\sigma},$$

$$= \frac{2\sigma}{\sqrt{2\pi}} \int_{0}^{\infty} e^{-\frac{1}{2}z^2} |z| dz \quad \text{(since the integral is even)}$$

$$= \sigma\sqrt{\frac{2}{\pi}} = \frac{4}{5}\sigma \quad \text{(approximately)}$$

4.26.11.1 Solved Examples

Example 4.81: A normal distribution has a mean 20 and a SD of 4. Find out the probability that a value of X lies between 20 and 24?

Solution: We have $\mu = 20$, and $\sigma = 4$ (given)

$$Z = \frac{x-\mu}{\sigma} = \frac{x-20}{4}$$

When

$$X = 20, \quad Z = \frac{20 - 20}{4} = 0$$

And when

$$X = 24, \quad Z = \frac{24 - 20}{4} = 1$$

Therefore Z lies between 20 and 24, the value of Z lies between 0 and 1.

Hence

$$P(0 \leq Z \leq 1) = 0.3413$$

Example 4.82: The mean and SD of a normal variable x are 50 and 4, respectively. Find the values of the corresponding standard normal variable when x is equal to 42, 54, and 84?

Solution: We have
$\mu = 50$, and $\sigma = 4$ (given)

$$Z = \frac{x - \mu}{\sigma} = \frac{x - 50}{4}$$

when

$$x = 42, \quad Z = \frac{42 - 50}{4} = -2$$

when

$$x = 54, \quad Z = \frac{54 - 50}{4} = 1$$

and when

$$x = 84, \quad Z = \frac{84 - 50}{4} = 8.5$$

Example 4.83: Find the area under the standard normal curve which lies

1. to the left of $Z = 1.73$.
2. to the right of $Z = -0.66$.
3. between $Z = 1.25$ and $Z = 1.67$.
4. between $Z = -1.45$ and $Z = 1.45$.

Solution:

1. Area to the left of $Z = 1.73$

$$\begin{aligned}&= 0.5 + \text{area between } Z = 0 \text{ and } Z = 1.73\\&= 0.5 + 0.4582\\&= 0.9582 \text{ (using the table)}\end{aligned}$$

2. Area to the right of $Z = -0.66$

$$\begin{aligned}&= \text{area from } Z = -0.66 \text{ to } Z = 0 + 0.5\\&= 0.2454 + 0.5\\&= 0.7454 \text{ (using the table)}\end{aligned}$$

3. Area between $Z = 1.25$ and 1.67

$$\begin{aligned}&= \text{Area between } Z = 0 \text{ and } 1.67\\&- (\text{Area between } Z = 0 \text{ and } 1.25)\\&= 0.4525 - 0.3944\\&= 0.0581 \text{ (using the table)}\end{aligned}$$

4. Area between $Z = -1.45$ and 1.45

$$\begin{aligned}&= \text{Area between } Z = 0 \text{ and } 1.45\\&- (\text{Area between } Z = 0 \text{ and } 1.45)\text{(by symmetry)}\\&= 2(\text{Area between } Z = 0 \text{ and } 1.45)\\&= 2(0.4265)\\&= 0.8530 \text{ (using the table)}\end{aligned}$$

Example 4.84: The ABC company uses a machine to fill boxes with soap powder. Assume that the net weight of the boxes of soap powder is normally distributed with mean 15 ounces and SD 8 ounces.

1. What proportion of boxes will have net weight of more than 14 ounces?
2. 25% of boxes will be heavier than a certain net weight ω and 75% of the boxes will be lighter than this net weight. Find ω?

   Solution: We have $\mu = 15$ ounce, $\sigma = 0.8$ ounce

   Let Z be the standard normal variable

$$Z = \frac{x - \mu}{\sigma} = \frac{x - 15}{0.8}$$

1. When $x = 14$,

$$Z = \frac{14 - 15}{0.8} = \frac{-1}{0.8} = -1.25$$

when

$$X = 14 \text{ the probability that } X \geq 14 = P(X \geq 14)$$
$$= P(Z > -1.25)$$
$$= P(-1.25 \leq Z \leq 0) + P(Z \geq 0)$$
$$= P(0 \leq Z \leq 1.25) + 0.5$$
$$= 0.3944 + 0.5$$
$$= 0.8944$$

89.44% boxes will have their net weight greater than 14.

2. Taking $x = \omega$, we have $Z = \frac{x - \mu}{\sigma} = \frac{\omega - 15}{0.8}$

Probability of Z being greater than $\frac{\omega - 15}{0.8}$ is 0.25.

Therefore $0.25 = 0.5 - P\left(0 \leq Z \leq \frac{\omega - 15}{0.8}\right)$

i.e., $P\left(0 \leq Z \leq \frac{\omega - 15}{0.8}\right) = 0.5 - 0.25 = 0.25$

The value of Z for which the probability is $0.25 = 0.675$

Therefore $\frac{\omega - 15}{0.8} = 0.694$

or
$$\omega = 0.675 \times 0.8 + 15$$
$$= 15.54 \text{ ounces}$$

Example 4.85: Let X denote the number of successes in test. If X is normally distributed with mean 100 and SD 15. Find the probability that x does not exceed 130?

Solution: We have $\mu = 100$ and $\sigma = 15$

Let

$$Z = \frac{x - \mu}{\sigma} = \frac{x - 100}{15}$$

Probability that x does not exceed $130 = P(X \leq 130)$.

$$= P$$
$$= P(Z \leq 2)$$
$$= P(-\infty < Z \leq 0) + P(0 \leq Z \leq 2)$$
$$= 0.5 + 0.4772$$
$$= 0.9772 \text{ (from the area table)}$$

Example 4.86: Assume that mean height of soldiers to be 68.22 in. with a variance of 10.8 in. How many soldiers in a regiment of 1000 would you expect to be over 6 ft. tall?

Solution: We have $\mu = 68.22$, $\sigma^2 = 10.8$, $\sigma = \sqrt{10.8} = 3.286$

$$X = 6 \text{ ft.} = 6 \times 12 = 72 \text{ in.}$$

when

$$X = 72, Z = \frac{x - \mu}{\sigma} = \frac{72 - 68.22}{3.286} = 1.1503$$

$$\begin{aligned} P(X > 72) &= P(Z > 1.1503) \\ &= 0.5 - P(0 < Z < 1.1503) \\ &= 0.5 - 0.3749 \\ &= 0.1251 \end{aligned}$$

Expected number of soldiers over 6 ft. tall in regiment of

$$\begin{aligned} 1000 &= 1000 \times 0.1251 \\ &= 125.1 \\ &= 125 \text{ soldiers.} \end{aligned}$$

Example 4.87: What is the probability that a standard normal variate Z will be lying between 1.25 and 2.75?

Solution:

$$\begin{aligned} P(1.25 < Z < 2.75) &= P(0 < Z < 2.75) - P(0 < Z < 1.25) \\ &= \text{Area between } Z = 0 \text{ and } Z = 2.75 \\ &- (\text{Area between } Z = 0 \text{ and } Z = 1.25) \\ &= 0.4970 - 0.3944 \\ &= 0.1026 \end{aligned}$$

Example 4.88: In a distribution exactly 7% of the items are under 35 and 89% of the items are under 63. What are the values of mean and SD of the distribution?

Solution: Let μ be the mean and σ be the SD of the distribution.

The area lying to the left of the ordinate at $x = 35$ is 7%, i.e., 0.07. The corresponding values of Z is negative.

The area lying to the right of the ordinate at $x = 63$ up to the mean μ is

$$= 0.5 - 0.07 = 0.43.$$

Hence the values of Z corresponding in the area 0.43 is 1.48.

i.e.,

$$\frac{35 - \mu}{\sigma} = -1.48 \text{ or } \mu - 1.48\sigma = 35 \tag{4.20}$$

Similarly the area lying to the left of the ordinate at $x = 63$ up to the mean is 0.39 (i.e., 39%).

The value of Z corresponding to the area 0.39 is 1.23.

i.e.,

$$\frac{63 - \mu}{\sigma} = 1.23 \text{ or } \mu + 1.23\sigma = 63 \tag{4.21}$$

From Eqs. (4.20) and (4.21), we get

$$\mu = 50.3; \quad \sigma = 1.33$$

Example 4.89: The income of a group of 10,000 persons was found to be normally distributed with mean equal to Rs. 750, and SD equal to Rs. 50. What was lowest income among the richest 250?

Solution: We have $\mu = 750$, $\sigma = 50$ (given)

Let x' be the lowest income among the richest 250 people.

Then

$$P(x \geq x) = \frac{250}{10,000} \text{(given)}$$

i.e.,

$$P\left(\frac{x - \mu}{\sigma} \geq \frac{x' - 750}{50}\right) = 0.025$$

or

$$P\left(Z \geq \frac{x' - 750}{50}\right) = 0.025$$

We obtain

$$\left(\frac{x' - 750}{50}\right) = 1.96$$

or

$$x' - 750 = 50 \times 1.96$$

or

$$x' = 750 + 98 = 848$$

Hence Rs. 848 is the lowest income among the richest 250 people.

Example 4.90: If $\log_e x$ is normally distributed with mean 1 and variance 4. Find $P\left(\frac{1}{2} < X < 2\right)$ given that $\log_e 2 = 0.693$?

Solution: Here we have $\mu = 1$, variance $= \sigma^2 = 4$ (given)

$$\therefore \sigma = \sqrt{4} = 2$$

Let

$$X = \log_e x$$

$$Z = \frac{x - \mu}{\sigma} = \frac{x - 1}{2} = \frac{\log_e x - 1}{2}$$

When

$$x = \frac{1}{2},$$

$$Z_1 = \frac{\log_e \frac{1}{2} - 1}{2} = \frac{\log_e 1 - \log_e 2 - 1}{2}$$

$$= \frac{-\log_e 2 - 1}{2} = \frac{0.693 - 1}{2}$$

$$= -0.8465$$

When $x = 2$,

$$Z_2 = \frac{\log_e 2 - 1}{2} = \frac{0.693 - 1}{2}$$

$$= -0.1535$$

$$P\left(\frac{1}{2} < X < 2\right) = P(Z_1 < Z < Z_2)$$

$$= P(-0.8465 < Z < -0.1535)$$

$$= P(0.1535 > Z > 0.8465)$$

$$= \text{Area between } Z = 0 \text{ and } Z = 0.8465$$

$$- (\text{Area between } Z = 0 \text{ and } Z = 0.1535)$$

$$= 0.2996 - 0.0596$$

$$= 0.24$$

Example 4.91: In a normal distribution, 31% of the items are under 45 and 8% are over 64. Find the mean and SD of the distribution?

Solution: Let X be the normal variable, μ denote the mean, and σ be the SD of X.

Let

$$Z = \frac{x - \mu}{\sigma}$$

It is given that

$$P(X > 64) = \frac{31}{100} = 0.31$$

And

$$P(X > 64) = \frac{8}{100} = 0.08$$

Since

$$P(X < 45) < 0.5$$

X lies to the left of $X = \mu$, therefore the value of Z is negative say Z_1.
i.e., when $x = 45$, let $Z = \frac{45 - \mu}{\sigma} = -Z_1, (Z_1 > 0)$(say)
Similarly Z is positive when $X = 64$
Let

$$Z = \frac{64 - \mu}{\sigma} = Z_2 (Z_2 > 0)$$

We have

$$P(X < 45) = 0.3$$

i.e.,

$$P(Z < -Z_1) = 0.31$$

or

$$P(Z > -Z_1) = 0.31$$

or

$$0.5 - P(0 \leq Z \leq Z_1) = 0.31$$

i.e.,

$$P(0 \leq Z \leq Z_1) = 0.19$$
$$Z_1 = 0.5 \text{ (from the area table)}$$

Therefore

$$\frac{45-\mu}{\sigma}=-0.5$$

or

$$45-\mu=-0.5\sigma \tag{4.22}$$

Also we have

$$P(X>64)=0.08$$

i.e.,

$$P(Z>Z_2)=0.08$$

or

$$0.5-P(0\leq Z\leq Z_2)=0.08$$
$$P(0\leq Z\leq Z_2)=0.42$$

Hence $Z_2=1.4$ (from the area table)
Therefore

$$\frac{64-\mu}{\sigma}=1.4$$

or

$$64-\mu=1.4\sigma \tag{4.23}$$

Solving Eqs. (4.22) and (4.23) we get

$$\mu=50 \text{ and } \sigma=10$$

Example 4.92: The mean inside diameter of a sample of 200 washers produced by a machine is 0.502 cm and that SD is 0.005 cm. The purpose for which these washers are intended allows a maximum tolerance is the diameter of 0.496–0.508 cm. Otherwise the washers are considered to be defective washers produced by machine, assuming the diameters are naturally distributed.

Solution: we have $\mu=0.502$ cm and $\sigma=0.005$ cm

Let z be the standard normal variable

$$Z=\frac{x-\mu}{\sigma}=\frac{x-0.502}{0.005}$$

Probability that a washer is not defective

$$\begin{aligned}
&= P(0.496 \leq X \leq 0.508) \\
&= P\left(\frac{0.496 - 0.502}{0.005} \leq \frac{X - 0.502}{0.005} \leq \frac{0.508 - 0.502}{0.005}\right) \\
&= P(-1.2 \leq X \leq 1.2) \\
&= 2 \cdot P(0 \leq X \leq 1.2)(\text{By symmetry}) \\
&= 2 \cdot (0.3849) = 0.7698
\end{aligned}$$

Probability that a washer is defective

$$\begin{aligned}
&= 1 - \text{Probability that a washer is not defective} \\
&= 1 - 0.7698 \\
&= 0.2302
\end{aligned}$$

Therefore the percentage of defective washers = 23.02.

Example 4.93: The mean of a normal distribution is 60 and 6% of the values are greater than 70. Find the SD of the distribution?

Solution: Let X denote the normal random variable

We have $P(X > 70) = \dfrac{6}{100} = 0.6$ (given)

Therefore

$$P\left(\frac{x - \mu}{\sigma} > \frac{70 - \mu}{\sigma}\right) = 0.6$$

i.e.,

$$P\left(Z > \frac{70 - 60}{\sigma}\right) = 0.6$$

or

$$P\left(Z > \frac{10}{\sigma}\right) = 0.6$$

Let $Z_1 = 10/\sigma$, then $\frac{10}{\sigma} = 1.56$ (from the table of standard normal probabilities)

We get

$$\sigma = \frac{10}{1.56} = \frac{1000}{156} = 6.4$$

Hence the SD is 6.4.

Example 4.94: Fit a normal curve to the following data:

x	6	7	8	9	10	11	12
$f(x)$	3	6	9	13	8	5	4

Solution: Mean of the distribution is $\mu = \frac{\sum x_i f_i}{\sum f_i}$

i.e., $$\mu = \frac{(6)(3)+(7)(6)+(8)(9)+(9)(13)+(10)(8)+(11)(5)+(12)(4)}{3+6+9+13+8+5+4}$$

$$= \frac{18+42+72+117+80+55+48}{48}$$

$$= \frac{432}{48} = 9$$

$\sigma =$ Standard deviation

$$= \sqrt{\frac{\sum f_i x_i^2}{\sum f_i} - \mu^2}$$

$$= \sqrt{\frac{(3)(36)+(6)(49)+(9)(64)+(13)(81)+(8)(100)+(5)(121)+(4)(144)}{48} - 9^2}$$

$$= \sqrt{2.5833} = 1.607$$

The required best fitting normal curve is

$$Y = \frac{N}{\sigma\sqrt{2\pi}} e^{-\frac{1}{2}\left[\frac{x-\mu}{\sigma}\right]^2}$$

$$= \frac{48}{(1.607)\sqrt{2\pi}} e^{-\left[\frac{(x-9)^2}{5.1666}\right]}$$

$$= 11.916.e^{-\left[\frac{(x-9)^2}{5.1666}\right]}$$

4.26.12 Fitting a Normal Distribution

There are two methods for fitting normal distribution, namely

1. Method of ordinates
2. Area method.

We explain these methods with the help of examples.

Example 4.95: Fit a normal distribution for the following data by the ordinates method:

Class interval	150–160	160–170	170–180	180–190	190–200	200–210	210–220	220–230	230–240
Frequency	9	24	51	66	72	48	21	6	3

Solution: In order to fit a normal distribution we compute the values of mean (σ) and SD (σ) of the distribution using the values of (of μ and σ) obtained, we compute

$$Z = \frac{x - \mu}{\sigma}$$

For the given values of x, after finding the values of Z and using the table of areas the values of $\varphi(Z)$ are obtained. The ordinate Z for any value of $\varphi(Z)$ is obtained by the relation

$$Z = \frac{N}{\sigma}\varphi(Z)$$

The theoretical frequencies can be calculated by using the above relation

Class interval	f	x_i(midpoint)	$f_i x_i$	$f_i x_i^2$
150–160	9	155	1345	216,225
160–170	24	165	3960	653,400
170–180	51	175	8925	156,187
180–190	66	185	12210	2,258,850
190–200	72	195	14040	2,737,800
200–210	48	205	9840	2,017,200
210–220	21	215	4515	970,725
220–230	6	225	1350	303,750
230–240	3	235	705	165,675
	300		56940	10,885,500

$$\text{Mean of the distribution is } \mu = \frac{\sum x_i f_i}{\sum f_i}$$

$$= \frac{56940}{300} = 189.8$$

$$\sigma = \text{Standard deviation} = \sqrt{\frac{\sum f_i x_i^2}{\sum f_i} - \mu^2}$$

$$= \sqrt{\frac{10885500}{300} - 189.8^2} = 16.15$$

Hence $\sigma = 16.15$

Using the values of μ and σ we fit the normal distribution for the given data as follows (where C denotes the length of the class interval).

Class interval	Midpoint, x	$\left\lvert Z = \frac{x-\mu}{\sigma} \right\rvert$	Value of $\frac{N}{\sigma}\varphi(Z)$	$\left[\frac{N}{\sigma}\varphi(Z)\right] \times C$	The ordinate $\varphi(Z)$
150–160	155	2.15	0.0369	0.7356	$7.35 \sim 7$
160–170	165	0.5356	0.1238	2.299	$28.99 \sim 23$
170–180	175	0.9164	0.2637	4.902	49
180–190	185	0.297	0.3825	7.1113	71
190–200	195	0.322	0.379	7.05	71
200–210	205	0.942	0.2565	4.77	48
210–220	215	1.562	0.1182	2.198	22
220–230	225	2.181	0.0371	0.689	7
230–240	235	2.80	0.0079	0.15	2

Hence the expected frequencies are 7, 23, 49, 71, 71, 48, 22, 7, and 2.

Example 4.96: Fit a normal distribution to the following data and calculate the theoretical frequencies by area method:

Class interval	59.5–62.5	62.5–65.5	65.5–68.5	68.5–72.5	71.5–74.5
No. of students	5	18	42	27	8

Solution:

Class interval	f	x_i (midpoint)	$f_i\, x_i$	$f_i\, x_i^2$
59.5–62.5	5	61	305	18,605
62.5–65.5	18	64	1152	73,728
65.5–68.5	42	67	2814	188,538
68.5–71.5	27	70	1890	132,300
71.5–74.5	8	73	584	42,632
	100		6745	455,803

We have

$$\sum f_i = N = 100$$

$$\sum f_i x_i = 6745$$

$$\sum f_i x_i^2 = 455,803$$

$$\text{Mean of the distribution is } \overline{x} = \frac{\sum x_i f_i}{\sum f_i} = \frac{6745}{100} = 67.45$$

$$\sigma = \text{Standard deviation} = \sqrt{\frac{\sum f_i x_i^2}{\sum f_i} - \mu^2}$$

$$= \sqrt{\frac{455803}{100} - (67.45)^2} = 2.92$$

Hence $\sigma = 2.92$

The expected frequencies can be computed as shown in the table given below. Hence the theoretical frequencies are 4, 21, 39, 28, and 7.

Lower limits	$Z = \frac{x-\mu}{\sigma}$	Area under the normal curve $\varphi(Z)$	Area included in each class $\Delta\varphi(Z)$	Expected frequency $N \cdot \Delta\varphi(Z)$
59.2	− 2.72	0.4967	0.0413	4.13 ~ 4
62.5	− 1.70	0.4554	0.2068	2068 ~ 21
65.5	− 0.67	0.2486	0.3892	38.92 ~ 39
68.5	0.36	0.1406	0.2771	27.71 ~ 28
71.5	1.39	0.4177	0.0743	7.43 ~ 7
74.5	2.4	0.4920		

Example 4.97: Find the points of inflection of a normal curve. Find the asymptote of the cure?

Solution: The equation of the normal curve is

$$y = f(x) = \frac{1}{\sigma\sqrt{2\pi}} e^{-\frac{1}{2}\left[\frac{x-\mu}{\sigma}\right]^2} \tag{4.24}$$

Differentiating Eq. (4.24) with respect to x, we get

$$f'(x) = \frac{dy}{dx} = \frac{1}{\sigma\sqrt{2\pi}} e^{-\frac{1}{2}\left[\frac{x-\mu}{\sigma}\right]^2} \left[-\frac{(x-\mu)}{\sigma^2}\right]$$

$$= -\frac{1}{\sigma\sqrt{2\pi}\sigma^2}(x-\mu).e^{-\frac{1}{2}\left[\frac{x-\mu}{\sigma}\right]^2}$$

Again differentiating we get

$$f''(x) = \frac{d^2y}{dx^2} = \frac{1}{\sigma\sqrt{2\pi}}\left[e^{-\frac{1}{2}\left[\frac{x-\mu}{\sigma}\right]^2}\left[-\frac{(x-\mu)}{\sigma^2}\right]^2 - \frac{1}{\sigma^2}e^{-\frac{1}{2}\left(\frac{x-\mu}{\sigma}\right)^2}\right] = 0$$

$$\frac{d^2y}{dx^2} = 0 \quad \text{when}\left(\frac{x-\mu}{\sigma}\right)^2 - 1 = 0$$

i.e.,

$$(x-\mu)^2 = \sigma^2$$

or

$$(x-\mu) = \pm\sigma$$

or

$$x = \mu \pm \sigma$$

Differentiating $\frac{d^2y}{dx^2}$ with respect to x, we get

$$\frac{d^3y}{dx^3} = \frac{1}{\sigma\sqrt{2\pi}}\left[e^{-\frac{1}{2}\left[\frac{x-\mu}{\sigma}\right]^2}\left[-\frac{(x-\mu)}{\sigma^2}\right]^3 + \frac{3(x-\mu)}{\sigma^2}e^{-\frac{1}{2}\left(\frac{x-\mu}{\sigma}\right)^2}\right]$$

Hence $\frac{d^3y}{dx^3} = \pm\frac{2}{\sigma^4\sqrt{2\pi}}e^{-\frac{1}{2}}$ when $\mu = x \pm \sigma$

i.e., $\frac{d^3y}{dx^3} \neq 0$ when $\mu = x \pm \sigma$

The values of $y = f(x)$ at $x = \mu \pm \sigma$ is $\frac{1}{\sigma\sqrt{2\pi}}e^{-\frac{1}{2}}$

Therefore the points of inflection are $\left(\mu \pm \sigma, \frac{1}{\sigma\sqrt{2\pi}}e^{-\frac{1}{2}}\right)$

The line $y = 0$, meets the curve at $x = \pm\infty$.
Hence the curve has x-axis for its asymptote.

Exercise 4.10

1. If X is a normal variable with mean 25 and SD 5. Find the probability that (a) $15 \leq X \leq 30$ and (b) $|x - 30| \geq 10$?

   Ans: (a) 0.8185; (b) 0.16

2. Find the probability that the standard normal variate lies between 0 and 1.5?

   Ans: 0.4327

3. X is normally distributed and the mean of X is 12 and the SD is 4. Find the probability of the following: (a) $x \geq 20$; (b) $0 \leq x \leq 12$?

   Ans: (a) 0.0228; (b) 0.9972

4. In a normal distribution 31% of the items are under 45 and 8% are over 64. Find the mean and the SD?

   Ans: $\mu = 50$, $\sigma = 10$

5. A large number of measurements are normally distributed with mean 65.5″ and SD of 6.2″. Find the percentage of measurement that fall between 54.8″ and 68.8″?

   Ans: 6690

6. In a distribution exactly normal 7% of the items are under 35% and 89% are under 63. What are the values of mean and SD of the distribution?

   Ans: 50.3, 10.33

7. The weekly wages of 1000 workers are normally distributed around a mean of Rs.70 and with a SD of Rs. 5. Estimate the number of workers whose weekly wages will be
   a. between Rs. 70 and Rs. 72
   b. between Rs. 68 and Rs. 72
   c. more than Rs. 75
   d. Less than Rs. 63
   e. more than Rs. 80

   Ans: (a) 155; (b) 235; (c) 159; (d) 81; (e) 23

8. Given the mean heights of students in a class is 159 cm with the SD 20 cm. Find how many students heights lie between 150 cm and 170 cm, if there are 100 students in the class?

   Ans: 38

9. A manufacturer knows from experience that the resistance of the resistors he produces is normal with mean = 100 ohms and SD 20 hours. What percentage of resistors will have resistance between 96 ohms and 104 ohms?

   Ans: 95

10. X is a normal variate with mean 30 and SD 5. Find the probability of (a) $26 \leq x \leq 40$; (b) $x \geq 45$?

    Ans: (a) 0.7653; (b) 0.0014

11. In a sample of 1000 cases, the mean of common list is 14 and the SD is 2.5. Assuming the distribution to be normal, find (a) how many students score between 12 and 15; (b) how many score above 18; (c) how many score below 8?

    Ans: (a) 0.4435; (b) 0.0547; (c) 0.0082

12. Students of a class were given mechanical application test. Their marks were found to be normally distributed with mean 60 and SD 5. What percent of students scored (a) more than 60 marks; (b) less than 56 marks; (c) between 45 and 65 marks?

    Ans: (a) 50; (b) 21.19; (c) 84

13. If the diameters of ball bearing are normally distributed with 0.614 and SD 0.0025 cm. Determine the percent of ball bearing with diameter (a) between 0.610 and 0.618 cm; (b) greater than 0.617 cm; (c) less than 0.608 cm.

Ans: (a) 89.04; (b) 11.51%; (c) 0.82%

14. The life of electronic tubes of certain types may be assumed to be normally distributed with mean 155 hours and SD 19 hours. What is the probability that the life of a randomly chosen tube is (a) less than 117 hours; (b) between 136 and 194 hours; (c) more than 395 hours?

Ans: (a) 0.0288; (b) 0.8211; (c) 0

15. Obtain the equation of the normal probability curve that may be filled to the following data:

x	4	6	8	10	12	14	16	18	20	22	24
f	1	7	15	22	25	43	38	20	13	15	1

Ans: $-\left(\dfrac{x-13.85}{29.34}\right)^2$

16. Fit a normal curve to the following data and obtain the expected frequencies:

x	8.60	8.59	8.58	8.57	8.56	8.55	8.54	8.53	8.52
f	2	3	4	9	10	8	4	1	1

Ans: $(9.8)\ e^{-0.163} e^{(x-8.563)^2}$, 0.99, 2.24, 5.63, 7.5, 9.5, 7.5, 5.63, 2.24, 0.79

17. Fit a normal curve to the following data:

x	1	2	3	4	5	6	7	8	9	10	11	12
f	2	10	19	25	40	44	41	28	25	15	5	1

Ans: $e^{-0.99(x-6.259)^2}$

18. Prove that for the normal distribution *s* the Quartile deviation, mean deviation, and SD are approximately in the ratio 10:12:15.

Ans: $\dfrac{9\sqrt{3}}{4\sqrt{2\pi}} e^{-3\left(\frac{x-5}{32}\right)_2}$

19. Find the equation to the best fitting normal curve to the following distribution:

x	1	3	5	7	9
f	1	2	3	2	1

Ans: $\dfrac{113}{\sqrt{2\pi}} e^{-0.099(x-6.259)^2}$

20. Fit a normal curve to the following data:

x	1	2	3	4	5	6	7	8	9	10	11	12
f	2	10	19	25	40	44	41	28	25	15	5	1

***Ans*:** 7, 23, 49, 71, 71, 48, 22, 7, 2

21. Fit a normal distribution to the following data by ordinates method:

Class interval	150–160	160–170	170–180	180–190	190–200	200–210	210–220	220–230	230–240
f	9	24	51	66	72	48	21	6	3

***Ans*:** 0.79, 2.24, 5.63, 7.5, 9.5, 7.5, 5.63, 2.24, 0.79

22. Find the theoretical frequencies for the distribution?

x	2	4	6	8	10
f	1	4	6	4	1

***Ans*:** 0.864, 3.872, 6.3824, 3.872, 0.864

23. Obtain the equation of the normal probability curve that may be filled to the following data and calculate the theoretical frequencies?

x	4	6	8	10	12	14	16	18	20	22	24
f	1	7	15	22	25	43	38	20	13	15	1

***Ans*:** $y = \frac{1}{3.83\sqrt{2\pi}} e^{-\left(\frac{(x-13.85)^2}{29.355}\right)}$

4.26.13 Linear Combination of Independent Normal Variables

Theorem 1:

If $\lambda_1, \lambda_2, \ldots, \lambda_n$ are independent normal variables, and $a_1, a_2, \ldots, a_n$ are real constants, then $a_1x_1 + a_2x_2 + \cdots + a_nx_n$ is also a normal variate.

Proof:

Since

$$M_x(t) = e^{t\mu + \frac{1}{2}t^2\sigma^2}$$

We have

$$M(t)_{a_ix_i} = M_X(a_it) = e^{a_i\mu_i + \frac{1}{2}a_i^2t^2\sigma_i^2}$$

$$= e^{t(a_i\mu_i) + \frac{1}{2}t^2(a_i\sigma_i)^2}$$

$$M(t)_{\sum a_i x_i} = M(t)_{a_1x_1+a_2x_2+\cdots+a_nx_n}$$

$$= M_{a_1x_1}(t)\cdot M_{a_2x_2}(t)\ldots M_{a_nx_n}(t)$$

$$= e^{t(a_1\mu_1)+\frac{1}{2}t^2(a_1\sigma_1)^2}\cdot e^{t(a_2\mu_2)+\frac{1}{2}t^2(a_2\sigma_2)^2}\ldots e^{t(a_n\mu_n)+\frac{1}{2}t^2(a_n\sigma_n)^2}$$

$$= e^{t\sum_1^n (a_i\mu_i)+\frac{1}{2}t^2\sum_1^n (a_i\sigma_i)^2}$$

is the mgf of a normal variate with mean $\sum a_i\mu_i$ and variance, $\sum a_i^2\sigma_i^2$.

Hence the theorem.

4.26.14 Fitting a Normal Distribution

If N denotes the sum of all frequencies of a normal distribution, the normal curve is given by $y = \frac{N}{\sigma\sqrt{2\pi}} e^{-\frac{1}{2}\left[\frac{x-\mu}{\sigma}\right]^2}$.

The area under the normal curve is equal to N.

To fit a normal curve to a distribution, there are two methods namely

1. Ordinate method
2. Area method.

These methods are explained with the help of few examples.

Example 4.98: Fit a normal curve to the distribution

x	1	3	5	7	9
f	1	2	3	2	1

Solution: Let μ denote the mean and σ denote the SD of the distribution.

We have,

$$N = \sum f_i = 1 + 2 + 3 + 2 + 1 = 9$$

Mean of the distribution is $\overline{x} = \frac{\sum x_i f_i}{\sum f_i} = \frac{1.1 + 3.2 + 5.3 + 7.2 + 9.1}{9}$

$$= \frac{45}{9} = 5$$

$$\sigma^2 = \text{variance} = \frac{\sum x_i^2 f_i}{\sum f_i} - \mu^2$$

$$= \frac{1^2 \cdot 1 + 3^2 \cdot 2 + 5^2 \cdot 3 + 7^2 \cdot 2 + 9^2 \cdot 1}{9} - 5^2$$

$$= \frac{1 + 18 + 75 + 98 + 81}{9} - 25$$

$$= \frac{273}{9} - 25$$

$$= \frac{273 - 225}{9} = \frac{48}{9}$$

$$\therefore \quad \sigma = \sqrt{\frac{48}{9}} = \frac{4\sqrt{3}}{3}$$

The equation of best fitting curve is

$$y = \frac{N}{\sigma\sqrt{2\pi}} e^{-\frac{1}{2}\left[\frac{x-\mu}{\sigma}\right]^2}$$

$$= \frac{9}{\frac{4\sqrt{3}}{3}\sqrt{2\pi}} e^{-\frac{1}{2}\left[\frac{x-5}{\frac{48}{9}}\right]^2}$$

$$= \frac{9\sqrt{3}}{4\sqrt{2\pi}} e^{-\frac{9}{96}[x-5]^2}$$

$$= \frac{9\sqrt{3}}{4\sqrt{2\pi}} e^{-\frac{3}{32}[x-5]^2}$$

Example 4.99: Fit a normal distribution to the following data by the method of ordinates:

Class interval	59.5–62.5	62.5–65.5	65.5–68.5	68.5–72.5	71.5–74.5
Frequency (f)	5	18	42	27	8

Solution:

Class interval	f	x_i (midpoint)	$f_i x_i$	$f_i x_i^2$
59.5–62.5	5	61	305	18,605
62.5–65.5	18	64	1152	73,728
65.5–68.5	42	67	2814	188,538
68.5–71.5	27	70	1890	132,300
71.5–74.5	8	73	584	42,632
	100		6745	455,803

We have

$$\sum f_i = N = 100$$

$$\sum f_i x_i = 6745$$

$$\sum f_i x_i^2 = 455,803$$

$$\text{Mean of the distribution is } \overline{x} = \frac{\sum x_i f_i}{\sum f_i} = \frac{6745}{100} = 67.45$$

$$\sigma = \text{Standard deviation} = \sqrt{\frac{\sum f_i x_i^2}{\sum f_i} - \mu^2}$$

$$= \sqrt{\frac{455803}{100} - (67.45)^2} = 2.92$$

Hence $\sigma = 2.92$

The ordinates are obtained at various distances from the mean, we use the formula

$$f(x) = \frac{1}{\sqrt{2\pi}} e^{-\frac{z^2}{2}}$$

where

$$Z = \frac{x - \mu}{\sigma}$$

We have the following table:

Class interval	x(Midpoint)	$Z = \frac{x-\mu}{\sigma}$	$f(x) = \frac{1}{\sqrt{2\pi}} e^{-\frac{z^2}{2}}$	$y = \frac{N}{\sigma} f(Z)$	$\frac{N}{\sigma} f(Z) \times c$
59.5–62.5	61	−2.21	0.0347	1.188	3.565
62.5–65.5	64	−1.18	0.1984	6.794	20.382
65.5–68.5	67	−0.15	0.3943	13.50	40.50
68.5–71.5	70	0.87	0.2725	9.333	27.999
71.5–74.5	73	1.90	0.0656	2.247	6.741

The expected (i.e., theoretical) frequencies are 3.565, 20.382, 40.50, 27.999, 6.741 i.e., 4, 20, 41, 28, and 7.

Example 4.100: Fit a normal distribution to the following data by the area method:

Class interval	60–65	65–70	70–75	75–80	80–85	85–90	90–95	95–100
Frequency(*f*)	3	21	150	335	326	135	26	4

Solution: We find the values of mean and SD as follows:

Class interval	F	Midpoint, x_i	$f_i\, x_i$	$f_i\, x_i^2$
60–65	3	62.5	187.5	11718.75
65–70	21	67.5	1417.5	95681.25
70–75	150	72.5	10875.0	788437.5
75–80	335	77.5	25962.5	2012093.75
81–85	326	82.5	26895.0	2218837.5
85–90	135	87.5	11812.5	1033593.75
90–95	26	92.5	2405.0	22462.5
95–100	4	97.5	390.0	38025.0
	N = 100		79945	64200850.0

We have

$$\sum f_i = N = 1000$$

$$\sum f_i x_i = 79945$$

$$\sum f_i x_i^2 = 64200850.0$$

$$\text{Mean of the distribution is } \mu = \frac{\sum x_i f_i}{\sum f_i} = \frac{79945}{1000} = 79.945$$

$$\sigma^2 = \text{variance} = \frac{\sum x_i^2 f_i}{\sum f_i} - \mu^2$$

$$= \frac{64200850}{1000} - (79.945)^2$$

$$= 6420.85 - 639120$$

$$= 29.65$$

$$\therefore \sigma = \sqrt{29.65} = 5.4451$$

Since the normal distribution takes values from, $-\infty$ to ∞ the given data can be modified as follows:

Class interval	f	Lower limit	$Z = \frac{X-\mu}{\sigma}$	Area $\varphi(Z)$	$\Delta\varphi(X) = \varphi(Z+1) - \varphi(Z)$	$N \cdot \Delta\varphi(Z)$
$-\infty$ to 60	0	$-\infty$	$-\infty$	0	0.0001	0.1
60–65	3	60	-3.6629	0.0001	0.003	3
65–70	21	62	-2.7447	0.0031	0.0313	31.3
70–75	150	70	-1.8264	0.0344	0.1497	149.7
75–80	335	75	-0.9081	0.1841	0.3199	319.9
80–85	326	80	0.0101	0.5040	0.3172	317.2
85–90	135	85	0.9283	0.8212	0.1459	145.9
90–95	26	90	1.8466	0.9671	0.03	30
95–100	4	95	2.7649	0.9671	0.0028	2.8
100		$-\infty$	∞ 100	3.6831	0.9999	

The expected frequencies are 0.1, 3, 31.3, 149.7, 319.9, 317.2, 145.9, 30, 2.8, i.e., 0, 3, 31, 150, 320, 317, 146, 30, 3.

4.26.15 Normal Approximation to Binomial Distribution

Normal can be used to approximate the binomial distribution. Consider a random variable X. If X follows binomial distribution then we have

$$P(X = x) = {}^nC_x p^x q^{n-x}$$

which is very large, then

$$P(X_1 < x < x_2) = \sum_{x=x_1}^{n=x_2} {}^nC_x p^x q^{n-x}$$

1. When $p = q = \frac{1}{2}$, n may or may not be large
   Let $Z = \frac{x-\mu}{\sigma}$

Where μ mean $= np$ and SD $= \sigma = \sqrt{npq}$

Then
$$P(X_1 < X < X_2) = P\left(\frac{X_1 - \mu}{\sigma} < \frac{X - \mu}{\sigma} < \frac{X_2 - \mu}{\sigma}\right)$$
$$= P(Z_1 < Z < Z_2) = \int_{z_1}^{z_2} \varphi(z)\mathrm{d}z$$

which can be obtained from the area table.

2. When $p \neq q$, and n is large

The real class interval is $\left(x - \frac{1}{2}, x + \frac{1}{2}\right)$

For X to lie in the interval, we consider the interval $\left(x - \frac{1}{2}, x + \frac{1}{2}\right)$

The corresponding values of Z are $\frac{x_1 - \frac{1}{2} - \mu}{2}$ and $\frac{x_2 - \frac{1}{2} - \mu}{2}$ where $\mu = np$ and $\sigma = \sqrt{npq}$.

The required probability is $\int_{z_1}^{z_2} \varphi(z)\mathrm{d}z$

Example 4.101: Find the probability of getting 1 or 3 or 4 or 5 in throwing a die 5–7 times among 9 trials?

Solution: We have $n = 9$

$$p = 4 \cdot \frac{1}{6} = \frac{2}{3}$$

$$q = 1 - p = 1 - \frac{2}{3} = \frac{1}{3}$$

Mean $\sigma = np = 9 \cdot \frac{2}{3} = 6$

$$\sigma^2 = \text{variance} = npq = 9 \cdot \frac{2}{3} \cdot \frac{1}{3} = 2$$

$$\therefore \sigma = \sqrt{2} = 1.414$$

$$x_1 = 5, x_2 = 7$$

$$Z_1 = \frac{x_1 - \frac{1}{2} - \mu}{\sigma} = \frac{5 - \frac{1}{2} - 6}{1.414} = \frac{4.5 - 6}{1.414}$$

$$= \frac{-1.5}{1.414} = -1.0608$$

$$Z_2 = \frac{x_2 - \frac{1}{2} - \mu}{\sigma} = \frac{7 - \frac{1}{2} - 6}{1.414} = \frac{7.5 - 6}{1.414}$$

$$= \frac{1.5}{1.414} = 1.0608$$

$$P(X_1 < X < X_2) = P\left(X_1 - \frac{1}{2} < X < X_2 + \frac{1}{2}\right)$$

$$\begin{aligned}\therefore P(4.5 < X < 7.5) &= P(-1.0608 < Z < 1.0608)\\ &= P(-1.0608 < Z < 0) + P(0 < Z < 1.0608)\\ &= 2[P(0 < Z < 1.0608)] = 2(0.3554)\\ &= 0.7108\end{aligned}$$

Example 4.102: Eight coins are tossed together, find the probability of getting 1–5 heads in a single toss?

Solution: We have $n = 8$

$$p = \frac{1}{2} \text{ and } q = \frac{1}{2}$$

Mean $\mu = np = 8 \cdot \frac{1}{2} = 4$

$$\sigma^2 = \text{variance} = npq = 8. \ \frac{1}{2} \cdot \frac{1}{2} = 2$$

$$\therefore \sigma = \sqrt{2} = 1.414$$

$$x_1 = 1, x_2 = 5$$

$$Z = \frac{x_1 - \frac{1}{2} - \mu}{\sigma} = \frac{1 - \frac{1}{2} - 4}{1.414}$$

$$= \frac{-3.5}{1.414} = -2.4752$$

$$Z_2 = \frac{x_2 - \frac{1}{2} - \mu}{\sigma} = \frac{5 - \frac{1}{2} - 4}{1.414}$$

$$= \frac{3.5}{1.414} = 2.4752$$

$$\therefore P(1 < X < 5) = P(-2.475 < Z < 2.475)$$
$$= 2[P(0 < Z < 2.475)] = 2(0.4932)$$
$$= 0.9864$$

Example 4.103: Random variable X is normally distributed with $m = 12$ and SD 2. Find $P(9.6 < x < 13.8)$?

Given that for $\frac{x}{\sigma} = 0.9$, $A = 0.3159$ and
for $\frac{x}{\sigma} = 1.2$ $A = 0.3849$

Solution: We have

$$P(9.6 < x < 13.8) = \int_{9.6}^{13.8} \frac{1}{\sigma\sqrt{2\pi}} e^{-\frac{1}{2}\left[\frac{x-\mu}{\sigma}\right]^2} dx \qquad (4.25)$$

$$\mu = 12, \quad \sigma = 2, \quad Z = \frac{x - \mu}{\sigma} = \frac{x - 12}{2}$$

When

$$X = 9.6, \quad Z = Z_1 = \frac{9.6 - 12}{2} = -1.2$$

and

$$X = 13.8, \quad Z = Z_2 = \frac{13.8 - 12}{2} = 0.9$$

Substituting in Eq. (4.25) we get

$$P(9.6 < x < 13.8) = \int_{9.6}^{13.8} \frac{1}{2\sqrt{2\pi}} e^{-\frac{1}{2}\left[\frac{x-12}{2.4}\right]^2} dx$$

$$= \frac{1}{\sqrt{2\pi}} \int_{-1.2}^{0.9} e^{-\frac{1}{2}z^2} dx$$

$$= \frac{1}{\sqrt{2\pi}} \int_{-1.2}^{0} e^{-\frac{1}{2}z^2} dx + \frac{1}{\sqrt{2\pi}} \int_{0}^{0.9} e^{-\frac{1}{2}z^2} dx$$

$$= 0.3849 + 0.3159 = 0.7008$$

4.27 CHARACTERISTIC FUNCTION

Let x be a continuous random variable. The characteristic function of x, for the continuous probability distribution with density function $f(x)$ is defined as

$$E[e^{itx}] = \int_{-\infty}^{\infty} e^{itx} f(x)\mathrm{d}x$$

$$= \int_{-\infty}^{\infty} e^{itx} \mathrm{d}F(x), (i = \sqrt{-1})$$

where t is a real parameter. The characteristic function of x is denoted by $\varphi_x(t)$ or $\varphi(t)$.

1. $\varphi(t)$ is continuous in t

2. $\varphi(0) = 1$

3. $\varphi(t)$ is defined in every finite t interval

4. $|\varphi(t)| \leq 1$

5. $\varphi(t)$ and $\varphi(-t)$ are conjugate.

4.28 GAMMA DISTRIBUTION

Let x be a continuous random variable. X is said to follow a Gamma distribution with parameter λ, if its probability function is given by

$$P(X = x) = P(x) = \frac{e^{-x} x^{\lambda - 1}}{\Gamma\lambda}, \quad \lambda > 0, \quad 0 < x < \infty$$

$$= 0, \qquad \text{otherwise}$$

The above distribution is known as Gamma distribution of first kind.

4.28.1 Mean and Variance of Gamma Distribution

Since

$$M_X(t) = (1 - t)^{-\lambda}$$

We get

$$M'_x(t) = \lambda(1 - t)^{-\lambda - 1}(-1)$$

$$= \lambda(1 - t)^{-\lambda - 1}$$

Hence

$$\mu'_1 = M'_x(0) = \lambda$$

Since

We get

Hence

Since

$$M'_x(t) = \lambda(1-t)^{-\lambda-1}, \quad \text{we get}$$

$$\begin{aligned} M''x(t) &= (-\lambda-1)\lambda(1-t)^{-\lambda-2}(-1) \\ &= (\lambda+1)\lambda(1-t)^{-\lambda-2} \\ \therefore\ \mu'_2 &= M''x(0) = \lambda(\lambda+1) \end{aligned}$$

$$\begin{aligned} \text{Variance} &= \mu'_2 - \mu'_2 \\ &= \lambda(\lambda+1) - \lambda^2 \\ &= \lambda^2 + \lambda - \lambda^2 = \lambda \end{aligned}$$

Hence mean of Gamma distribution = variance of gamma distribution = λ

The first four central moments of Gamma distribution of first kind are

$$\mu_1 = 0, \quad \mu_2 = \lambda, \quad \mu_3 = 2\lambda \quad \text{and} \quad \mu_4 = 3\lambda(\lambda+2)$$

4.28.2 Gamma Distribution of Second Kind

A random variable x is said to follow Gamma distribution of second kind if it has the following density function:

$$\begin{aligned} f(x) = P(X=x) &= \frac{a^\lambda}{\Gamma\lambda} e^{-ax} x^{\lambda-1}, \quad && a>0,\ \lambda>0,\ 0<x<\infty \\ &= 0, && \text{otherwise} \end{aligned}$$

4.29 BETA DISTRIBUTION OF FIRST KIND

Let X be a continuous random variable. X is said to follow a beta distribution of first kind if its pdf is given by

$$f(x) = \frac{1}{\beta(m,n)} \cdot x^{m-1}(1-x)^{n-1}, \quad m, n>0, \quad 0<x<1$$

where m, n are parameters of the distribution and we have

$$\begin{aligned} \beta(m,n) &= \frac{\Gamma m \Gamma n}{\Gamma(m+n)} \\ &= \frac{\beta(m+r,n)}{\beta(m,n)} \end{aligned}$$

4.29.1 Beta Distribution of Second Kind

A continuous random variable x is said to follow beta distribution of second kind if its pdf is given by

$$f(x) = \frac{1}{\beta(m,n)} \cdot \frac{x^{m-1}}{(1+x)^{m+n}}, \quad m,n>0, \ 0<x<\infty$$
$$= 0, \quad \text{otherwise}$$

4.30 WEIBULL DISTRIBUTION

The random variable X has a Weibull distribution if its pdf has the form

$$f(x) = \frac{\beta}{\alpha}\left(\frac{x-\nu}{\alpha}\right)^{\beta-1} e^{-\left(\frac{x-\nu}{\alpha}\right)^{\beta}}, \quad x \geq \nu$$
$$= 0, \quad \text{otherwise}$$

The three parameters of Weibull's distribution are $\nu(-\infty<\nu<\infty)$, α and β, where $\alpha>0$ and $\beta>0$

ν is called location parameter; α is called scale parameter; β is called the shape parameter.

When $\nu=0$, the Weibulls pdf becomes

$$f(x) = \frac{\beta}{\alpha}\left(\frac{x}{\alpha}\right)^{\beta-1} e^{-\left(\frac{x}{\alpha}\right)^{\beta}}, \quad x \geq 0$$
$$= 0, \quad \text{otherwise}$$

When $\nu=0$, and $\beta=1$ the Weibull distribution reduces to

$$f(x) = \frac{1}{\alpha} e^{-x/a}, \quad x \geq 0$$
$$= 0, \quad \text{otherwise}$$

which is an exponential distribution with parameter $\lambda = 1/\alpha$.

The mean and variance of the Weibull distribution are given by the following expressions:

$$E(X) = \nu + \alpha\Gamma\left(\frac{1}{\beta}+1\right)$$

$$V[X] = \alpha^2\left[\Gamma\left(\frac{2}{\beta}+1\right) - \left[\Gamma\left(\frac{1}{\beta}+1\right)\right]^2\right]$$

Thus the location parameters has no effect in the variance, however the mean is increased or decreased by ν.

The cumulative distribution function of the Weibull distribution is given by

$$F(x) = 0, \qquad x < \nu$$
$$= 1 - e^{-\left(\frac{x-\nu}{\alpha}\right)^{\beta}}, \quad x \geq \nu$$

Example 4.104: The time it takes for an aircraft to clear the runway at a major international airport has a Weibull distribution with $\nu = 1.34$ minutes. $\beta = 0.5$ and $\alpha = 0.04$ minutes. Determine the probability that an incoming airplane will take more than 1.5 minutes to land and clear the runway.

Solution: we have $P(X \leq 1.5) = F(1.5) = 1 - e^{-\left(\frac{1.5-1.34}{0.04}\right)^{0.5}}$
$$= 1 - e^{-2} = 1 - 0.135 = 0.865$$

$P(X > 1.5)$ = Probability then an incoming airplane will take more than 1.5 minutes $= 1 - 0.865 = 0.135$.

Exercise 4.11

1. Define the terms (a) Uniform distribution; (b) Hypergeometric distribution; (c) Beta distribution of second kind.
2. The number of cars passing a busy road between 5 p.m. and 6 p.m. is normally distributed with mean 352 and a variance of 961. What percent of time will there be more than 400 cars between and 6 p.m?

   Ans: 6.06
3. Define: (a) Geometric distribution, (b) Gamma distribution.
4. If X is a binomial random variable with $n = 6$, satisfying $9P(X = 4) = P(X = 2)$. What is the value of p?

   Ans: 0.25
5. The following table shows the distribution of the number of vacancies in transportation company occurring per year during the years 1837–1932. Fit Poisson distribution.

Vacancies	0	1	2	3
Frequency	59	27	9	1

   Ans: 58, 29, 8, 1
6. Trains arrive at a station 15 minutes interval starting at 4:00 a.m. If a passenger arrives at the station at a particular time that is uniformly

distributed between 9:00 and 9:30. Find the probability that he/she has to wait for the train (a) for less than 6 minutes; (b) more than 10 minutes?

Ans: (a) $\frac{2}{5}$; (b) $\frac{1}{3}$

7. The number of fatal accidents in electric trains during a week has Poisson distribution such that the probability of two fatal accidents is the same as that of three accidents. What is the mean number of accidents?

Ans: 3

8. In a construction place, 4 lorries arrive per hour to unload building materials. What is the probability that at most 18 lorries will arrive during 6 hour day?

Ans: 0.128

CHAPTER 5

Curve Fitting

5.1 INTRODUCTION

The process of constructing a curve or mathematical function that has the best fit to a series of data points is known as curve fitting. For a given set of data, the fitting curves of a given type are generally not unique. The best fitting curve with minimum deviations from all data points can be obtained by the method of least squares.

5.2 THE METHOD OF LEAST SQUARES

The method of least squares assumes that the best fit curve of a given type is the curve that has the minimal sum of deviations, i.e., least square error from a given set of data.

Suppose that the data points are $(x_1, y_1), (x_2, y_2), \ldots, (x_n, y_n)$ where x is the independent variable and y is the dependent variable. The fitting curve $f(x)$ has the deviation (error) e_i from each data point, as follows:

$$\begin{aligned} e_1 &= y_1 - f(x_1), \\ e_2 &= y_2 - f(x_2), \\ &\vdots \\ e_n &= y_n - f(x_n) \end{aligned}$$

According to the method of least squares, the best fitting curve has the property that $\sum_1^n e_i^2 = \sum_1^n [y_i - f(x_i)]^2$ is minimum.

We now introduce the method of least squares using polynomials in the following sections.

5.3 THE LEAST-SQUARES LINE

The least-squares line uses a straight line

$$y = a + bx$$

To approximate the given set of data, $(x_1, y_1), (x_2, y_2), \ldots, (x_n, y_n)$ where $n \geq 2$. The best fitting curve $y = f(x)$ has the least square error. Let

Statistical Techniques for Transportation Engineering.
DOI: http://dx.doi.org/10.1016/B978-0-12-811555-8.00005-2

E denote the error where E is a function of two variables a and b. Then $E = \sum_1^n e_i^2 = \sum_1^n [y_i - f(x_i)]^2 = \sum_1^n [y_i - (a + bx_i)]^2$ is minimum, where a and b are unknown coefficients. The necessary conditions for E to be minimum are

$$\frac{\partial E}{\partial a} = 0 \text{ and } \frac{\partial E}{\partial b} = 0$$

i.e.,

$$\frac{\partial E}{\partial b} = 2\sum_1^n (-x_i)[y_i - (a + b_{x_i})] = 0$$

Expanding the above equations we have:

$$\sum_1^n y_i = a\sum_1^n 1 + b\sum_1^n x_i$$

$$\sum_1^n x_i y_i = a\sum_1^n x_i + b\sum_1^n x_i^2$$

or

$$\sum_1^n x_i y_i = \sum_1^n x_i a + b\sum_1^n x_i^2 \tag{5.1}$$

$$\sum_1^n x_i y_i = a\sum_1^n x_i + b\sum_1^n x_1^2 \tag{5.2}$$

The Eqs. (5.1) and (5.2) are known as the normal equations. Solving the normal equations we get:

$$a = \frac{\sum y_i \sum x_i^2 - \sum x_i \sum x_i y_i}{n\sum x_i - (\sum x_i)^2}$$

and

$$b = \frac{n\sum x_i y_i - \sum x_i \sum y_i}{n\sum x_i - (\sum x_i)^2}$$

Remark:

The normal equations for fitting the least square line $y = ax + b$ are:

$$\sum_1^n y_i = nb + a\sum_1^n x_i$$

$$\sum_1^n x_i y_i = b\sum_1^n x_i + a\sum_1^n x_i^2$$

Example 5.1: Fit the straight line $y = ax + b$ for the data given below:

x	5	10	15	20	25
y	16	19	23	26	30

Solution: We have the following table:

x_i	y_i	x_i^2	$x_i . y_i$
5	16	25	80
10	19	100	190
15	23	225	345
20	26	400	520
25	30	625	750
$\sum x_i = 75$	$\sum y_i = 114$	$\sum x_i^2 = 1375$	$\sum x_i y_i = 1885$

From the table we have

$$n = 4, \ \sum x_i = 75, \ \sum y_i = 114, \ \sum x_i^2 = 1375, \text{ and } \sum x_i y_i = 1885$$

The normal equations for fitting the least square line $y = ax + b$ are

$$\sum_1^n y_i = nb + a\sum_1^n x_i = \sum_1^n y_i$$

$$\sum_1^n x_i y_i = b\sum_1^n x_i + a\sum_1^n x_i^2 = \sum_1^n x_i y_i$$

Substituting the values in normal equations, we get

$$75a + 5b = 114$$

and

$$1375a + 75b = 1885$$

Solving these equations, we get

$$a = 0.7$$

$$b = 12.3$$

Therefore the required least square line is

$$y = 0.7x + 12.3$$

Example 5.2: Find the least square line $y = a + b \cdot x$ for the data points (−1, 10), (0, 9), (1, 7), (2, 5), (3, 4), (4, 3), (5, 0), and (6, −1)?

Solution: Here

x_i	y_i	x_i^2	$x_i\, y_i$
−1	10	1	−10
0	9	0	0
1	7	1	7
2	5	4	10
3	4	9	12
4	3	16	12
5	0	25	0
6	−1	36	−6
$\sum x_i = 20$	$\sum y_i = 37$	$\sum x_i^2 = 92$	$\sum x_i y_i = 25$

From the table we have

$$n = 8,\ \sum x_i = 20,\ \sum y_i = 37,\ \sum x_i^2 = 92, \text{ and } \sum x_i y_i = 25$$

The normal equations are

$$na + b\sum_1^n x_i = \sum_1^n y_i \tag{5.3}$$

$$\sum_1^n x_i = b\sum_1^n x_i^2 = \sum_1^n x_i y_i \tag{5.4}$$

Putting the values in normal equations

$$8a + 20b = 37$$

and

$$20a + 97b = 25$$

and solving these equations, we get

$$a = 8.6428571$$

$$b = -1.6071429$$

hence the required least square line is

$$y = 6428571 + (-1)\ 1.6071429x$$

i.e.,

$$y = -1.6071429x + 8.6428571$$

5.4 FITTING A PARABOLA BY THE METHOD OF LEAST SQUARES

Let $(x_1, y_1), (x_2, y_2), \ldots, (x_n, y_n)$, (where $n \geq 2$) denote a set of n observations of two variables of two variables x and y. The best fitting parabola $y = a + bx + cx^2$ has the least error. Let E denote the error where E is a function of two variables a, b, and c. Then

$$E = \sum_1^n e_i^2 = \sum_1^n [y_i - f(x_i)]^2 = \sum_1^n [y_i - (a + bx_i + cx_i^2)]^2$$

is minimum, where a, b, and c are unknown coefficients. The necessary conditions for E to be minimum are

$$\frac{\partial E}{\partial a} = 0 \text{ and } \frac{\partial E}{\partial b} = 0, \quad \frac{\partial E}{\partial c} = 0$$

i.e., we have

$$\frac{\partial E}{\partial a} = \sum_1^n 2[y_i - (a + bx_i + cx_i^2)](-1) = 0$$

On simplifying we get

$$na + b\sum x_i + c\sum x_i^2 = \sum y_i \tag{5.5}$$

$$\frac{\partial E}{\partial b} = \sum_1^n 2y_i - (a + bx_i + cx_i^2)(-x_i) = 0$$

On simplifying we get

$$\sum x_i + b\sum x_i^2 + c\sum x_i^3 = \sum x_i y_i \tag{5.6}$$

$$\frac{\partial E}{\partial c} = \sum_1^n 2y_i - (a + bx_i + cx_i^2)(-x_i^2) = 0$$

On simplifying we get

$$a\sum x_i^2 + b\sum x_i^3 + c\sum x_i^4 = \sum x_i^2 y_i \tag{5.7}$$

Hence the normal equations for fitting the least square curve $y = a + bx + c \cdot x^2$ (Parabola) are

$$na + b\sum x_i + c\sum x_i^2 = \sum y_i$$
$$a\sum x_i + b\sum x_i^2 + \sum x_i^3 = \sum x_i y_i$$
$$a\sum x_i^2 + b\sum x_i^3 + c\sum x_i^4 = \sum x_i^2 y_i$$

Dropping the suffix we can write

$$na + b\sum x + c\sum x^2 = \sum y$$
$$a\sum x + b\sum x^2 + c\sum x^3 = \sum xy$$
$$a\sum x^2 + b\sum x^3 + c\sum x^4 = \sum x^2 y$$

We can find the values of a, b, and c by solving these equations. Substituting the values of a, b, and c in $y = a + bx + cx^2$ we get the required equation of the curve.

Example 5.3: Fit a least square parabola $y = a + bx + cx^2$ by the method of least squares for the data given below:

x	0	1	2	3	4
y	1	1.8	1.3	2.5	2.3

Solution: We have the following table:

x	y	xy	x^2y	x^2	x^3	x^4
0	1	0	0	0	0	0
1	1.8	1.8	1.8	1	1	1
2	1.3	2.6	5.2	4	8	16
3	2.5	7.5	22.5	9	27	81
4	2.3	9.2	36.8	16	64	256
$\sum x = 10$	$\sum y = 8.9$	$\sum xy = 21.1$	$\sum x^2y = 66.3$	$\sum x^2 = 30$	$\sum x^3 = 100$	$\sum x^4 = 354$

The required normal equations are

$$5a + 10b + 30c = 8.9$$
$$10a + 30b + 100c = 21.1$$
$$30a + 100b + 354c = 66.3$$

Solving these equations we get $a = 1.078$, $b = 0.414$, and $c = -0.021$ therefore the equation of the curve is $y = 1.078 + 0414x - 0\ 021x^2$

Example 5.4: Find a formula for the line of the form $y = a + bx + cx^2$ that will fit the following data:

x	0	0.1	0.2	0.3	0.4	0.5	0.6	0.7	0.8	0.9
y	3.1950	3.2299	3.2532	3.2611	3.2516	3.2282	3.1807	3.1266	3.0594	2.9759

Solution: The normal equations are

$$10a + 4.5b + 2.85c = 31.7616$$

$$4.5a + 2.85b + 2.025c = 14.0896$$

$$2.85a + 2.0256 + 1.5333c = 8.82881$$

and solving these equations we obtain

$$a = 3.1951,\ b = 0.44254,\ c = -0.76531.$$

The required equation is

$$y = 3.1951 + 0.44254x - 0.76531x^2$$

Example 5.5: Find a second degree $y = ax^2 + b\ x + c$ in the least square sense for the following data:

x	1	2	3	4	5
y	10	12	13	16	19

Solution: Let $U = x-3,\ V = y-14$
We have the following table:

x	y	U	V	UV	U^2V	U^2	U^3	U^4
1	10	−2	−4	8	−16	4	−8	16
2	12	−1	−2	2	−2	1	−1	1
3	13	0	−1	0	0	0	0	0
4	16	1	2	2	2	1	1	1
5	19	2	5	10	20	4	8	16
Total	–	0	0	22	4	10	0	34

The normal equations are

$$na + b\sum U + c\sum U^2 = \sum V$$

$$a\sum U + b\sum U^2 + c\sum U^3 = \sum UV$$

$$a\sum U^2 + b\sum U^3 + c\sum U^4 = \sum U^2 V$$

Substituting the values of a, b, and c, the normal equations can be written as

$$10a + 5c = 0$$

$$10b = 22$$

$$34a + 10c = 4$$

Solving the above equations we get $c = 0.29$, $b = 2.2$, and $a = -0.58$
The equation the parabola is $V = a + b\,U + c\,U^2$, i.e.,

$$y - 14 = a + b(x - 3) + c(x - 3)^2$$

or

$$y - 14 = -0.58 + 2.2(x - 3) + (0.29)(x - 3)^2$$

Hence the required equation of the parabola is

$$y = 9.43 + 0.46\,x + 0.29x^2$$

5.5 FITTING THE EXPONENTIAL CURVE OF THE FORM $y = a\,e^{bx}$

We explain the method of fitting the curve of the form $y = a\,e^{bx}$ with the help of the following example:

Example 5.6: Fit a least square curve of the form $y = a\,e^{bx}$ $(a > 0)$ to the data given below:

x	1	2	3	4
y	1.65	2.70	4.50	7.35

Solution: Consider $y = a\,e^{bx}$
Applying logarithms (with base 10) on both sides, we get

$$\log_{10}y = \log_{10}a + bx\,\log_{10}e \tag{5.8}$$

taking $\log_{10} y = Y$, the Eq. (5.8) can be written as

$$Y = A + B\,X \tag{5.9}$$

where $Y = \log_{10} y$ $A = \log_{10} a$, $B = \log_{10} e$
Eq. (5.9) is a linear equation in X and Y, the normal equations are

$$nA + B\sum x = \sum Y$$

x	y	$Y = \log_{10} y$	xY	x^2
1	1.65	0.2175	0.2175	1
2	2.70	0.4314	0.8628	4
3	4.50	0.6532	1.9596	9
4	7.35	0.8663	3.4652	16
Total 10	–	2.1684	6.5051	30

We have

$$4A + 10B = 2.1684 \tag{5.10}$$

$$10A + 30B = 6.5051. \tag{5.11}$$

Solving the Eqs. (5.10) and (5.11), we get
Now

$$A = 0.0001, \quad B = 0.2168$$

$$A = 0.0001 \Rightarrow \log_{10} a = 0.0001 \Rightarrow A = 1.0002$$

$$B = 0.218 \Rightarrow b \ \log_{10} e = 0.2168$$

or

$$b = \frac{0.2168}{\log_{10} e} = \frac{0.2169}{0.4343}$$

or

$$b = 0.4992$$

Therefore the required curve is $Y = (1.0002)\ e^{0.4992x}$.

Remark:
We can also take $A = \log_e a$, $B = \log_e e = 1$ and apply the method of least squares. The method is explained below:

Example 5.7: Fit the curve $y = a\ e^{bx}$ for the following data:

x	0	2	4
y	8.12	10	31.82

Solution: Consider the equation $y = a\ e^{bx}$
Taking logarithms on both sides we get $\log_e y = (\log_e a) b \log_e e$
we have

$$A = \log_e a, \quad \log_e e = 1$$

the normal equations are

$$nA + b\sum x = \sum Y$$

$$A\sum x + b\sum x^2 = \sum xY$$

We have the following table:

x	y	$Y = \log_e y$	xY	x^2
0	8.12	2.09	0	0
2	10	2.30	4.60	4
4	31.82	3.46	13.84	16
$\sum x = 6$	–	$\sum Y = 7.85$	$\sum XY = 18.44$	$\sum x^2 = 30$

The normal equations are

$$3A + 6b = 7.85,$$
$$6A + 20b = 18.44$$

Solving we get

$$A = 7.85$$
$$a = e^A = 6.903, \ b = 0.3425$$

Hence the required equation of the curve is $Y = 6.903 \cdot e^{0.3425x}$ where $Y = \log_e y$

Exercise 5.1

1. Find the least square line $y = a + bx$ for the following data:

x	−4	−2	0	2	4
y	1.2	2.8	6.2	7.8	13.2

   Ans: $y = 6.24 + 1.45x$

2. Find the least square line $y = a_0 + a_1x$ for the following data:

x	1	2	3	4
y	0	1	1	2

   Ans: $y = -\frac{1}{2} + \frac{3}{5}x$

3. Find the least-squares parabola for the points (−3, 3), (0, 1), (2, 1), (4, 3)?

   Ans: $y = 0.850519 - 0.192495x + 0.178462x^2$

4. Find the least-squares parabolic fit $y = ax^2 + bx + c$, for the following data:

x	−3	−1	1	3
y	15	5	1	5

   Ans: $y = \frac{7}{8}x^2 - \frac{17}{10}x + \frac{17}{8}$

5. The pressure and volume of a gas are related by the equation $PV^{\lambda} = k$ (λ and k are constants). Fit this equation for the data given below:

P	0.5	1.0	1.5	2.0	2.5	3.0
V	1.62	1.00	0.75	0.62	0.52	0.046

Ans: $PV^{1.\ 4225} = 0.997$

6. Using the method of least-squares fit a curve of the form $y = ab^x$ to the following data:

x	1	2	3	4
y	4	11	35	100

Ans: $y = (1.3268) \cdot (2.948)^x$

7. Using the method of least squares, fit a relation of the form $y = ab^x$ to the following data:

x	2	3	4	5	6
y	1.44	172.8	207.4	248.8	298.5

Ans: $y = 9.986x^{1.2}$

8. Fit a least square curve of the form $y = a\ e^{bx}$ $(a > 0)$ to the data given below:

x	1	2	3	4
y	1.65	2.70	4.50	7.35

Ans: $y = (1.0002)\ e^{0.4992x}$

9. Find the least-squares parabolic fit $y = a + bx + cx^2$?

x	−3	−1	1	3
y	15	5	1	5

Ans: $y = 2.125 - 1.70x + 0.875x^2$

10. Fit a least square curve of the form $y = ax^b$ for the following data where a and b are constants:

x	61	26	7	2.6
y	350	400	500	600

Ans: $y = (701.94)\ x^{-0.1708}$

11. The observations from an experiment are as given below:

x	2	10	26	61
y	600	500	400	350

It is known that a relation of type $y = a\, e^{bx}$ exists.
Find the best possible values of a and b?
Ans: $y = 43.12777\, e^{-0.0057056x}$

12. If P is the Pull required to lift a load W by means of a pulley block, find a liner law of the form $P = mW + c$, connecting P and W, using the data?

P	12	15	21	25
W	50	70	100	120

where P and W are taken in kg-wt. Compute P when $W = 150$ kg.
Ans: $2.2759 + 0.1879\, W$, 30.4635 kg

CHAPTER 6

Correlation and Regression

6.1 INTRODUCTION

In this chapter we introduce the concepts of correlation and regression. In many problems of science we may be dealing with two variables which may have relationship or association between them. Correlation measures the strength of relationship between two variables. The correlation in a sample is measured by the sample coefficient of correlation that is denoted by r and the coefficient of correlation in a population is measured by population coefficient correlation that is denoted by ρ (Rho). The technique of correlation is used to test the statistical significance of association. Correlation does not necessarily demonstrate a casual relationship.

Regression measures the nature and extent of correlation. Regression is used for prediction of one variable with respect to another variable. It is used to examine the relationship between one dependent variable and one independent variable. Regression analysis is used to describe the relationship precisely by means of an equation that has predictive value.

6.2 CORRELATION

Correlation is a statistical measure for finding out degree of association between two or more variables.

Definition 6.1: The relationship between two variables such that a change in one variable results in the positive or negative change in the other. Also a greater change in one variable results in corresponding greater or smaller change in the other variable is known as correlation.

Statistical Techniques for Transportation Engineering.
DOI: http://dx.doi.org/10.1016/B978-0-12-811555-8.00006-4

6.2.1 Types of Correlation

Correlation is classified into many types:

1. *Positive or Negative*: If two variables tend to move together in the same direction then the correlation is called positive correlation. When there is a negative correlation as the value of one variable increases, the value of other variable decreases and vice versa.

   Examples: Height and weight, demand and supply.
2. *Simple and Multiple:* When we study only two variables say X and Y, then the relationship is described as simple correlation.

   If we study more than two variables simultaneously then the correlation is called multiple correlations.

   Examples: Relationship of price, demand, and supply.
3. *Partial and Total:* If the study of the variables excluding some other variables is called partial correlation.

   If all the facts are taken into account then the correlation is called the amount of change in the other variables is called nonlinear correlation.

6.3 COEFFICIENT OF CORRELATION

The extent or degree of relationship between two variables measured in terms of another parameter is called the coefficient of correlation.

It is denoted by r.

Let X and Y denote two variables and r denote the coefficient of correlation between X and Y. Depending on the value of r, we can classify correlation as follows:

1. If $r = 1$, both the variables X and Y increase or decrease in the same proportion. In this case we say that there is a perfect positive correlation.
2. If $r = -1$, both the variables X and Y are inversely proportional to each other. We say that there is a perfect negative correlation.
3. If $r = 0$, we say that there is no relation between X and Y.
4. If $0 < r < 1$, there is moderate (partial) positive correlation between X and Y.
5. If $-1 < r < 0$, there is moderate (partial) negative correlation between X and Y.

6.3.1 Properties of Coefficient of Correlation

1. The correlation is a measure of relationship between two variables.
2. The value of coefficient of correlation lies between -1 and 1, i.e., $-1 \leq r \leq 1$.

3. If $r = 0$, the variables are said to be independent.
4. If $r = \pm 1$, there is a perfect correlation coefficient.

6.4 METHODS OF FINDING COEFFICIENT OF CORRELATION

Coefficient of correlation may be computed by any one or more of the following methods:

1. Scatter diagram
2. Correlation diagram
3. Direct method (Karl Pearson's method)
4. Two way frequency table method
5. Concurrent deviation method

In this chapter we discuss some of these methods.

6.5 SCATTER DIAGRAM

When the pair of values $(x_1, y_1), (x_2, y_2), (x_3, y_3), \ldots, (x_n, y_n)$ are plotted on a graph paper, the points show the pattern in which they lie, such a diagram is called a scatter diagram.

Consider the points $(x_1, y_1), (x_2, y_2), (x_3, y_3), \ldots, (x_n, y_n)$. In a scatter diagram the variable x is shown along the x-axis (horizontal axis) and the variable y is shown along the y-axis (vertical axis) and all the pairs of values of x and y are shown by points (or dots) on the graph paper. The scatter diagram of these points reveals the nature and strength of correlation between these two variables x and y. We observe the following:

If the points plotted lie on a straight line rising from lower left to upper right, then there is a perfect positive correlation between the variables x and y (Fig. 6.1A). If all the points do not lie on a straight line, but their tendency is to rise from lower left to upper right then there is a positive correlation between the variables x and y (Fig. 6.1B). In these cases the two variables x and y are in the same direction and the association between the variables is direct.

If the movements of the variables x and y are opposite in direction and the scatter diagram is a straight line, the correlation is said to be negative, i.e., association between the variables is said to be indirect.

Example 6.1: Draw a scatter diagram for the following data:

x	3	5	7	9	11	13	15
y	5	8	11	13	15	17	19

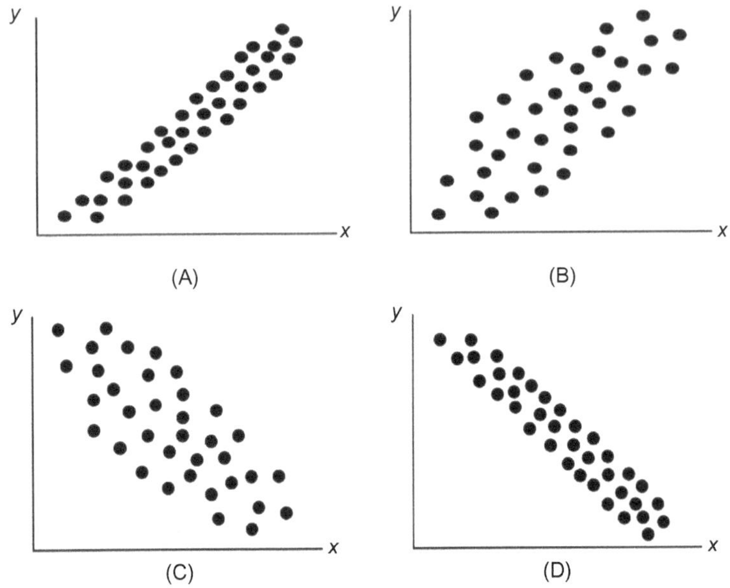

Figure 6.1 (A) Perfect positive correlation ($r = 1$). (B) Positive correlation. (C) Negative correlation. (D) Perfect negative correlation.

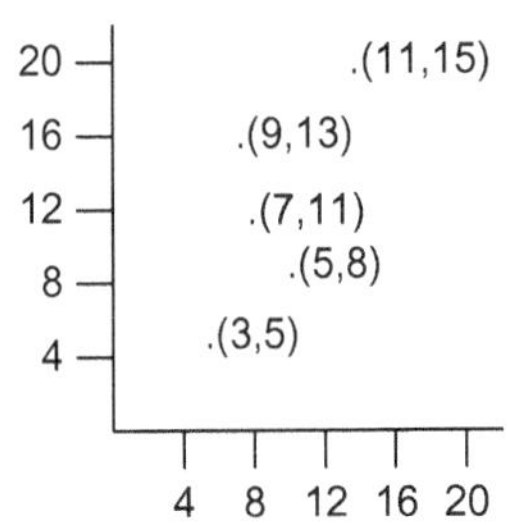

6.6 DIRECT METHOD

In this method the coefficient of correlation is between two variables x and y is given by

$$r = \frac{\sum \mathrm{d}x\, \mathrm{d}y}{\sqrt{\sum \mathrm{d}x^2 \sum \mathrm{d}y^2}}$$

where

$$\mathrm{d}x = x - \bar{x}$$

$$\mathrm{d}y = y - \bar{y}$$

$$dx^2 = (x-\overline{x})^2$$
$$dy^2 = (y-\overline{y})^2$$

$\overline{x}$ and $\overline{y}$ are the means of the variates X and Y, respectively.

The above formula can also be written as

$$r = \frac{\sum dxdy}{n\sigma_x\sigma_y}$$

or

$$\sigma_x = \text{Standard deviation of } X = \sqrt{\frac{\sum X^2}{n} - (\overline{X})^2}$$

$$\sigma_y = \text{Standard deviation of } Y = \sqrt{\frac{\sum Y^2}{n} - (\overline{Y})^2}$$

n = Number of observations in X or Y series.

Or

$$r = \frac{\text{Covariance }(X, Y)}{\sigma_X\sigma_Y}$$

where covariance $(X, Y) = 1/n(\sum XY - \overline{XY})$

This method is known as Karl Pearson's method. It is a popular mathematical method.

Example 6.2: Find the coefficient of correlation between industrial production (x) and export (y) from the following data:

Production (x)	55	56	58	59	60	60	62
Export (y)	35	38	37	39	44	43	44

Solution: Let 58 and 40 be two arbitrary numbers taken such that $X = x-58$, $Y = y-40$

x	y	$X = x-58$	$Y = y-40$	X^2	Y^2	XY
55	35	−3	−5	9	25	15
56	38	−2	−2	4	4	4
58	37	0	−3	0	9	0
59	39	1	−1	1	1	−1
60	44	2	4	4	16	8
60	43	2	3	4	9	6
62	44	4	4	16	16	16
		4	0	38	80	48

We have

$$\overline{X} = \frac{\sum x}{n} = \frac{4}{7} = 0.5714, \quad \overline{Y} = \frac{\sum y}{n} = \frac{0}{5} = 0$$

$$\text{Covariance}\,(X, Y) = \frac{1}{n}\Sigma XY - \overline{XY} = \frac{48}{7} - 0 = 6.857$$

$$\sigma_X = \sqrt{\frac{\sum X^2}{n} - (\overline{X})^2} = \sqrt{\frac{38}{7} - (0.5714)^2} = 2.2588$$

$$\sigma_Y = \sqrt{\frac{\sum Y^2}{n} - (\overline{Y})^2} = \sqrt{\frac{80}{7} - 0} = 3.38$$

$$x = \frac{\text{Covariance}\,(X, Y)}{\sigma_X \sigma_Y} = \frac{6.857}{2.258 \times 3.38} = 0.898$$

Example 6.3: Find the coefficient of correlation between X and Y for the following data:

X	1	2	3	4	5	6	7	8	9
Y	10	11	12	14	13	15	16	17	18

Solution: We have

$$\overline{x} = \frac{\sum x}{n} = \frac{1+2+3+4+5+6+7+8+9}{9} = \frac{45}{9} = 5$$

$$\overline{y} = \frac{\sum y}{n} = \frac{10+11+12+13+14+15+16+17+18}{9} = \frac{126}{9} = 14$$

Computation table

X	$dx = x - 5$	dx^2	Y	$dy = y - 14$	dy^2	$dxdy$
1	−4	16	10	-4	16	16
2	−3	9	11	−3	9	9
3	−2	4	12	−2	4	4
4	−1	1	14	0	0	0
5	0	0	13	−1	1	0
6	1	1	15	1	1	1
7	2	4	16	2	4	4
8	3	9	17	3	9	9
9	4	16	18	4	16	16
$\sum x = 45$	0	$\sum dx^2 = 60$	$\sum Y = 136$	0	$\sum dy^2 = 60$	$\sum dxdy = 59$

Coefficient of correlation $r = \dfrac{\sum dxdy}{\sqrt{\sum dx^2 \sum dy^2}} = \dfrac{59}{\sqrt{60 \times 60}} = \dfrac{59}{60}$

$= 0.9833$

Example 6.4: Calculate the coefficient of correlation between the variables X and Y from the following data:

X	65	66	67	67	68	69	70	72
Y	67	68	65	68	72	72	69	71

Solution: We have

$$\bar{x} = \frac{\sum x}{n} = \frac{65 + 66 + 67 + 67 + 68 + 69 + 70 + 72}{8} = \frac{544}{8} = 68$$

$$\bar{y} = \frac{\sum y}{n} = \frac{67 + 68 + 65 + 68 + 72 + 72 + 69 + 71}{8} = \frac{552}{8} = 69$$

X	$dx = x - 5$	dx^2	Y	$dy = y - 14$	dy^2	$dxdy$
65	−3	9	67	−2	4	6
66	−2	4	68	−1	1	2
67	−1	1	65	−4	16	4
67	−1	1	68	−1	1	1
68	0	0	72	3	9	0
69	1	1	72	3	9	3
70	2	4	69	0	0	0
72	4	16	71	4	16	8
544		36	552		44	24

$$r = \frac{\sum dxdy}{\sqrt{\sum dx^2 \sum dy^2}} = \frac{24}{\sqrt{36 \times 44}} = 0.603$$

Example 6.5: Find the coefficient of correlation between the height of fathers and sons from the following data:

Height of father	64	65	66	67	68	69	70
Height of son	66	67	68	69	70	71	72

Solution: Let X denote the height of father and Y denote the height of the son

We have

$$\bar{x} = \frac{\sum x}{n} = \frac{64 + 65 + 66 + 67 + 68 + 69 + 70}{7} = \frac{469}{7} = 67$$

$$\bar{y} = \frac{\sum y}{n} = \frac{66 + 67 + 68 + 69 + 70 + 71 + 72}{7} = \frac{483}{7} = 69$$

X	dx = x − 5	dx²	Y	dy = y − 14	dy²	dxdy
64	−3	9	66	−3	9	9
65	−2	4	67	−2	4	4
66	−1	1	68	−1	1	1
67	0	0	69	0	0	0
68	1	1	70	1	1	1
69	2	4	71	2	4	4
70	3	9	72	3	9	9
469		28	483		28	28

$$r = \frac{\sum dxdy}{\sqrt{\sum dx^2 \sum dy^2}} = \frac{28}{\sqrt{28 \times 28}} = 1$$

Example 6.6: Find coefficient of correlation for the following data and comment:

Fertilizers used (X)	15	18	20	24	30	35	40	50
Productivity (Y)	85	93	95	105	120	130	150	160

Solution: Let X denote the fertilizer used and Y denote the productivity.

Computation of coefficient of correlation.

Assume $A = 29$, and $B = 119$

Fertilizer used				Productivity		
X	dx = x − 5	dx²	Y	dy = y − 14	dy²	dxdy
15	−14	196	85	−34	1156	476
18	−11	121	93	−26	676	286
20	−9	81	95	−24	576	216
24	−5	25	105	−14	196	70
30	1	1	120	1	1	1
35	6	36	130	11	121	66
40	11	121	150	31	961	341
50	21	441	160	41	1681	861
	0	1022		−14	5368	2317

$$\text{Coefficient of correlation, } r = \frac{\sum dxdy - \frac{\sum dx \times \sum dy}{N}}{\sqrt{\sum dx^2 - \frac{(\sum dx)^2}{N}}\sqrt{\sum dy^2 - \frac{(\sum dy)^2}{N}}}$$

$$= \frac{2317 - \frac{0 \times (-14)}{8}}{\sqrt{1022 - \frac{0^2}{8}}\sqrt{5368 - \frac{(-14)^2}{8}}}$$

$$= \frac{2317}{\sqrt{1022}\sqrt{5368 - \frac{196}{8}}}$$

$$= \frac{2317}{\sqrt{1032 \times 5343.51}} = 0.99$$

There is a high degree of correlation between fertilizer used and productivity.

6.7 SPEARMAN'S RANK CORRELATION COEFFICIENT

This method is based on rank and is useful in measuring characteristics such as beauty, intelligence, characters, etc. It is applicable only to individual observations. It is defined as follows:

$$r = 1 - \frac{6\sum D^2}{n(n^2 - 1)}$$

where n = number of paired observations, D^2 = sum of squares of the difference of two ranks, r = rank correlation coefficient, The rank correlation coefficient is also denoted by ρ.

Example 6.7: A random sample of five college students is selected and their grades in mathematics and statistics are found to be

Mathematics	85	60	73	40	90
Statistics	93	75	65	50	80

Calculate rank correlation coefficient.

Solution: Calculation of rank correlation coefficient:

Marks in mathematics, X	Ranks of X	Marks in statistics, Y	Ranks of Y	Rank difference, D	D^2
85	2	93	1	1	1
60	4	75	3	1	1
73	3	65	4	−1	1
40	5	50	5	0	0
90	1	80	2	−1	1

$$r = 1 - \frac{6\sum D^2}{n(n^2-1)} = 1 - \frac{6.4}{5(5^2-1)} = \frac{4}{5} = 0.8$$

6.7.1 Rank Correlation Coefficient When the Ranks Are Tied

When two or more values are equal it is customary that values are given the average of the ranks they would have received. In this case the formula for computing rank correlation coefficient takes the form

$$r = 1 - \frac{6(\sum D^2 + T_x + T_y)}{n(n^2-1)}$$

$$T_x = \sum \frac{1}{12}(t_i^3 - t_i)$$

i denotes the ith tie, t_i denotes the number of ranks tied in the ith tie in the variable x.

$$T_y = \sum \frac{1}{12}(t_j^3 - t_j)$$

j denotes the jth tie, t_j denotes the number of ranks tied in the jth tie in the variable y.

Example 6.8: Compute the rank correlation coefficient for the following data:

X	68	64	75	50	64	80	75	40	55	64
Y	62	58	68	45	81	60	68	48	50	70

Solution:

X	Rank of X	Y	Rank of Y	$D=X-Y$	D^2
68	4	62	5	−1	1
64	6	58	7	−1	1
75	2.5	68	3.5	−1	1
50	9	45	10	−1	1
64	6	81	1	5	25
80	1	60	6	−5	25
75	2.5	68	3.5	−1	1
40	10	48	9	1	1
55	8	50	8	0	0
64	6	70	2	4	16

$$r = 1 - \frac{6\left(\sum D^2 + \frac{6(t^3-t)}{12} + \cdots\right)}{n(n^2-1)}$$

$$= 1 - \frac{6\left[72 + \frac{1}{12}(2^3-2) + \frac{1}{12}(3^3-3) + \frac{1}{12}(2^3-2)\right]}{10(10^2-1)}$$

$$= 1 - \frac{6[72+0.5+2+0.5]}{990} = 0.545$$

Exercise 6.1

1. Find the coefficient of correlation for the data:

Height of father	66	67	67	68	69	71	73
Height of son	68	64	68	72	70	69	70

Ans: $r=0.472$

2. Calculate the coefficient of correlation from the following data:

X	12	9	8	10	11	13	7
Y	14	8	6	9	11	12	3

Ans: $r=0.95$

3. Ten students got the following percentage of marks in mathematics and physics. Find the coefficient of rank correlation:

Mathematics, X	8	36	98	25	75	82	92	62	35
Physics, Y	84	51	91	60	68	62	86	58	49

Ans: 0.4551

4. Find the coefficient of correlation between x and y and interpret.

X	1	3	4	5	7	8	10
Y	2	6	8	10	14	16	20

Ans: Perfect positive correlation

5. Find the correlation for the data given below:

X	78	36	98	25	75	82	90	62	35	39
Y	84	51	91	60	68	62	86	58	53	47

Ans: $r = 0.78$

6. Compute the coefficient of correlation between A and B.

A	5	10	5	11	12	4	3	2	7	1
B	1	6	2	8	5	1	4	6	5	2

Ans: $r = +0.58$

7. Find the coefficient of correlation.

X	1	2	3	4	5	6	7	8	9
Y	9	8	10	12	11	13	14	16	15

Ans: $r = 0.95$

8. Calculate the Pearson's coefficient of correlation from the following data using 44 and 26 as the origins of X and Y, respectively.

X	43	44	46	40	44	42	45	42	38	40	42	57
Y	29	31	19	18	19	27	27	29	41	30	26	10

Ans: $r = 0.85$

9. Calculate the Karl Pearson's correlation coefficient between X and Y for the following data:

X	43	54	59	68	76
Y	105	98	84	63	50

Ans: $r = -1$

10. Calculate the coefficient of correlation between X and Y.

X	92	89	87	86	83	77	71	63	53	50
Y	86	83	91	77	68	85	52	82	37	57

Ans: $r = 0.73$

11. Calculate the coefficient of correlation between X and Y.

X	1	2	3	4	5	6	7	8	9
Y	12	11	13	15	14	17	16	19	10

Ans: $r = 0.933$

6.8 CALCULATION OF *r* (CORRELATION COEFFICIENT) (KARL PEARSON'S FORMULA)

If $(x_1, y_1), (x_2, y_2), (x_3, y_3), \ldots, (x_n, y_n)$ be n paired observation, then

$$r = \frac{\sum\left[(x_i - \overline{x})(y_i - \overline{y})\right]}{n\sigma_x\sigma_y}$$

or simply

$$r = \frac{\sum\left[(x_i - \overline{x})(y_i - \overline{y})\right]}{\sqrt{\sum(x-\overline{x})^2}\sqrt{\sum(y-\overline{y})^2}}$$

where σ_x = Standard deviation of $x_1, x_2, \ldots, x_n$, σ_y = Standard deviation of $y_1, y_2, \ldots, y_n$, $\overline{x} = (\sum x_i)/n,$, $\overline{y} = (\sum y_i)/n$.

And

$$\sigma_x = \sqrt{\frac{\sum x_i}{n}}, \ \sigma_y = \sqrt{\frac{y_i}{n}}$$

If $x_i = x_i - \overline{x}$ and $y_i = y_i - \overline{y}$, then

$$r = \frac{\sum x_i y_i}{n\sigma_x\sigma_y}$$

$$= \frac{\sum x_i y_i}{\sqrt{\sum x_i^2}\sqrt{\sum y_i^2}}$$

If A and B denote the assumed means, then

$$r = \frac{\sum x_i y_i - \frac{\left(\sum x_i\right)\left(\sum y_i\right)}{n}}{\sqrt{\sum x_i^2 - \frac{\left(\sum x_i\right)^2}{n}}\sqrt{\sum y_i^2 - \frac{\left(\sum y_i\right)^2}{n}}}$$

Karl Pearson's formula is a direct method of computing r. It can be proved mathematically that $-1 \leq r \leq 1$. The Karl Pearson's coefficient r, is also denoted by ρ (rho) and is also called Karl Pearson's moment correlation coefficient.

6.9 REGRESSION

Correlation methods are used to know, how two or more variables are interrelated. Correlation cannot be used to estimate or perfect the most likely values of one variable for specified values of other variable. The term regression was coined by Sir Francis Galton (while studying the linear relation between two variables).

The independent variable in regression analysis is called "prediction" or "regressor" and the dependent variable is called regressed variable.

Definition 6.2: Regression is the measure of the average relationship between two or more variables in terms of the original units of data.

6.10 REGRESSION EQUATION

The functional relationship of a dependent variable with one or more independent variables is called a regression equation. It is also called prediction equation (or estimating equation).

6.11 CURVE OF REGRESSION

The graph of the regression equation is called the curve of regression. If the curve is a straight line, then it is called the line of regression.

6.12 TYPES OF REGRESSION

If there are only two variables under consideration, then the regression is called simple regression.

Example:

1. Study of regression between heights and age for a group of persons.
2. The study of regression between "income" and "expenditure" for a group of persons.

In this case the relationship is linear.

If there are more than two variables under consideration then the regression is called multiple regressions.

If there are more than two variables under consideration and relation is called multiple regression.

If there are more than two variables under consideration and relation between only two variables is established, after excluding the effect of the remaining variables, then the regression is called partial regression.

If the relationship between x and y is nonlinear, then the regression is curvilinear regression. In some cases polynomials are selected to predict or estimate, which is called polynomial regression.

6.13 REGRESSION EQUATIONS (LINEAR FIT)

6.13.1 Linear Regression Equation of *y* on *x*

In linear regression if we fit a straight line of the form $y = a + bx$ to the given data by the method of least squares, we obtain the regression of y on x.

Let $(x_1, y_1), (x_2, y_2), (x_3, y_3), \ldots, (x_n, y_n)$ denote n pairs of observations and let the corresponding straight line to be fitted, to these data points be

$$y = a + bx \tag{6.1}$$

Applying the method of least squares, we get the following normal equations:

$$na + \sum bx_i = \sum y_i \tag{6.2}$$

$$a\sum x_i + b\sum x_i^2 = \sum x_i y_i \tag{6.3}$$

Dividing Eq. (6.2) by n, we get

$$a + b\frac{\sum x_i}{n} = \frac{\sum y_i}{n}$$

or

$$a + b \cdot \overline{x} = \overline{y} \tag{6.4}$$

Subtracting Eq. (6.4) from Eq. (6.1), we obtain

$$(y_i - \overline{y}) = b(x_i - \overline{x}) \tag{6.5}$$

Multiplying Eq. (6.2) by $\sum x_i$ and Eq. (6.3) by n and then subtracting, we get

$$\sum x_i y_i - n\sum x_i y_i = b\left(\sum x_i\right) - nb\sum x_i^2$$

or

$$b\left[n\sum x_i^2 - \left(\sum x_i\right)^2\right] = n\sum x_i y_i - \sum x_i \sum y_i$$

or

$$b = \frac{n\sum x_i y_i - \sum x_i \sum y_i}{\left[n\sum x_i^2 - \left(\sum x_i\right)^2\right]}$$

or

$$b = \frac{\frac{\sum x_i y_i}{n} - \frac{\sum x_i}{n} \cdot \frac{\sum y_i}{n}}{\frac{\sum x_i^2}{n} - \frac{(\sum x_i)^2}{n}} \tag{6.6}$$

Replacing b by b_{yx} in Eq. (6.5), we get

$$(y_i - \overline{y}) = b_{yx}(x_i - \overline{x}) \tag{6.7}$$

where

$$b_{yx} = \frac{\frac{\sum x_i y_i}{n} - \frac{\sum x_i}{n} \cdot \frac{\sum y_i}{n}}{\frac{\sum x_i^2}{n} - \frac{(\sum x_i)^2}{n}}$$

Eq. (6.7) is called regression equation of y on x, and is used to estimate the values of y for given values of $n.b_{yx}$ is also given by $b_{yx} = r(\sigma_y/\sigma_x)$.

And it is called the regression coefficient of y on x.

Regression coefficient is an absolute figure. It explains that the decrease in one variable is associated with the increase in the other variable. The regression coefficients b_{yx} and b_{xy} will have the same sign and are independent of change of origin but not of scale.

6.13.2 Regression Equation of *x* and *y*

It is the best equation of best fitted line of the type

$$x = a' + b'y$$

to the given data.

Applying the principle of least squares, we get the following two normal equations:

$$na' + b'\sum y = \sum x \tag{6.8}$$

$$a'\sum y + b'\sum y^2 = \sum xy \tag{6.9}$$

Solving Eqs. (6.8) and (6.9) for and proceeding as before, we obtain the regression equation of x on y as follows:

$$(x_i - \overline{x}) = b_{xy}(y_i - \overline{y})$$

where

$$b_{xy} = \frac{\frac{\sum x_i y_i}{n} - \frac{\sum x_i}{n} \cdot \frac{\sum y_i}{n}}{\frac{\sum y_i^2}{n} - \frac{\left(\sum y_i\right)^2}{n}}$$

b_{xy} is also given by

$$b_{xy} = r\frac{\sigma_x}{\sigma_y} \tag{6.10}$$

and is called the regression coefficient of x on y.

Since $b_{yx} = r(\sigma_y/\sigma_x)$ and $b_{xy} = r(\sigma_x/\sigma_y)$

We have

$$b_{yx} \cdot b_{xy} = r^2 \tag{6.11}$$

Note:

1. If $x_i = x_i - \overline{x}$ and $y_i = y_i - \overline{y}$, then

$$b_{yx} = \frac{\frac{\sum x_i y_i}{n}}{\frac{\sum x_i^2}{n}} = \frac{\sum x_i y_i}{\sum x_i^2} \quad \left(\because \sum x_i = 0, \ \sum y_i = 0\right).$$

Similarly $b_{xy} = \dfrac{\sum x_i y_i}{\sum y_i^2}$

The two regression lines are $(x_i - \overline{x}) = b_{xy}(y_i - \overline{y})$ and $(y_i - \overline{y}) = b_{yx}(x_i - \overline{x})$ are identical if $b_{yx} \cdot b_{xy} = 1$ or $b_{yx} = 1/b_{xy}$ or $r^2 = 1$, i.e., the lines Eqs. (6.10) and (6.11) are identical if $r^2 = 1$, i.e., $r = \pm 1$.

2. The two regression lines always intersect at $(\overline{x}, \overline{y})$.

6.14 ANGLE BETWEEN TWO LINES OF REGRESSION

Consider the regression lines

$$(y_i - \overline{y}) = b_{yx}(x_i - \overline{x}) \tag{6.12}$$

$$(x_i - \overline{x}) = b_{xy}(y_i - \overline{y}) \tag{6.13}$$

Eq. (6.13) can be written as

$$(y_i - \overline{y}) = \frac{1}{b_{xy}}(x_i - \overline{x}) \tag{6.14}$$

Let θ be the angle between the regression lines, then the slopes of the lines Eqs. (6.12) and (6.14) are:

$$m_1 = b_{yx} = r\frac{\sigma_y}{\sigma_x}$$

and

$$m_2 = \frac{1}{b_{xy}} = \frac{\sigma_y}{r\sigma_x}$$

We have

$$\begin{aligned}
m_2 &= \frac{1}{b_{xy}} = \frac{\sigma_y}{r\sigma_x} \\
&= \pm \frac{\dfrac{\sigma_y}{r\sigma_x} - \dfrac{r\sigma_y}{\sigma_x}}{1 + \dfrac{\sigma_y}{r\sigma_x}\cdot\dfrac{r\sigma_x}{\sigma_y}} = \pm \frac{\dfrac{\sigma_y}{\sigma_x}\left(\dfrac{1}{r} - r\right)}{1 + \dfrac{\sigma_y^2}{\sigma_x^2}} \\
&= \pm \left(\frac{1-r^2}{r}\right)\frac{\sigma_y}{\sigma_x}\cdot\frac{\sigma_x^2}{\sigma_x^2+\sigma_y^2} \\
&= \pm \left[\left(\frac{1-r^2}{r}\right)\frac{\sigma_x\sigma_y}{\sigma_x^2+\sigma_y^2}\right]
\end{aligned}$$

Since $r^2 \leq 1$ and σ_x, σ_y are positive, the positive sign gives the acute angle between the lines and the negative sign gives the obtuse angle between the lines.

If θ_1 denotes acute angle and θ_2 denotes the obtuse angle between the regression lines, then

$$\theta_1 = \tan^{-1}\left[\left(\frac{1-r^2}{r}\right)\frac{\sigma_x\sigma_y}{\sigma_x^2+\sigma_y^2}\right]$$

and

$$\theta_2 = \tan^{-1}\left[\left(\frac{r^2-1}{r}\right)\frac{\sigma_x\sigma_y}{\sigma_x^2+\sigma_y^2}\right].$$

If $r=0$ then $\tan\theta = \infty$ and $\theta = \pi/2$. In this case x and y are uncorrelated and lines of regression are perpendicular to each other.

If $r = \pm 1$, then $\tan\theta = 0$, and $\theta = 0$ in this case there is a perfect correlation (positive or negative) between x and y. The two lines of regression coincide, but are not parallel since the lines pass through the points $(\overline{x}, \overline{y})$.

Note: If $r = 0$, then from the equation of lines of regression, we have

$$(y, \overline{y}) = r\frac{\sigma_y}{\sigma_x}(x_i - \overline{x}) = 0$$

$$(y, \overline{y}) = 0 \text{ or } y = \overline{y}$$

and

$$(x - \overline{x}) = r\frac{\sigma_x}{\sigma_y}(y - \overline{y}) = 0$$

$$x - \overline{x} = 0 \quad \text{or} \quad x = \overline{x}$$

When $r = 0$, the equations of lines of regression are $x = \overline{x}$ and $y = \overline{y}$, which are the equations of the lines parallel to the axis.

6.15 COEFFICIENT OF DETERMINATION

If r denotes the coefficient of correlation, then the square of r (i.e., r^2), is called the coefficient of determination. It gives the proportion of the variance (fluctuation) of one variable that is predictable from the other variable.

The coefficient of determination is a measure that allows us to determine how certain one can be in making predictions from a certain model graph. The *coefficient of determination* is such that $0 \leq r^2 \leq 1$, and denotes the strength of the linear association between x and y.

The coefficient determination is also defined as follows:

$$\text{Coefficient of determination} = \frac{\text{Explained variance}}{\text{Total variance}}$$

6.16 COEFFICIENT NONDETERMINATION

$$\text{The coefficient of nondetermination} = \frac{\text{Unexplained variance}}{\text{Total variance}}$$

It is denoted by k^2, where $k^2 = 1 - r^2$ the coefficient of nondetermination is the percent of variation that is unexplained by the regression.

6.17 COEFFICIENT OF ALIENATION

The square root of coefficient of nondetermination is called coefficient of Alienation.

6.17.1 Solved Examples

Example 6.9: For the following data, find the regression line (by applying the method of least squares):

x	5	10	15	20	25
y	20	40	30	60	50

Solution: We have

x	y	x^2	y^2	xy
5	20	25	400	100
10	30	100	900	400
15	40	225	1600	450
20	60	400	3600	1200
25	50	625	2500	1250
75	200	1375	9000	3400

$$\therefore \sum x_i = 75, \sum y_i = 200, \sum x_i^2 = 1375, \sum y_i^2 = 9000, \sum x_i y_i = 3400$$

Regression of y on x

The normal equations are

$$na + b\sum x_i = \sum y_i$$

$$a\sum x_i + b\sum x_i^2 = \sum x_i y_i$$

i.e.,

$$na + b\sum x_i = \sum y_i \quad (6.15)$$

$$75a + 375b = 3400 \quad (6.16)$$

Solving these equations we get $a = 16$ and $b = 1.6$. The regression equation of y on x is

$$y = 16 + 1.6x$$

Regression equation of x on y

The normal equations are

$$na' + b'\sum y_i = \sum x_i$$

$$a'\sum y_i + b'\sum y_i^2 = \sum x_i y_i$$

i.e.,

$$5a + 200b = 75 \quad (6.17)$$

$$200a + 9000b = 3400 \quad (6.18)$$

Solving Eqs. (6.17) and (6.18) we get

$$a = -1 \text{ and } b = 0.4$$

The regression equation of x on y is

$$x = -1 + 0.4y$$

The regression equations are

$$y = 16 + 1.6x$$

and

$$x = -1 + 0.4y$$

Example 6.10: For the following data, find the regression line of y on x:

x	1	2	3	4	5	8	10
y	9	8	10	12	14	16	15

Solution: We have

x	y	xy	x^2
1	9	9	1
2	8	16	4
3	10	30	9
4	12	48	16
5	14	70	25
8	16	128	64
10	15	150	100

Total:

$$\sum x_i = 33, \quad \sum y_i = 84, \quad \sum x_i^2 = 219, \quad \sum x_i y_i = 451$$

$$\overline{x} = \frac{\sum x_i}{n} = \frac{33}{7} = 4.714$$

$$\bar{y} = \frac{\sum y_i}{n} = \frac{84}{7} = 12$$

and

$$b = \frac{n\sum x_i y_i - \sum x_i \sum y_i}{\left[n\sum x_i^2 - \left(\sum x_i\right)^2\right]} = \frac{7(451) - (33)(84)}{7(2119) - (33)^2} = 0.867$$

The regression equation of y on x is

$$(y_i - \bar{y}) = b_{yx}(x_i - \bar{x})$$

i.e.,

$$y - 12 = 0.867(x - 4.714)$$

or

$$y = 0.867x + 7.9129$$

Example 6.11: From the following data, fit two regression equations by finding actual means (of x and y), i.e., by actual means method.

x	1	2	3	4	5	6	7
y	2	4	7	6	5	6	5

Solution: We change the origin and find the regression equations as follows:

We have

$$\bar{x} = \frac{\sum x_i}{n} = \frac{1+2+3+4+5+6+7}{7} = \frac{28}{7} = 4$$

$$\bar{y} = \frac{\sum y_i}{n} = \frac{2+4+7+6+5+6+5}{7} = \frac{35}{7} = 5$$

x	y	$X = x - \bar{x}$	$Y = y - \bar{y}$	X^2	Y^2	XY
1	2	−3	−3	9	9	9
2	4	−2	−1	4	4	2
3	7	−1	2	1	1	−2
4	6	0	1	0	0	0
5	5	1	0	1	1	0
6	6	2	1	4	4	2
7	5	3	0	9	9	0
28	35	0	0	28	16	11

We have

$$\sum x_i = 28, \sum y_i = 35, \sum = 0, \quad \sum Y_i = 0, \quad \sum X_i^2 = 28,$$
$$\sum Y_i^2 = 16, \quad \sum X_i Y_i = 11$$

$$b_{yx} = \frac{\sum X_i Y_i}{\sum X_i^2} = \frac{11}{28} = 0.3928 = 0.393 \text{ (Approximately)}$$

$$b_{xy} = \frac{\sum X_i Y_i}{\sum Y_i^2} = \frac{11}{16} = 0.6875 = 0.688 \text{ (Approximately)}$$

The regression equation of y on x is

$$(y_i - \overline{y}) = b_{yx}(x_i - \overline{x})$$

$$(y - 5) = 0.393(x - 4)$$

or

$$y = 0.393x + 3.428$$

And the regression equation of x on y is

$$(x_i - \overline{x}) = b_{xy}(y_i - \overline{y})$$

$$(x_i - 4) = 0.688(y_i - 5)$$

or

$$x = 0.688y + 0.56$$

The required equations are

$$y = 0.393x + 3.428$$

$$x = 0.688y + 0.56$$

Example 6.12: From the following results obtain the two regression equations and estimate the yield of crops when the rainfall is 29 cm and the rainfall when the yield is 600 kg.

	Y (yield in kg)	*X* (rainfall in cm)
Mean	508.4	26.7
Standard deviation	36.8	4.6

Coefficient of correlation between yield and rainfall is 0.52.

Solution: We have

$$\overline{x} = 26.7, \quad \overline{y} = 508.4$$

$$\sigma_x = 4.6, \quad \sigma_y = 36.8$$

and $r = 0.52$

$$b_{yx} = r\frac{\sigma_y}{\sigma_x} = (0.52)\frac{36.8}{4.6} = 4.16$$

$$b_{xy} = r\frac{\sigma_x}{\sigma_y} = (0.52)\frac{4.6}{36.8} = 0.065$$

Regression equation of y on x

$$(y_i - \overline{y}) = b_{yx}(x_i - \overline{x})$$

i.e.,

$$y = 508.4 = 4.16\,(x - 26.7)$$

or

$$y = 397.328 + 4.16x$$

when

$$x = 29,$$

we have

$$y = 397.328 + 4.16\,(29) = 517.968 \text{ kg}$$

Regression equation of x on y

$$(x_i - \overline{x}) = b_{xy}(y_i - \overline{y})$$

or

$$x - 26.7 = -0.065\,(y - 508.4)$$

or

$$x = -6.346 + 0.065x$$

When $y = 600$ kg.

$$x = -6.346 + 0.065\,(-600)$$
$$= 32.654 \text{ cm}$$

The regression equations are

$$y = 397.328 + 4.16x$$

$$x = -6.346 + 0.065y$$

When the rainfall is 29 cm the yield of the crop is 517.968 kg and when the yield is 600 kg the rainfall is 32.654 cm.

Example 6.13: Find the most likely price of a commodity in Bombay corresponding to the price of Rs. 70 at Calcutta from the following:

	Calcutta	Bombay
Average prize	65	67
Standard deviation	2.5	3.5

Correlation coefficient between the prices of commodity in the two cities is 0.8.

Solution: we have

$$\overline{x} = 65, \quad \overline{y} = 67$$

$$\sigma_x = 2.5, \quad \sigma_y = 3.5$$

and $r = 0.8$

$$(y, \overline{y}) = r\frac{\sigma_y}{\sigma_x}(x_i - \overline{x}) = 0$$

$$\Rightarrow \quad y - 67 = (0.8).\left(\frac{3.5}{2.5}\right)(x - 65)$$

$$\Rightarrow \quad y = 67 + 1.12x - 72.8$$

$$\Rightarrow \quad y = -5.8 + 1.12x$$

when $x = 70$,

$$\Rightarrow \quad y = -5.8 + 1.12 \times 70 = -5.8 + 78.4$$

$$\Rightarrow \quad y = 72.60$$

The price of the commodity in Bombay corresponding to Rs. 70 at Calcutta is Rs. 72.60.

Example 6.14: The regression equation calculated from a given set of observations

$$x = -0.4y + 6.4$$

and

$$y = -0.6x + 4.6$$

Calculate $\overline{x}$, $\overline{y}$ and r_{xy}

Solution: We have

$$x = -0.4y + 6.4 \tag{6.19}$$

and

$$y = -0.6x + 4.6 \quad (6.20)$$

From Eq. (6.20), we have $y = -0.6(-0.4y + 6.4) + 4.6$

$$\Rightarrow \quad y = 0.24y - 3.84 + 4.6$$

$$\Rightarrow \quad 0.76y = 0.76$$

$$\Rightarrow \quad y = 1$$

From Eq. (6.19), we have $x = -0.4 \times 1 + 6.4 = 6.0$

But $(\overline{x}, \overline{y})$ in the point of intersection of Eqs. (6.19) and (6.20)

Hence

$$(\overline{x}, \overline{y}) = (1, 6)$$

$$\overline{x} = 1, \quad \overline{y} = 6$$

Clearly, Eq. (6.19) in the regression equation x on y and Eq. (6.20) is the regression equation y on x.

We have

$$b_{yx} = -0.6, b_{xy} = -0.4$$

and

$$r^2 = (-0.4)(-0.6) = 0.24$$

$$r = r_{xy} = \pm\sqrt{0.24}$$

Since b_{yx} and b_{yx} are both negative $r = r_{xy}$ is negative

$$r_{xy} = \pm\sqrt{0.24}$$

Example 6.15: Show that the coefficient of correlation is the Geometric mean (GM) of the coefficient of regression.

Solution: The coefficients of regression are $r(\sigma_x/\sigma_y)$ and $r(\sigma_y/\sigma_x)$.

GM of the regression coefficients is

$$= \sqrt{r\frac{\sigma_x}{\sigma_y} \cdot r\frac{\sigma_y}{\sigma_x}} = \sqrt{r^2} = r$$

$$= \text{correlation of correlation}$$

Example 6.16: In a partially destroyed laboratory record of an analysis of correlation data, the following results are legible:

Variance of $x = 9$

Regression equation: $8x - 10y + 66 = 0, 40x - 18y = 214$

What were

i. The mean values of x and y
ii. The standard deviation of y
iii. The coefficient of correlation between x and y.

Solution: Variance of $x = 9$
i.e., $\sigma_x^2 = 9 \Rightarrow \sigma_x = 3$
Solving the regression equations

$$8x - 10y + 66 = 0 \tag{6.21}$$

$$40x - 18y = 214 \tag{6.22}$$

We obtain

$$x = 13, \quad y = 17$$

Since the pair of intersection of the regression lines is $(\overline{x}, \overline{y})$, we have

$$(\overline{x}, \overline{y}) = (x, y) = (13, 17)$$

Therefore $\overline{x} = 13, \quad \overline{y} = 17$
The regression Eqs. (6.21) and (6.22) can be written as

$$y = 0.8x + 6.6 \tag{6.23}$$

$$x = 0.45y + 5.35 \tag{6.24}$$

The regression coefficient of y on x is

$$r\frac{\sigma_y}{\sigma_x} = 0.8 \tag{6.25}$$

And the regression coefficient of x on y is

$$r\frac{\sigma_x}{\sigma_y} = 0.45 \tag{6.26}$$

Multiplying Eqs. (6.25) and (6.26), we get

$$r^2 = 0.45 \times 0.8$$

$$\Rightarrow \quad r^2 = 0.36$$

$$\Rightarrow \quad r = 0.6$$

Putting the values of r and σ_x in Eq. (6.25), we get the value of σ_y as follows:

$$r\frac{\sigma_y}{\sigma_x} = 0.8$$

$$(0.6)\frac{\sigma_y}{3} = 0.8$$

$$\sigma_y = \frac{0.8}{0.2} = 4$$

Example 6.17: If one of the regression coefficients is greater than unity. Show that the other regression coefficient is less than unity.

Solution: Let one of the regression coefficient, say $b_{yx} > 1$

Then

$$b_{yx} > 1 \Rightarrow b_{yx} < 1$$

Since,

$$b_{yx} \cdot b_{xy} = r^2 \leq 1$$

We have

$$b_{xy} \leq \frac{1}{b_{yx}}$$

$$b_{xy} < 1 \left(\because \frac{1}{b_{yx}} < 1 \right)$$

Example 6.18: Show that the Arithmetic mean of the regression coefficients is greater than the correlation coefficient.

Solution: We have to show that $\frac{b_{xy} + b_{yx}}{2} > r$

Consider

$$(\sigma_y - \sigma_x)^2 > 0$$

Clearly

$$(\sigma_y - \sigma_x)^2 > 0$$

(Since square of two real qualities is always >0)

$$\sigma_x^2 + \sigma_y^2 - 2\sigma_x\sigma_y > 0$$

or

$$\frac{\sigma_x^2}{\sigma_y\sigma_x}+\frac{\sigma_y^2}{\sigma_y\sigma_x}>2$$

or

$$\frac{\sigma_x}{\sigma_y}+\frac{\sigma_y}{\sigma_x}>2$$

or

$$r\frac{\sigma_x}{\sigma_y}+r\frac{\sigma_y}{\sigma_x}>2$$

or

$$b_{xy}+b_{yx}>2r$$

or

$$\frac{b_{xy}+b_{yx}}{2}>r$$

Hence proved.

Example 6.19: Given that $x=4y+5$ and $y=kx+4$ are two lines of regression. Show that $0\leq k\leq 1/4$. If $k=1/8$, find the means of the variables and ratio of their variables.

Solution: $x=4y+5\Rightarrow b_{xy}=4$

$$y=kx+4\Rightarrow b_{yx}=k$$

$$r^2=b_{xy}\cdot b_{yx}=4\cdot k$$

but

$$-1\leq r\leq 1$$

$$\Rightarrow\quad 0\leq r^2\leq 1$$

$$\Rightarrow\quad 4\leq 4k\leq 1$$

$$\Rightarrow\quad 0\leq k\leq\frac{1}{4}$$

when $k=\frac{1}{8}$, $r^2=4\cdot\frac{1}{8}=\frac{1}{2}\Rightarrow r=0.7071$

$$y=kx+4$$

or

$$y=\frac{1}{8}x+4$$

or

$$8y = x + 32$$

or

$$8y = 4y + 5 + 32$$

or

$$4y = 37 \Rightarrow y = 9.25$$

Now

$$x = 4y + 5 \Rightarrow x = 4\,(9.25) + 5$$

or

$$x = 42$$

Therefore $(\bar{x}, \bar{y}) = (x, y)$ (the point of intersection is $(\bar{x}, \bar{y})$)

$$= (42,\ 9.25)$$

i.e.,

$$\bar{x} = 42, \quad \bar{y} = 9.25$$

Hence

$$\frac{b_{xy}}{b_{yx}} = \frac{r\dfrac{\sigma_x}{\sigma_y}}{r\dfrac{\sigma_y}{\sigma_x}} = \frac{\sigma_x \cdot \sigma_x}{\sigma_y \cdot \sigma_y} = \frac{4}{\left(\dfrac{1}{8}\right)} = 32$$

i.e.,

$$\frac{(\sigma_x)^2}{(\sigma_y)^2} = 32$$

The ratio of the variances is 32:1.

Multiple correlation is a statistical technique that predicts value of one variable on the basis of two or more other variables.

Regression analysis involves identifying the relationship between a dependent variable and one or more independent variables. In simple linear regression, the model is used to describe the relationship between a single dependent variable y and a single independent variable x. Multiple regression is concerned with the collection and interpretation of quantitative data and the use of probability theory to estimate population parameters multiple correlation coefficients is denoted by R. It is never negative $ab =$ nd there are several ways to compute its value. R lies between 0 and 1, and it tells the only strength of association between the dependent and independent variables.

If $R = 0$ then there is no linear association relationship, if $R = 1$ there is stronger linear association, i.e., perfect linear relationship between x and y. Correlation and regression analysis are related in the sense that both deal with relationships among variables.

6.18 MULTILINEAR REGRESSION

In some cases the value of a variate may not depend only on a single variable. It may happen that there are several variables, which when taken jointly, will serve as a satisfactory basis for estimating the desired variable. If $x_1, x_2, \ldots, x_k$ represent the independent variables, is the variable which is to be predicted, and represents the regression equation.

$$y' = a_0 + a_1x_1 + a_2x_2 + \cdots + a_kx_k$$

The unknown coefficients a_0, a_1, $a_2, \ldots$, a_k will be estimated by the method of least squares. To obtain the values of the variables, we have n set of values of $(k + 1)$ variables. Geometrically, the problem is one of finding the equation of the plane which best fits in the sense of least squares of n points in $(k + 1)$th dimension. The normal equations are

$$na_0 + a_1\sum x_1 + a_2\sum x_2 + \cdots + a_k\sum x_k = \sum y$$

$$a_0\sum x_1 + a_1\sum x_1^2 + a_2\sum x_1x_2 + \cdots + a_k\sum x_1x_k = \sum x_1y$$

$$\vdots$$

$$a_0\sum x_k + a_1\sum x_1x_k + a_2\sum x_kx_2 + \cdots + a_k\sum x_k^2 = \sum x_ky$$

If there are two independent variables say x_1 and x_2 the normal equations are

$$na_0 + a_1\sum x_1 + a_2\sum x_2 = \sum y$$

$$a_0\sum x_1 + a_1\sum x_1^2 + a_2\sum x_1x_2 = \sum x_1y$$

$$a_0\sum x_2 + a_1\sum x_1x_2 + a_2\sum x_2^2 = \sum x_2y$$

And the regression equation is

$$y = a_0 + a_1x_1 + a_2x_2$$

6.19 USES OF REGRESSION ANALYSIS

There are many uses of regression analysis. In many situations, the dependent variable y is such that it cannot be measured directly. In such cases, with the help of some auxiliary variables, which are taken as independent variable in a regression, the value of y is estimated. Regression equation is often used as a prediction equation. Regressional analysis is used in predicting yield of a crop, for different doses of a fertilizer, and in predicting future demand for food. Regression analysis is also used to estimate the height of a person at a given age, by finding the regression of height on age.

Exercise 6.2

1. Heights of fathers and sons are given below in inches:

Height of father	65	66	67	67	68	69	71	73
Height of son	67	68	64	68	72	70	69	70

Form the lines of regression and calculate the average height of the son when the height of father is 67.5 in.

***Ans*:** When father's height is 67.5 in. the son's height is 68.19 in.

2. For the following data, determine the regression lines:

x	6	2	10	4	8
y	9	11	5	8	7

***Ans*:** $y = 11.9 - 0.65x$, $x = 16.4 - 1.3y$

3. Find the regression equations for the following data:

Age of husband(x)	36	23	27	28	28	29	30	31	33	35
Age of wife(y)	29	18	20	22	27	21	29	27	29	28

***Ans*:** $y = -1.739 + 0.8913x$, $x = 11.25 + 0.75y$

4. By the method of least squares find the regression of y and x, find the value of y when $x = 4$, also find the regression equation of x on y, and find the value of x when $y = 24$. Use the table given below:

x	1	3	5	7	9
y	15	18	21	23	22

***Ans*:** $y = 15.05 + 0.95x$, the value of y when $x = 4$ is 18.85.
$x = -12.58 + 0.88y$, the value of x when $y = 24$ is 8.73.

5. Using the method of least squares find the two regression equation for the data given below:

x	5	10	15	20	25
y	20	40	30	60	50

***Ans*:** $y = 16 + 1.6x$, $x = -1 + 0.4y$

6. Define regression and find the regression equation of y on x for the data given below:

x	2	6	4	3	2	2	8	4
y	7	2	1	1	2	3	2	6

Ans: $y = 4.16 - 0.3x$

7. From the following data, obtain the two regression equations:

Sales	91	97	108	121	67	124	51
Purchase	71	75	69	97	70	91	39

Ans: $y = 15.998 + 0.607x, \quad x = 0.081 + 1.286y$

8. Find the equation of regression lines for the following pairs for the variables x and y: (1, 2), (2, 5), (3, 3), (4, 8), (5, 7).
Ans: $y = 1.1 + 1.3x, \quad x = 0.5 + 0.5y$

9. From the following data, find the yield of wheat in kilogram per unit area when the rainfall is 9 in.

	Means	SD
Yield of wheat (kg)	10	8
Annual rainfall (in.)	8	2

Ans: 12 kg

10. Show that the regression coefficients are independent of the change of origin but not the change of scale.

11. Given $\sum x_i = 60, \sum y_i = 40, \sum x_i^2 = 4160, \sum y_i^2 = 1, \sum x_i y_i = 1150, x = 10$. Find the regression equation of x or y and find r.
Ans: $x = 3.68 + 0.58y, \quad r = 0.37$

12. Using the data given below find the demand when the price of the quantity is Rs.12.50:

	Price	Demand
Means	10	35
Standard deviation	2	5

Coefficient of correlation $(r) = 0.8$
Ans: $y = 15 + 2x$, demand $= 40,000$ units

13. Find the means of x_i and y_i, also find the coefficient of correlation given

$$2y - x - 50 = 0$$
$$2y - 2x - 10 = 0$$

Ans: $\overline{x} = 130, \overline{y} = 90, \quad r = 0.866$

14. From the following data obtain the two regression equations and calculate the correlation coefficient:

X	1	2	3	4	5	6	7	8	9
Y	9	8	10	12	11	13	14	16	15

Ans: $x = -6.4 + 0.95y, \quad y = 7.25 + 0.95x$

15. From the following regression equations: $8X - 10Y = -66$, $40X - 18Y = 214$

Find

a. Average values of X and Y

b. Correlation coefficient between the two variables X and Y

c. Standard deviation of Y

Ans: (a) 13, 17 (b) $b_{yx} = 0.8, b_{xy} = 0.45$ (c) $\sigma_y = 4$

16. Find the most likely price in Bombay (X), corresponding to the price of Rs.70 at Calcutta (Y) from the following data:

	Bombay	Calcutta
Average prize	67	65
Standard deviation	3.5	2.5

Ans: 72.6

17. The following data based on 450 students are given for marks in statistics and economics at a certain examination:

	Statistics	Economics
Means marks	40	48
Standard deviation	12	16

Sum of the products of deviation of marks from their respective mean is 42.075.

Give the equations to the two lines of regression.

Estimate the average marks of regression.

Ans: $x = 22.24 + 0.37y, y = 22 + 0.65x, y = 54.5$

18. From the following information calculate the line of regression of y on x:

	x	y
Means	40	60
Standard deviation	10	15
Correlation coefficient	0.7	

Ans: $y = 1.05x + 18$

19. The correlation coefficient between two variables x and y is $r = 0.6$. If $\sigma_x = 1.5, \sigma_y = 2.0, \overline{x} = 10, \overline{y} = 20$. Find the regression lines of (a) y on x and (b) x on y.

Ans: $x = 0.45y + 1, y = 0.8x + 12$

20. The following table gives the age of cars for a certain make and the maintenance costs. Obtain the regression equation for costs related to age. Also estimate maintenance cost for a 10-year-old car.

Age of car (in years)	2	4	6	8
Maintenance cost (in Rs. hundred)	10	20	25	30

Ans: $y = 5 + 3.25x$, Rs.37.50

21. Two brands of tyres are tested for their life and the following results were obtained:

Length of life	20–25	25–30	30–35	35–40	40–45
No. of tyres, X	1	22	64	10	3
No. of tyres, Y	3	21	74	1	1

If consistency is the criterion, which brand of tyres would you prefer?

Ans: Y brand of tyres

22. Calculate the coefficient of correlation between the age of cars and annual maintenance cost and comment

Age of cars (years)	2	4	6	7	8	10	12
Annual maintenance cost (Rs.)	1600	1500	1800	1900	1700	2100	2000

Ans: 0.836, There is a high degree of positive correlation between output age cost of an automotive, of cars, and maintenance cost.

23. Find the coefficient of correlation between output and cost of an automotive factory from the following data:

Output of cars (in 1000s)	3.5	4.2	5.6	6.5	7.0	8.2	8.8	9	9.7	10.0
Cost per car (in 1000s of Rs.)	9.8	9.0	8.8	8.4	8.3	8.2	8.2	8.0	8.0	8.1

Ans: $r = -0.938$

24. A factory produces two types of tyres. In an experiment in the working life of these tyres, the following results were obtained:

Length of life(in 100 h)	15–17	17–19	19–21	21–23	23–25
Type A	50	110	260	100	80
Type B	40	300	120	80	60

State which type of the tyre is more stable?

Ans: Type A

Partial correlation: Partial correlation is a method used to describe the relationship between two variables whilst taking away the effects of another variable, or several other variables, on this relationship.

A partial correlation coefficient is a measure of the linear dependence of a pair of random variables from a collection of random variables in the case where the influence of the remaining variables are eliminated.

A measure of the strength of association between a dependent variable and an independent variable when the effect of all other independent variables is removed; equal to the square root of the partial coefficient of determination. Multiple correlation is a statistical technique that predicts value of one variable on the basis of two or more other variables. Population parameters multiple correlation coefficient is denoted by R. It is never negative and lies between 0 and 1. R tells the only strength of association between the dependent and independent variables.

If $R=0$, then there is no linear association relationship; if $R=1$ there is stronger linear association, i.e., perfect linear relationship between x and y.

Multiple regression involves one continuous criterion (dependent) variable and two or more predictors (independent variables). The equation for a line of best fit is derived in such a way as to minimize the sums of the squared deviations from the line. Although there are multiple predictors, there is only one predicted Y value, and the correlation between the observed and predicted Y values is called Multiple R. The value of Multiple R will range from 0 to 1. In the case of bivariate correlation, a regression analysis will yield a value of Multiple R, i.e., the absolute value of the Pearson product moment correlation coefficient between X and Y. The multiple linear regression equation will take the following general form:

$$\widehat{Y} = b_0 + b_1X_1 + b_2X_2 + \cdots + b_kX_k$$

The Partial Correlation procedure computes partial correlation coefficients that describes the linear relationship between two variables while controlling for the effects of one or more additional variables. Correlations are measures of linear association. Two variables can be perfectly related, but if the relationship is not linear, a correlation coefficient is not an appropriate statistic for measuring their association.

Partial correlation is the correlation of two variables while controlling for a third or more variables.

Partial correlation is a method used to describe the relationship between two variables whilst taking away the effects of another variable, or several other variables, on this relationship in simple correlation, we measure the strength of the linear relationship between two variables, without taking into consideration of the fact that both these variables may be influenced by a third variable.

Partial correlation aims at eliminating effects of other variables. The partial correlation coefficient of two variables X_1 and X_2, partialling X_3 is denoted by $r_{12.3}$. It is given by

$$r_{12.3} = \frac{r_{12} - r_{13}r_{23}}{\sqrt{1 - r_{13}{}^2}\sqrt{1 - r_{23}{}^2}}$$

where r_{12} is the correlation coefficient between X_1 and X_2 ignoring altogether any influence of X_3.

r_{13} is the correlation coefficient between X_1 and X_3 ignoring altogether any influence of X_2 is the correlation coefficient between X_2 and X_3 ignoring altogether any influence of X_1.

$r_{12.3}$ lies between -1 and 1.

Example 6.20: If $r_{12} = 0.41$, $r_{13} = 0.71$, and $r_{23} = 0.50$ then $r_{12.3} = 0.09$.

Example 6.21: If $r_{12} = 0.65$, $r_{13} = 0.60$, and $r_{23} = 0.90$ then $r_{12.3} = 0.32$.

Probable error: If r denotes the coefficient of correlation, and n denotes the number of pairs of observations, then the probable error (PE) is given by $\frac{2}{3}\left[\frac{1 - r^2}{\sqrt{n}}\right]$, where $\left[\frac{1 - r^2}{\sqrt{n}}\right]$ is the standard error. The probable error is also given by $PE = 0.6745\left[\frac{1 - r^2}{\sqrt{n}}\right]$.

If $\frac{r}{PE} \geq 6$ then r is significant.

Example 6.22: If $n = 100$, $r = 0.4$, then $PE = 0.6745$ $\left[\frac{1 - r^2}{\sqrt{n}}\right] = 0.6745 \cdot \left[\frac{1 - (0.4)^2}{\sqrt{100}}\right] = 0.06$ $\frac{r}{PE} = \frac{0.4}{0.06} = 6.66 > 6$, hence r is significant.

Example 6.23: If the value of Karl Pearson's coefficient of correlation $r = 0.9$, $PE = 0.04$, then find n?

Solution: We have $PE = 0.6745\left[\frac{1 - r^2}{\sqrt{n}}\right]$

$$PE = 0.6745\left[\frac{1 - (0.9)^2}{\sqrt{n}}\right] \Rightarrow \sqrt{n} = 3.204 \Rightarrow n = 10.287$$

Hence $n = 10$

CHAPTER 7
Sampling

7.1 INTRODUCTION

In this chapter we briefly introduce sampling theory. The aim of the theory is to get as much information as possible, ideally the whole of the information about the population from which the sample has been drawn. We begin our discussion by defining the term population.

7.2 POPULATION

The aggregate of all units pertaining to a study is called population or universe.

It is a collection of items or individuals or observations having common fundamental characteristics.

If a population consists of finite or fixed number of items or values then it is called a finite population.

A population which is not finite is called an infinite population. It contains endless succession of values.

7.3 SAMPLE

A part of the population is called a sample. It is a subset of a population.

A population may be finite or infinite according to the number of observations in it. The number of observations included in a finite sample is called the size of a sample.

If the size of the sample is less than or equal to 30 (i.e., $n \leq 30$) then the sample is called small sample.

If $n > 30$, the sample is known as a large sample.

A sample must be representative of the population from which it is selected. It should be free from any influence that causes any difference between the sample value and the population value. The sample yield precise estimates. A good sample must be adequate in size in order to be reliable. Hence the requirement of a good sample is representativeness, adequacy, and avoiding bias.

Statistical Techniques for Transportation Engineering.
DOI: http://dx.doi.org/10.1016/B978-0-12-811555-8.00007-6

7.4 SAMPLING

The process of drawing a sample from a population is called sampling.

Sampling reduces the time and, cost of a study and it saves labor. Sampling demands thorough knowledge of sampling methods and procedures. Sampling methods may be classified into two types:

1. Probability or random sampling.
2. Nonprobability or nonrandom sampling.

Types of sampling methods: There are various methods of sampling. We briefly explain some of these methods in the following sections.

7.5 RANDOM SAMPLING

Sampling in which every member of a parent population has an equal chance of being included is called a random sampling. According to Moser and Alton "a random method of selection is one which gives each of the *N* units in the population to be covered a calculable probability of being selected."

Under random sampling, the universe is clearly defined, every element in the population has an equal chance of being represented and the scope for bias is limited. To avoid bias one can use lottery method or random numbers.

7.6 SIMPLE RANDOM SAMPLING

In this method each member of the population has an equal chance of being selected as subject. The entire process of sampling is done in a single step with each subject selected independently of the other members of the population.

We can define simple random sample as follows: When a sample size of *n* is drawn from a population size of *N* in such a way that every possible sample size *n* has the same chance (probability) of being selected, then the sample is known as simple random sampling.

There are many methods to proceed with simple random sampling. The most primitive and mechanical would be the lottery method. We can also use the table of random numbers. One of the best things about simple random sampling is the ease of assembling the sample. It is also considered as a fair way of selecting a sample from a given population. Another key feature of simple random sampling is its representativeness of the population. If the sampling is done with replacement, then there are N^n samples of size n. If the sampling is done without replacement, then the number of samples size n is N_{C_r}.

7.7 STRATIFIED SAMPLING

Stratification is the process of dividing the members of the population into homogeneous subgroups called "Strata" before sampling. The strata should be mutually exclusive such that every element of the population must be assigned only one stratum. The strata should also be collectively exhaustive. From among the strata, the sample is drawn according to the size determined. The sample can be drawn randomly. The stratified sampling allows the use of smaller sample than does simple random sampling with greater precision and consequent saving in time and money.

7.8 SYSTEMATIC SAMPLING

It is a method of selecting sample members from a larger population according to a random starting point and a fixed periodic interval. This method is applied when complete list of the population is available. The most common form of systematic sampling is an equal-probability method. In this approach progression through the list is treated circularly, with a return to the top, once the end of the list is passed. The sampling starts by selecting an element from the list at random and then every kth element in the frame is selected, where k, the sampling interval, which is calculated as:

$$k = \frac{N}{n}$$

where n is the sample size and N is the population size.

Systematic sampling is applied only when the population from which the sample selected is logically homogeneous.

This makes systematic sampling functionally similar to simple random sampling. Systematic sampling is to be applied only if the given population is logically homogeneous. The main advantage of the systematic sampling is its simplicity. The disadvantage of this method is that the process of selecting the sample can interact with a hidden periodic trait to each other. The population is divided into various "clusters" or groups and from each group of cluster, sample is selected randomly. Ideally the clusters chosen should be dissimilar so that the sample is as representative of the population as possible. Cluster is used when "natural" but relatively homogeneous groupings are evident in a population, the total clustering saves traveling time consequently reducing the cost. It is useful for surveying employees in a particular industry, where individual companies can form clusters. Clustering has some disadvantages. The units close to each other may be similar and so less likely to represent the whole population. Clustering has larger sampling error than simple random sampling.

One of the most common quantities used to summarize a set of data is its center. The center is a single value, chosen in such a way that it gives a reasonable approximation of normality. There are many ways to approximate the center of a set of data. One of the most familiar and useful measures of center is the mean, however, using only the mean to approximate normality can often be misleading. To obtain a better understanding of what is considered normal, other measures of central tendency such as the median, the trimmed mean, and the trimean may be utilized in addition to the mean.

Parameters: The statistical constants of the population viz., the mean, the variance, etc., are known as the parameters.

Statistics: The statistical concepts of the sample, computed from the members of the sample, to determine the parameters of the population from which the sample has been drawn, are known as statistics.

7.9 SAMPLE SIZE DETERMINATION

Sample size determination is the act of choosing the number of observations included in a statistical sample. The sample size is an important feature of any empirical study in which the goal is to make inferences about a population from a sample. In practice the sample size used in a study is determined based on the expense of data collection and the need to have sufficient statistical power.

Determining sample size is a very important issue because samples that are too large may waste time, resources, and money, while samples that are too small may lead to inaccurate results. In many cases, we can easily determine the minimum sample size needed to estimate a process parameter, such as the population mean μ.

When a sample data is collected and the sample mean $\overline{x}$ is calculated, that sample mean is typically different from the population mean μ.

This difference between the sample and population means can be thought of as an error. The margin of error E is the maximum difference between the observed sample mean $\overline{x}$ and the true value of the population mean μ

$$E = z_{\alpha/2} \cdot \frac{\sigma}{\sqrt{n}}$$

where $z_{\alpha/2}$ is known as the critical value, the positive z value that is at the vertical boundary for the area of $\alpha/2$ in the right tail of the standard normal distribution; σ is the population standard deviation; n is the sample size.

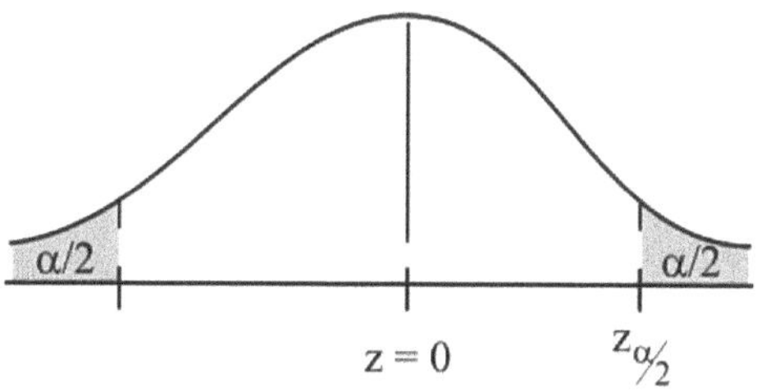

Rearranging this formula, we can solve for the sample size necessary to produce results accurate to a specified confidence and margin of error.

$$n = \left[\frac{z_{\alpha/2}\sigma}{E}\right]^2$$

This formula can be used when you know σ and want to determine the sample size necessary to establish, with a confidence of $1 - \alpha$, the mean value μ to within $\pm$ E. You can still use this formula if you do not know your population standard deviation σ and you have a small sample size. Although it is unlikely that you know σ when the population mean is not known, you may be able to determine σ from a similar process.

Sampling theory is based on sampling. It deals with statistical inferences drawn from sampling results and are of the following types:

1. Statistical estimation
2. Tests of significance
3. Statistical inference

Statistical information is the procedure by which we reach a conclusion about a population on the basis of the information contained in a sample drawn from that population.

7.10 SAMPLING DISTRIBUTION

The distribution of all possible values that can be assumed by some statistic, computed by some statistic, computed from samples of the same size randomly, is called the sampling distribution population.

To construct sampling distribution we randomly draw all possible samples of size n from a finite population of size N and compute the statistic of interest for each sample. We then list in one column the varied distinct observed values of the statistic and in another column list the corresponding frequency of occurrence of each distinct observed value of the statistic. We usually are interested in knowing the mean, variance, and the functional form of the sampling distribution.

When a population element can be selected more than one time, the sampling is known as sampling with replacement and when a population element can be selected only one time, the sampling is known as sampling without replacement.

In general when sampling is with replacement, the number of possible samples is equal to N^n.

And when the sampling is without replacement, the number of possible samples is N_{C_n}.

Standard error: The square root of the variance of the sampling distribution, i.e., σ/n is called the standard error of mean or simply standard error.

Example 7.1: Consider the five numbers 6, 8, 10, 12, and 14 representing a population of size $N = 5$.

The mean μ of the population is

$$\mu = \frac{\sum x_i}{N} = \frac{6 + 8 + 10 + 12 + 14}{5} = \frac{50}{5} = 10$$

$$\sigma^2 = \sum \frac{(x_i - \mu)^2}{N} = \frac{(-4)^2 + (-2)^2 + (0)^2 + (2)^2 + (4)^2}{5} = \frac{40}{5} = 8$$

$$S^2 = \sum \frac{(x_i - \mu)^2}{N - 1} = \frac{(-4)^2 + (-2)^2 + (0)^2 + (2)^2 + (4)^2}{4} = \frac{40}{4} = 10$$

All possible samples of size 2 from the population with replacement are shown below:

(6, 6)	(6, 8)	(6, 10)	(6, 12)	(6, 14)
(8, 6)	(8, 8)	(8, 10)	(8, 12)	(8, 14)
(10, 6)	(10, 8)	(10, 10)	(10, 12)	(10, 14)
(12, 6)	(12, 8)	(12, 10)	(12, 12)	(12, 14)
(14, 6)	(14, 8)	(14, 10)	(14, 12)	(14, 14)

Number of samples with replacement $= N^n = 5^2 = 25$

The sample means of the above samples are shown below:

6	7	8	9	10
7	8	9	10	11
8	9	10	11	12
9	10	11	12	13
10	11	12	13	14

The table above is known as sampling distribution of means.

The mean of the sampling distribution of means

$$\mu_{\overline{x}} = \frac{\sum \overline{x}_1}{N^n} = \frac{6+7+7+\cdots+14}{N^n} = \frac{250}{25} = 10$$

We observe that the mean of the population

$$= \frac{\text{The mean of sampling distribution of means}}{\text{The variance of the sampling distribution of means}}$$

$$\sigma^2_{\overline{x}} = \frac{(6-10)^2+(7-10)^2+\cdots+(14-10)^2}{25} = \frac{100}{25} = 4$$

We now consider all samples without replacement that can be drawn from the given population:

(6, 8)	(6, 10)	(6, 12)	(6, 14)
(8, 10)	(8, 12)	(8, 14)	
(10, 12)	(10, 14)		
(12, 14)			

The variance of this sampling distribution is

$$\sigma^2_{\overline{x}} = \frac{\sum(\overline{x}_1-\mu_{\overline{x}})^2}{N_{c_n}} = \frac{30}{10} = 3$$

When sampling is without replacement from a finite population, the sampling distribution of $\overline{x}$ will have mean μ and variance

$$\sigma_{\overline{x}}^2 = \frac{\sigma^2}{n} \cdot \frac{N-n}{N-1}$$

where the factor $(N-n)/(N-1)$ is called finite correction factor, which can be ignored when the sample is small compared to the population size.

Sampling from normally distributed population: When sampling is from normally distributed population, the distribution of the sample mean will be normal. The mean $\mu_{\overline{x}}$ of the distribution of $\overline{x}$ will be equal to the mean of the population from which the samples were drawn, and the variance $\sigma_{\overline{x}}^2$ of the distribution of $\overline{x}$ will be equal to the variance of the population divided by the sample size.

Central limit theorem: When sampling is from a normally distributed population we refer to an important theorem, which is known as the central limit theorem. The theorem is stated as follows:

"Given a population of any nonnormal functional form with mean population μ and finite variance σ^2, the sampling distribution of $\overline{x}$, computed from samples of size n from this population, will have the mean μ and variance σ^2/n and will be approximately normally distributed when the sample size is large."

The mathematical formulation of the central theorem is that the distribution of $(\overline{x}-\mu)/(\sigma/\sqrt{n})$ approaches a normal distribution with mean 0 and variance 1 as $n \rightarrow \infty$.

Exercise 7.1

1. Define the terms
   - a. Population
   - b. Sample
2. Define
   - a. Random sampling
   - b. Stratified sampling
3. Define
   - a. Systematic Sampling
   - b. Sampling with replacement
   - c. Sampling without repetition
4. Explain central theorem
5. What is sampling distribution?
6. Consider the population consisting of the numbers 3, 5, 7, 9, and 11. Find all the samples of size 2 and show that the mean of the population is equal to the mean of sampling distribution of mean?

CHAPTER 8

Hypothesis Testing

8.1 INTRODUCTION

In this chapter we shall deal with the problem of hypothesis testing, which enables us to make statement about population parameters. Many experiments are carried out with deliberate object of testing hypothesis. Hypothesis testing is a branch of statistics which helps us in arriving at criterion of decision making. We now explain the meaning of hypothesis and some of the basic concepts essential for better understanding of the hypothesis testing methods.

8.2 HYPOTHESIS

Hypothesis is an assumption or statement, which may or may not be true. It is the conclusion that is tentatively drawn as logical basis. It is defined as follows:

Definition 8.1: A hypothesis is a statement about the population parameter.

A hypothesis is an assumption made in order to arrive at a decision regarding the population through a sample of the population. A hypothesis should be specific, clear, and precise. It should state as far as possible in most single term so that the same is easily understood by all concerning. It should state the relationship between variables.

A hypothesis should be capable of testing.

8.3 HYPOTHESIS TESTING

It is the process of testing significance, which concerns with the testing of same hypothesis regarding parameter of the population on the basis of statistic from the population.

Statistical Techniques for Transportation Engineering.
DOI: http://dx.doi.org/10.1016/B978-0-12-811555-8.00008-8

8.4 TYPES OF HYPOTHESIS

To verify our assumption that is based on sample study we collect data and find out the difference between sample value and population value. If there is no difference, or if the difference is very small then our hypothesized value is correct. In general two types of hypothesis are constructed namely null hypothesis and alternative hypothesis.

8.4.1 Null Hypothesis

In the process of statistical test the hypothesis concerning a population is rejected or accepted based on the sample drawn from the population, the statistician test the hypothesis through observations and makes a probability statement. The hypothesis (i.e., statement) we have assumed is called null hypothesis. It is defined as follows:

Definition 8.2: The hypothesis formulated for the purpose of its rejection under the assumption that it is true is called the null hypothesis.

Null hypothesis is denoted by H_0.

8.4.2 Alternative Hypothesis

The negation of null hypothesis is called alternative hypothesis. It is denoted by H_1. If the null hypothesis is rejected, then the alternative hypothesis is accepted, the acceptance or rejection of null hypothesis is based on sample study. Suppose we want to test the hypothesis that the population mean is equal to the hypothesized mean 60. Symbolically they can be expressed as

$$H_0\colon \mu = \mu_0 = 60$$

$$\text{where } \mu_0 = \mu_{H_0}$$

The possible alternative hypothesis can be stated in one of the following form:

$$H_1\colon \mu \neq \mu_0, \text{ i.e., } \mu > \mu_0 \text{ or } \mu < \mu_0 \text{ (two-tailed test)}$$

$$H_1\colon \mu > \mu_0 \text{ (one-tailed test)}$$

$$H_1\colon \mu < \mu_0 \text{ (one-tailed test)}$$

In hypothesis testing, we usually proceed on the basis of null hypothesis, the probability of rejecting null hypothesis when it is true or the level of significance which is very small.

8.5 COMPUTATION OF TEST STATISTIC

Test statistic is computed after stating the null hypothesis. It is based on the appropriate probability distribution. The test statistic is used to test whether the null hypothesis H_0 should be accepted or rejected.

8.6 LEVEL OF SIGNIFICANCE

The level of significance is an important concept in hypothesis testing. It is always same percentage, it should be chosen with great care and thought and reason. The level of significance is maximum probability of rejecting null hypothesis when it is true and is denoted by α, the probability of making a correct decision is $1-\alpha$. The level of significance may be taken as 1% or 5% or 10% (i.e., $\alpha = 0.01$ or 0.05 or 0.1).

If we fix the level of significance at 5%, then the probability of making type I error is 0.05. This also means that we are 95% confident of making a correct decision.

When no level of significance is maintained it is taken as $\alpha = 0.05$.

8.7 CRITICAL REGION

A region corresponding to a statistic that amounts to rejection of null hypothesis H_0 is known as critical region. It is also called as region of rejection.

The critical region is the region of the standard normal curve corresponding to a predetermined level of significance α. The region under the normal curve that is not covered by normal curve is known as the acceptance region.

8.8 ONE-TAILED TEST AND TWO-TAILED TEST

8.8.1 One-Tailed Test

A test of any statistical hypothesis where alternative hypothesis is one sided such as is called one-tailed test.

$H_0: \mu = \mu_0$	$H_0: \mu = \mu_0$	$H_0: \mu \geq \mu_0$	$H_0: \mu \leq \mu_0$
or	or	or	or
$H_1: \mu > \mu_0$	$H_1: \mu < \mu_0$	$H_1: \mu > \mu_0$	$H_1: \mu < \mu_0$

where μ is the parameter to be tested and H_0 is the hypothesized value. When alternative hypothesis is one sided, i.e., one tailed, we apply one-tailed test. There are two types of one-tailed tests. When it is desired to see whether the population mean is as large as some specified value of the mean, we apply the right tailed test. In the right tailed test, the region of rejection (critical region) lies entirely on the right tail on the normal curve (Fig. 8.1).

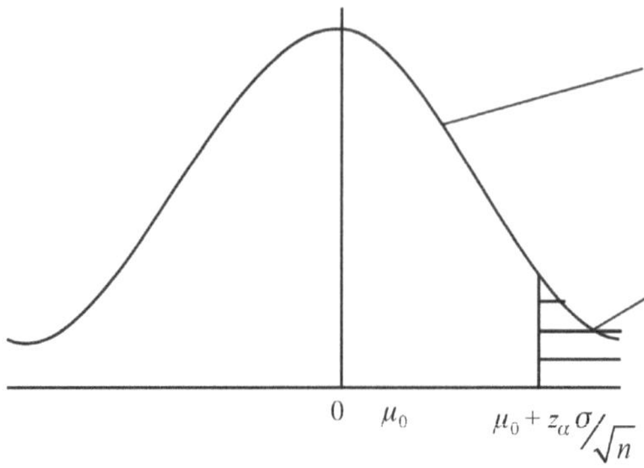

Figure 8.1 Right-tailed test: rejection and acceptance regions.

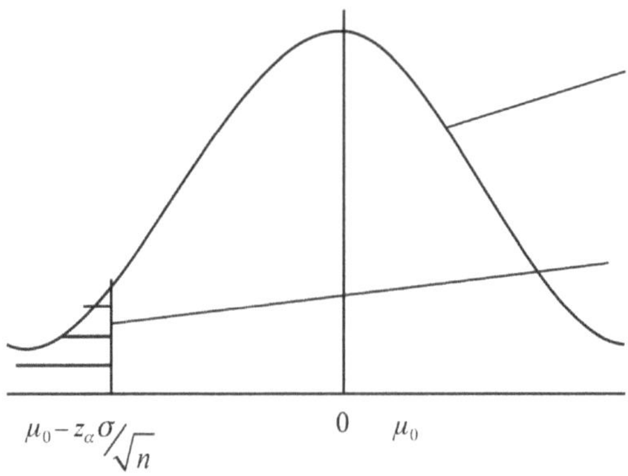

Figure 8.2 Left-tailed test: rejection and acceptance region.

When it is desired to see whether the population mean is at least as small as some specified value of the mean, we apply the left-tailed test.

In the left-tailed test the region of rejection (critical region) lies entirely on the left tail on the normal curve Fig. 8.2.

8.8.2 Two-Tailed Test

A test of any statistical hypothesis where the alternative hypothesis is two sided such as is called a two-tailed test.

$$H_0: \mu = \mu_0$$

$$H_1: \mu \neq \mu_0$$

where μ is the parameter to be tested and μ_0 is the hypothesized value. In a two-tailed test, there are two rejection regions, one an each tail of

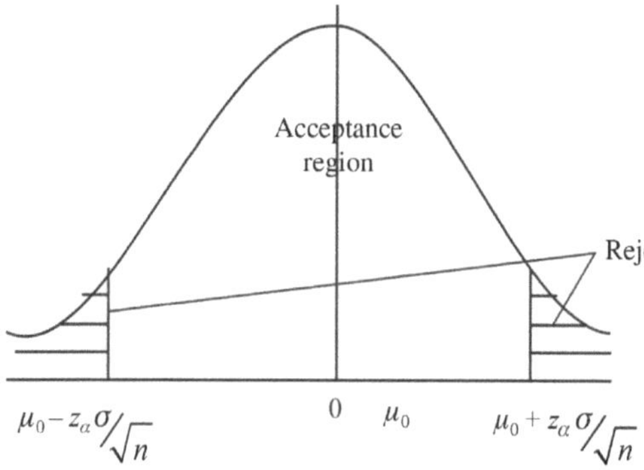

Figure 8.3 Two-tailed test: rejection and acceptance region.

the curve (Fig. 8.3). If the significance level is 5% and the test applied is a two-tailed test, the probability of the region area will be 0.05. It is equally splitted on both sides of the curve as 0.025. The region of acceptance in this case is 0.95.

Decision of rejecting Null hypothesis or not rejecting Null hypothesis: The statistical inference made on the basis of the sample mean is called a decision. If the sample mean lies in the region of rejection, null hypothesis is rejected while the alternative hypothesis is accepted, i.e., H_0 is rejected and H_1 is accepted. If the sample mean lies in the region of acceptance, H_0 is accepted and H_1 is rejected. In this case the area in the tails is equal to the level of significance α. In the case of one-tailed test α appears is one tail and for two-tailed test $\alpha/2$ appears in each tail of the curve. Instead of calculating the region of acceptance and rejection, we can directly calculate the value of the test statistic using the value of the sample mean calculated from the sample data.

Thus we have the following decision rules:

1. H_0: $\mu = \mu_0$ Accept H_0 if $\dfrac{|\overline{x} - \mu_0|}{SE(\overline{x})} \leq$ table value of $Z_{\alpha/2}$ or $t_{\alpha/2}$

   H_1: $\mu \neq \mu_0$ Reject H_0 if $\dfrac{|\overline{x} - \mu_0|}{SE(\overline{x})} >$ table value of $Z_{\alpha/2}$ or $t_{\alpha/2}$

2. H_0: $\mu \leq \mu_0$ Accept H_0 if $\dfrac{|\overline{x} - \mu_0|}{SE(\overline{x})} \leq$ table value of Z_α or t_α

   H_1: $\mu > \mu_0$ Reject H_0 if $\dfrac{|\overline{x} - \mu_0|}{SE(\overline{x})} >$ table value of Z_α or t_α

3. H_0: $\mu \geq \mu_0$ Accept H_0 if $\dfrac{|\overline{x} - \mu_0|}{SE(\overline{x})} \leq$ table value of Z_α or t_α

   H_1: $\mu < \mu_0$ Reject H_0 if $\dfrac{|\overline{x} - \mu_0|}{SE(\overline{x})} >$ table value of Z_α or t_α

Remarks:

1. When the size of the sample is $n \geq 30$, i.e., the sample is large, we make use of the unbiased estimate of σ.

   If the population standard deviation, σ, is known and n denotes the size of the sample the test statistic

$$Z = \frac{\overline{x} - \mu}{\sigma/\sqrt{n}} \text{(normal)}$$

2. When σ is unknown, and $n < 30$ (i.e., small sample) the statistic is $t = (\overline{x} - \mu)/(\hat{\sigma}/\sqrt{n})$, ($\hat{\sigma}$ is the unbiased estimate of σ).

8.9 ERRORS

The decision regarding a null hypothesis may or may not be correct. There are two types of errors we can make:

	Accept H_0	Reject H_0
H_0 (True)	Correct decision	Wrong decision Type I error
H_0 (False)	Wrong decision Type II error	Correct decision

The probability of rejecting H_0 when H_0 is true is denoted by α and the probability of accepting H_0 when H_1 is true is denoted by β.

i.e.,

Prob (Type I error) = α

Prob (Type II error) = β

An increase is the sample size n will reduce α.

The probability that we will reject a false null hypothesis is given by $1 - \beta$. The quantity $1 - \beta$ is called the power of a test. For a given α we may specify any number of possible values of the parameter of interest and for each compute the value of $1 - \beta$. The result is called a power function and the corresponding curve is called a power curve.

By adjusting the values of α, we can reduce the size of the rejection region. Probability of making one type of error can be reduced by allowing an increase in the probability of the other type of error. Hence a decrease in the probability of the other.

Remarks:

Of the two hypotheses, one is true and the other is false unfortunately we do not know which is which. Therefore we design a test and apply it and make a judgement. The problem is that there is no guarantee that we are correct in our decision or judgement.

8.10 PROCEDURE FOR HYPOTHESIS TESTING

The main question in hypothesis testing is whether to accept the null hypothesis or not to accept the null hypothesis. The following tests are involved in hypothesis testing:

1. *Stating of hypothesis*: The null hypothesis H_0, and the alternative hypothesis H_1 are constructed in this step. It is of the form right tailed.

Right tailed	**Left tailed**	**Two tailed**
H_0: $\mu \leq \mu_0$	H_0: $\mu \geq \mu_0$	H_0: $\mu = \mu_0$
or	or	or
H_1: $\mu > \mu_0$	H_1: $\mu < \mu_0$	H_1: $\mu \neq \mu_0$

   where μ is the population mean and μ_0 is its specified value.

2. *Identification of test statistic*: For large samples ($n \geq 30$), when population n standard deviation (SD) is known. The test statistic is $Z = (\overline{x} - \mu)/(\sigma/\sqrt{n})$.

   The corresponding distribution is normal for small samples (i.e., $n < 30$) we use *t*-test.

3. *Specifying the level of significance*: Very important concept in hypothesis testing is the level of significance. It is always some percentage. It may be 1% or 5% or 10%. The level of significance is the maximum value of the probability of rejecting the null hypothesis when it is true and is usually determined in advance before testing the hypothesis.
4. Determine the value of the test statistic.
5. Check whether the probability value is less than or equal to α and accordingly reject the null hypothesis, or accept it.

8.11 IMPORTANT TESTS OF HYPOTHESIS

For the purpose of testing hypothesis, several tests of hypothesis were developed. They can be classified as:

1. Parametric test.
2. Nonparametric test.

Parametric tests are also known as standard the distribution-free test of hypothesis. In this section, we consider only the parametric tests.

The important parametric tests are:

1. Z-Test
2. t-Test
3. χ^2-Test
4. F-Test

The parametric tests are based on the assumption of normality.

Z-Test is used for comparing the mean of a sample to some hypothesized mean for the population in case of a large sample. It is also used when the population variance is known. *Z*-Test is also used for comparing the sample proportion to a theoretical value of population proportion.

t-Test is used in the case of small samples. It is based on *t*-distribution and is considered an appropriate test for judging the significance of a sample mean or for judging the significance of difference between the means of two samples.

Chi-square test is used for comparing a sample variance to a theoretical population variance. It is based on chi-square distribution.

F-Test is used to compare the variance of the two independent samples. It is based on *F*-distribution.

8.12 CRITICAL VALUES

The value of the test statistic that separates the critical region and the acceptance region is called critical value.

It is also called significant value and is dependent on

1. the level of significance and
2. the alternative hypothesis.

For large samples corresponding to the statistic *t*, the variable $Z = (t - E(t)/(SE(t)))$ is normally distributed with mean 0 and variance 1. If α is the level of significance. The value Z_α of the test statistic for a two-tailed test is given by

$$P(|Z| > Z_\alpha) = \infty$$

i.e., Z_α is the value of Z so that the total area of the critical region and the two tails is α. Since the normal curve is a symmetrical curve.

$$P(|Z| > Z_\alpha) = \alpha$$

$$P(Z > Z_\alpha) + P(Z < -Z_\alpha) = \alpha$$

or

$$2P(Z > Z_\alpha) = \alpha$$

i.e.,

$$P(Z > Z_\alpha) = \frac{\alpha}{2}$$

Hence the area in each tail is $\alpha/2$.

The following table gives us the critical values of Z:

Critical value (Z_α)	Level of significance		
	1%	5%	10%
Two-tailed test	$\lvert Z_\alpha \rvert = 2.58$	$\lvert Z_\alpha \rvert = 1.96$	$\lvert Z_\alpha \rvert = 1.645$
Right-tailed test	$Z_\alpha = 2.33$	$Z_\alpha = 1.645$	$Z_\alpha = 1.28$
Left-tailed test	$Z_\alpha = -2.33$	$Z_\alpha = -1.645$	$Z_\alpha = -1.28$

8.13 TEST OF SIGNIFICANCE—LARGE SAMPLES

8.13.1 Test of Significance for Single Mean

Let $x_1, x_2, \ldots, x_n$ be a random sample of size n drawn from a large population with mean μ and variance σ^2. Let $\overline{x}$ denote the mean of sample and S^2 denote the variance of the sample.

We know that $\overline{x} \sim N\left(\mu, \dfrac{\sigma^2}{n}\right)$

The standard normal variate corresponding to $\overline{x}$

$$Z = \frac{\overline{x} - \mu}{SE(\overline{x})}$$

where

$$SE(\overline{x}) \frac{\sigma}{\sqrt{n}}$$

We set up the null hypothesis that there is no difference between the sample mean and the population mean. The test statistic is

$Z = \dfrac{\overline{x} - \mu}{\sigma/\sqrt{n}}$ (σi known and $SE(\overline{x}) = \sigma/\sqrt{n}$).

If σ is not known

$$Z = \frac{\overline{x} - \mu}{S/\sqrt{n}}$$

where S is the SD of the sample.

8.13.1.1 Solved Examples

Test of significance for single means:

Example 8.1: The heights of college students in a city are normally distributed with SD 6 cm; a sample of 100 students have a mean height of 158 cm. Test the hypothesis that the mean height of college students in the college is 160 cm.

Solution: We have
$\overline{x} = 158$ (Mean of the sample)
$\mu = 160$ (Mean of the population)
$\sigma = 6$ (SD)
$n = 100$ (Size of the sample)
Level of significance: 5% and 1%
$H_0{:}\mu = 160$, i.e., difference is not significant
$H_1{:}\mu \neq 160$
We apply two-tailed test
Test statistic is

$$Z = \frac{\overline{x} - \mu}{\sigma/\sqrt{n}} = \frac{158 - 160}{6/\sqrt{100}} = \frac{-2}{6/10} = -3.333$$

$$\therefore \quad |Z| = 3.333$$

Table value of Z at 5% level of significance = 1.96

Since the calculated value of Z at 5% level of significance is greater than the table value of Z, we reject H_0 at 5% level of significance.

Table value of Z at 1% level of significance is 2.58.

Calculated value at 1% level of significance is greater than the table value of Z, hence we reject the null hypothesis H_0 at 1%, i.e., H_0 is rejected both at 1% and 5% level of significance.

Example 8.2: A sample of 400 items is taken from a population whose SD is 10. The mean of the sample is 40. Test whether the sample has come from the population with mean 38. Also calculate 95% confidence interval for the population.

Solution: Here we have

$$\overline{x} = 40 \text{ (Mean of the sample)}$$

$$\mu = 38 \text{ (Mean of the population)}$$

$$\sigma = 10 \text{ (SD)}$$

$$n = 400$$

Null hypothesis: H_0: $\mu = 38$
i.e., the sample is from the population whose mean = 38
Alternative hypothesis: H_1: $\mu \neq 38$
Level of significance: $\alpha = 0.05$
Table value of $Z = Z_\alpha = 1.96$

Test statistic:

$$Z = \frac{\overline{x} - \mu}{\sigma/\sqrt{n}} = \frac{40 - 38}{10/\sqrt{400}} = \frac{2}{10/20} = \frac{40}{10} = 4$$

Calculated value of $Z = 4$ is greater than the table value of Z at 0.05 level of significance.

Hence, null hypothesis is rejected, i.e., confidence interval for the population whose mean is 38.

95% confidence interval for the population is given by

$$\overline{x} \pm 1.96 \cdot \frac{\sigma}{\sqrt{n}} = 40 \pm 1.96 \cdot \frac{10}{\sqrt{400}} = 40 \pm 0.98$$

or

$$[40 - 0.98, \quad 40 + 0.98] = [39.02, \quad 40.98]$$

Example 8.3: The mean and SD of a population are 11,795 and 14,054, respectively. If $n = 50$ find 95% confidence interval for the mean?

Solution: It is given that $\overline{x} = 11,795$, $\sigma = 14{,}054$, $n = 50$

Therefore

$$\frac{\sigma}{\sqrt{n}} = \frac{14054}{\sqrt{50}} = \frac{14054}{7.071} = 1987.55$$

and

$$1.96 \cdot \frac{\sigma}{\sqrt{n}} = (1.96)(1987.55) = 3895.607$$

Hence the 95% confidence interval is

$$\overline{x} \pm 1.96 \cdot \frac{\sigma}{\sqrt{n}} = [11795 - 3895.607, \quad 1175 + 3895.607]$$

$$= [7899.39, \quad 15690.60]$$

Example 8.4: It is claimed that a random sample of 49 types has a mean life of 15,200 km. This sample has drawn from a population whose mean is 15,150 km and a SD of 1200 km. Test the significance at 0.05 levels.

Solution: We have $\overline{x} = 15,200$, $\mu = 15{,}150$, $\sigma = 1200$, and $n = 49$

Null hypothesis: H_0: $\mu = 15{,}150$

Alternative hypothesis: H_1: $\mu \neq 15{,}150$ (two-tailed test)

Level of significance $= \alpha = 0.05$

Table value of $Z = 1.96$

Test statistic is

$$Z = \frac{\bar{x} - \mu}{\sigma/\sqrt{n}} = \frac{15200 - 15150}{1200/\sqrt{49}} = \frac{50}{171.42} = 0.29168$$

Since the calculated value of Z is less than the table value at 5% level, we accept null hypothesis.

Example 8.5: A sample size of 400 was drawn and the sample mean was found to be 98. Test whether this sample could have come from a normal population with mean 100 and SD 8 at 5% level of significance.

Solution: Here we have $\bar{x} = 98$, $\mu = 100$, $\sigma = 8$, and $n = 400$

Null hypothesis: H_0: $\mu = 100$

i.e., The sample comes from a normal population with mean 100.

Alternative hypothesis H_1: $\mu \neq 100$ (two-tailed test)

Level of significance = $\alpha = 0.05$

Table value of Z at 5% level of significance = 1.96

Test statistic:

$$Z = \frac{\bar{x} - \mu}{\sigma/\sqrt{n}} = \frac{98 - 100}{8/\sqrt{400}} = \frac{-2}{8/20} = \frac{-10}{2} = -5$$

$$|Z| = |-5| = 5 > 1.96$$

Since the calculated value of Z is greater than the table value of Z at 5% level of significance, we reject H_0, i.e., sample was not drawn from the normal population with mean 100 and SD 8.

Example 8.6: The average marks in mathematics of a sample of 100 students was 51 with a SD of 6 marks, could this have been a random sample from a population with average marks of 50?

Solution: Here we have $\bar{x} = 51$, $S = 6$, $\mu = 50$, $n = 100$

Null hypothesis: H_0: 50, The sample is from a population with an average of 50 marks.

i.e.,

$$\mu = 50$$

Alternate hypothesis: H_1: $\mu \neq 50$

Let the level of significance be taken as 5%

Table value of Z for 5% level of significance is $|Z| = 1.96$

$$Z = \frac{\bar{x} - \mu}{\sigma/\sqrt{n}} = \frac{51 - 50}{6/100} = \frac{1}{6/10} = \frac{10}{6} = 1.666$$

Since the calculated value of Z is less than the table value of Z at 5% level of significance, we accept H_0, i.e., sample is drawn from a population with a mean of 50 marks. The difference is not significant.

Example 8.7: The mean of certain production process is known to be 50 with a SD of 2.5. The production manager may welcome any change in mean value toward higher side but would like to safe guard against decreasing values of mean. He takes a sample of 12 items that gives a mean value of 46.5 what inference should the manager take for production process on the basis of sample results. Use 5% level of significance for the purpose.

Solution: $\bar{x} = 46.5$, $\mu = 50$, $\sigma = 2.5$

Null hypothesis H_0: $\mu = 50$

Alternative hypothesis H_1: $\mu < 50$ (one-sided test)

Level of significance $\alpha = 5\%$

Table value of $Z = 1.645$

$$Z = \frac{\bar{x} - \mu}{\sigma/\sqrt{n}} = \frac{46.5 - 50}{2.5/\sqrt{12}} = \frac{-3.5}{2.5/3.464} = \frac{-3.5}{0.721} = -4.854$$

$$|Z| = |-4.854| = 4.854$$

i.e., calculated value of Z is greater than the table value of Z.

Hence H_0 is rejected.

Therefore the production process shows a mean that is significantly less than the population mean and this calls for taking same corrective measures concerning the production process.

Example 8.8: A random sample of 100 students gave a mean height of 58 kg, with a standard deviation (SD) 4 kg. Test the hypothesis that the mean weight in the population is 60 kg.

Solution: Here $\bar{x} = 58$, $S = 4$, $\mu = 60$, $n = 100$

$$H_0: \mu = 60 \text{ kg}$$
$$H_1: \mu \neq 60 \text{ kg}$$

Level of significance $\alpha = 5\%$

Table value of $Z_\alpha = 1.96$

$$Z = \frac{\bar{x} - \mu}{S/\sqrt{n}} = \frac{58 - 60}{4/\sqrt{100}} = -5$$

$$|Z| = |-5| = 5$$

Since the calculated value of Z is greater than the table value at 5% level of significance, H_0 is rejected.

$$\therefore \quad \mu \neq 60$$

8.13.2 Test of Significance for Difference of Means of Two Large Samples

Let $\overline{x}_1$ be the mean of independent random sample of size n_1 from a population with mean μ_1 and variance σ_1^2 and let $\overline{x}_2$ be the mean of independent random sample of size n_2 from a population with mean μ_2 and variance σ_2^2, where n_1 and n_2 are large.

Clearly,

$$\overline{x}_1 \sim N\left(\mu_1, \frac{\sigma_1^2}{n_1}\right) \quad \text{and} \quad \overline{x}_2 \sim N\left(\mu_2, \frac{\sigma_2^2}{n_2}\right)$$

and

$$Z = \frac{(\overline{x}_1 - \overline{x}_2) - E(\overline{X}_1 - \overline{X}_2)}{S.E.(\overline{x}_1 - \overline{x}_2)} \sim N(0,1)$$

i.e., Z is the standard normal variate.

Setting up the null hypothesis, we have

$$E(\overline{x}_1 - \overline{x}_2) = E(\overline{x}_1) - E(\overline{x}_2)$$
$$= \mu_1 - \mu_2 = 0$$

Since $\overline{x}_1$ and $\overline{x}_2$ are independent, the covariance terms vanish. We have

$$\text{Var}(\overline{x}_1 - \overline{x}_2) = \text{Var}(\overline{x}_1) + \text{Var}(\overline{x}_2)$$
$$= \frac{\sigma_1^2}{n_1} + \frac{\sigma_2^2}{n_2}$$

Hence under the null hypothesis H_0: $\mu_1 = \mu_2$, the test statistic is

$$Z = \frac{\overline{x}_1 - \overline{x}_2}{\sqrt{\frac{\sigma_1^2}{n_1} + \frac{\sigma_2^2}{n_2}}} \tag{8.1}$$

If $\sigma_1 = \sigma_2 = \sigma$, then the test statistic is

$$Z = \frac{\overline{x}_1 - \overline{x}_2}{\sqrt{\sigma^2\left(\frac{1}{n_1} + \frac{1}{n_2}\right)}} = \frac{\overline{x}_1 - \overline{x}_2}{\sigma\sqrt{\left(\frac{1}{n_1} + \frac{1}{n_2}\right)}} \tag{8.2}$$

If σ_1 and σ_2 are not known then the test statistic is

$$Z = \frac{\overline{x}_1 - \overline{x}_2}{\sqrt{\frac{S_1^2}{n_1} + \frac{S_2^2}{n_2}}}$$

where

$$\sigma_1 \neq \sigma_2$$

If $\sigma_1 = \sigma_2$ and σ is not known, we compute σ^2 by using the formula

$$\sigma^2 = \frac{n_1 S_1^2 + n_2 S_2^2}{n_1 + n_2}$$

In this case the test statistic is

$$Z = \frac{\overline{x}_1 - \overline{x}_2}{\sqrt{\frac{n_1 S_1^2 + n_2 S_2^2}{n_1 + n_2}\left(\frac{1}{n_1} + \frac{1}{n_2}\right)}}$$

substitute S_1 for and S_2 for provided n_1 and n_2 are large.
i.e.,

$$Z = \frac{\overline{x}_1 - \overline{x}_2}{\sqrt{\frac{n_1 S_1^2 + n_2 S_2^2}{n_1 n_2}}} \tag{8.3}$$

If we want to test the null hypothesis $\mu_1 - \mu_2 = \delta$ where δ is a specified constant against one of the alternatives

$$\mu_1 - \mu_2 \neq \delta$$
$$\mu_1 - \mu_2 < \delta$$

or

$$\mu_1 - \mu_2 > \delta$$

We apply likelihood ratio technique and at the test based on $\overline{x}_1 - \overline{x}_2$. The corresponding critical regions for the test can be written as

$$|Z| \geq Z_{\alpha/2}, \quad Z \geq Z_\alpha$$

and

$$Z \leq Z_\alpha$$

where

$$Z = \frac{\overline{X}_1 - \overline{X}_2 - \delta}{\sqrt{\frac{\sigma_1^2}{n_1} + \frac{\sigma_2^2}{n_2}}}$$

When we deal with independent random samples drawn from populations with unknown variances, that may not be normal. We substitute S_1 for σ_1 and S_2 for σ_2 provided n_1 and n_2 are large.

When the sizes of the independent random samples, i.e., n_1 and n_2 are small and σ_1 and σ_2 are unknown the above test cannot be used. For independent random samples from two normal population having the same unknown variable, we use the test

$$Z = \frac{\overline{X}_1 - \overline{X}_2 - \delta}{S\sqrt{\frac{1}{n_1} + \frac{1}{n_2}}}$$

where

$$S_1 = \frac{(n_1 - 1)S_1^2 + (n_2 - 1)S_2^2}{n_1 + n_2 - 2}$$

(*Test of significance of difference between two means*).

Example 8.9: Random samples drawn from two places gave the following data relating to the heights of children:

	Place A	Place B
Mean height	68.50	68.58
Standard deviation (SD)	2.5	3.0
Number of items in the sample	1200	1500

Test at 5% level that the mean height is the same for the children at two places.

Solution: Here we have

$$\overline{x}_1 = 68.50, \quad \overline{x}_2 = 68.58$$

$$n_1 = 1200, \quad n_2 = 1500$$

$$\sigma_1 = 2.5, \quad \sigma_2 = 3.0$$

Level of significance = 0.05

$$Z_{0.05} = 1.96$$

$$SE\,(\overline{x}_1 - \overline{x}_2) = \sqrt{\frac{\sigma_1^2}{n_1} + \frac{\sigma_2^2}{n_2}}$$

$$= \sqrt{\frac{(2.5)^2}{1200} + \frac{3^2}{1500}}$$

$$= \sqrt{\frac{6.25}{1200} + \frac{9}{1500}} = 0.1058$$

Null hypothesis H_0: $\mu_1 = \mu_2$
Alternative hypothesis H_1: $\mu_1 \neq \mu_2$ (Two-tailed test)

$$Z_{0.05} = 1.96 \text{ (table value)}$$

$$Z = \frac{\overline{X}_1 - \overline{X}_2}{SE(\overline{x_1} - \overline{x_2})} = \frac{68.50 - 68.58}{0.1058} = -0.756$$

$|Z| = 0.756 < 1.96$, i.e., calculated value of Z is less than the table value. Hence null hypothesis is accepted.

Example 8.10: The mean produce of a sample of 100 fields is 200 lb, per acre with a standard deviation (SD) of 10 lb. Another sample of 150 fields gives the mean of 220 lb, with a SD 12 lb. Can the two samples be considered to have been taken from the same population whose SD is 11 lb. Use 5% level of significance.

Solution: It is given that

$$\overline{x}_1 - 200 \text{ lb}, \quad \overline{x}_2 = 220 \text{ lb}$$

$$n_1 = 100, \quad n_2 = 150$$

$$\sigma_1 = 10 \text{ lb}, \quad \sigma_2 = 12 \text{ lb}$$

$$\sigma = \text{SD of the population} = 11 \text{ lb}$$

$$H_0\text{: } \mu_1 = \mu_2$$

$$H_1\text{: } \mu_1 \neq \mu_2 \text{ (two-tailed test)}$$

Level of significance = 5%

$$Z_\alpha = 1.96$$

Test statistic:

$$Z = \frac{\overline{x}_1 - \overline{x}_2}{\sqrt{\sigma^2\left(\frac{1}{n_1} + \frac{1}{n_2}\right)}} = \frac{200 - 220}{\sqrt{(11)^2\left(\frac{1}{100} + \frac{1}{150}\right)}}$$

$$= \frac{-20}{\sqrt{(121)\left(\frac{250}{(100)(150)}\right)}} = \frac{-20}{1.42} = -14.08$$

$$\therefore \quad |Z| = |-14.08| = 14.08$$

The calculated value of Z is greater than the table value at 5% level of significance.

Hence we reject null hypothesis H_0.

Therefore the difference between the means of the two samples is significant and it is not due to sampling fluctuations.

Example 8.11: Two sets of 100 students each were taught to read by two different methods. After the instructions were over, a reading test given to them revealed $\overline{x}_1 = 73.4$, $\overline{x}_2 = 70.3$, $S_1 = 8$, and $S_2 = 10$. Test the hypothesis that $\mu_1 = \mu_2$.

Solution: It is given that

$$\overline{x}_1 = 73.4, \quad \overline{x}_2 = 70.3$$
$$S_1 = 8, S_2 = 10$$
$$n_1 = n_2 = 100$$

Null hypothesis H_0: $\mu_1 = \mu_2$

Alternative hypothesis: H_1: $\mu_1 \neq \mu_2$

$$Z = \frac{\overline{x}_1 - \overline{x}_2}{\sqrt{\frac{S_1^2}{n_1} + \frac{S_2^2}{n_2}}} = \frac{73.4 - 70.3}{\sqrt{\frac{64}{100} + \frac{100}{100}}}$$

$$= \frac{3.1}{\sqrt{164}} \times \sqrt{100} = \frac{31}{12.806} = 2.4207$$

$|Z| = 2.4207 > 1.96$, therefore at 5% level of significance H_0 is rejected.

$|Z| = 2.4207 < 2.58$, therefore at 1% Level of significance H_0 is accepted.

Example 8.12: Samples of students were drawn from two univariates and from their weights in kilograms and standard deviations (SDs) are calculated. Make a large sample test the significance of the difference between the means.

Solution: We have

$$\overline{x}_1 = 55, \quad \overline{x}_2 = 57$$
$$\sigma_1 = 10, \quad \sigma_2 = 15$$
$$n_1 = 400, \quad n_2 = 100$$

Null hypothesis $H_0{:}\mu_1 = \mu_2$
Alternative hypothesis: $H_1{:}\mu_1 \neq \mu_2$
Level of significance = 0.05
Table value of $Z = 1.96$

Test statistic:

$$Z = \frac{\overline{X_1} - \overline{X_2}}{\sqrt{\frac{\sigma_1^2}{n_1} + \frac{\sigma_2^2}{n_2}}} = \frac{55 - 57}{\sqrt{\frac{100}{400} + \frac{225}{100}}}$$

$$= \frac{-2}{\sqrt{\frac{100 + 900}{400}}} = \frac{-4}{\sqrt{10}} = -1.26$$

$$\therefore \quad |Z| = 1.26 < 1.96$$

i.e., calculated value of Z is less than the table value of Z at 5% level of significance. Hence H_0 is accepted.

Example 8.13: In a certain factory there are two independent processes for manufacturing the same item. The average from one process is formed to be 120 gm with a standard deviation (SD) of 400 items. Is the difference between the mean weights significant at 10% level of significance?

Solution: The given values are

$$\overline{x}_1 = 120, \quad \overline{x}_2 = 124$$

$$S_1 = 12, \quad S_2 = 14$$

$$n_1 = 250, \quad n_2 = 400$$

Level of significance = 10%
Null hypothesis H_0: $\mu_1 = \mu_2$
Alternative hypothesis: H_1: $\mu_1 \neq \mu_2$ (two-tailed test)
Z-value = 1.645 at 10% level

Test statistic is

$$Z = \frac{\overline{x}_1 - \overline{x}_2}{\sqrt{\frac{S_1^2}{n_1} + \frac{S_2^2}{n_2}}} = \frac{120 - 124}{\sqrt{\frac{12^2}{250} + \frac{14^2}{400}}}$$

$$= \frac{-4}{1.0325} = -3.87$$

$$|Z| = 3.87 > 1.645 \text{ (i.e., } Z \text{ at 10\% level)}$$

Since the calculated value of Z is greater than the table value of Z, we reject null hypothesis. Therefore there is a significance difference between the two sample mean weights.

8.13.3 Test of Significance for the Difference of SDs of Two Large Samples

If S_1 and S_2 are the SD of two independent samples and H_0: $\sigma_1 = \sigma_2$ is the null hypothesis, the test statistic is

$$Z = \frac{S_1 - S_2}{S.E(S_1 - S_2)} \sim N(0, 1)$$

If σ_1, σ_2 are known, then for large samples

$$Z = \frac{S_1 - S_2}{\sqrt{\frac{\sigma_1^2}{2n_1} + \frac{\sigma_2^2}{2n_2}}}$$

If σ_1, σ_2 are not known, the test statistic is

$$Z = \frac{S_1 - S_2}{\sqrt{\frac{S_1^2}{2n_1} + \frac{S_2^2}{2n_2}}}$$

where

$$\sqrt{\frac{S_1^2}{2n_1} + \frac{S_2^2}{2n_2}} \text{ is the SE } (S_1 - S_2)$$

Example 8.14: Intelligence test of two groups of boys and girls gave the following results:

Girls	$\overline{x}_1 = 84$	SD of $S_1 = 10$	$n_1 = 121$
Boys	$\overline{x}_2 = 80$	SD of $S_2 = 14$	$n_2 = 81$

Is the difference between the standard deviations (SDs) significant?

Solution: Null hypothesis H_0: $\sigma_1 = \sigma_2$
Alternative hypothesis: H_1: $\sigma_1 \neq \sigma_2$ (two-tailed test)
Level of significance = 5%
$Z = 1.96$ at 5% level of significance
We have

$$Z = \frac{S_1 - S_2}{\sqrt{\frac{S_1^2}{2n_1} + \frac{S_2^2}{2n_2}}} = \frac{10 - 14}{\sqrt{\frac{10^2}{242} + \frac{14^2}{162}}}$$

$$= \frac{-4}{\sqrt{0.4132 + 1.209}} = \frac{-4}{\sqrt{1.623}}$$

$$= \frac{-4}{1.274} = -3.139$$

$$\therefore \quad |Z| = |-3.139| = 3.139 > 1.96$$

Since calculated value of Z is greater than the table value of Z at 5% level of significance H_0 is rejected.

Therefore the difference between two standard deviations (SDs) is significant.

Example 8.15: The mean yield of two sets of plots and their variability are as given below. Examine whether the difference in the variables in yields significant?

	Set of 40 plots	Set of 60 plots
Mean yield per plot (kg)	1258	1243
SD per plot	34	28

Solution: We have

$$\bar{x}_1 = 1258 \text{ kg}, \qquad \bar{x}_2 = 1248 \text{ kg}$$

$$S_1 = 34, \qquad S_2 = 28$$

$$n_1 = 40, \qquad n_2 = 60$$

σ_1, σ_2 are unknown.

Null hypothesis H_0: $\sigma_1 = \sigma_2$
Alternative hypothesis: H_1: $\sigma_1 \neq \sigma_2$ (two-tailed test)
Level of significance $\alpha = 5\%$

$$Z_\alpha = 1.96$$

Test statistic:

$$Z = \frac{S_1 - S_2}{\sqrt{\frac{S_1^2}{2n_1} + \frac{S_2^2}{2n_2}}} = \frac{34 - 28}{\sqrt{\frac{34^2}{80} + \frac{28^2}{120}}}$$

$$= \frac{6}{\sqrt{4.45 + 6.63}} = \frac{6}{\sqrt{11.08}} = 1.8025$$

$Z = 1.8025 < 1.96$, i.e., table value of Z at 5% level of significance. Hence we accept null hypothesis, i.e., there is significant difference between two sample mean weights.

Example 8.16: In a study of the effect of chemicals on the laborers a chemical unit, the following results were obtained on their systolic blood pressures. Examine whether differences in blood pressures are significant?

	Males		Females	
	Exposed group	**Controlled group**	**Exposed group**	**Controlled group**
Number	250	55	105	50
Mean	117.5	121.0	111	112
SD	10.50	11	9.4	9.5

Solution: For males, we have

$$\bar{x}_1 = 117.5, \quad \bar{x}_2 = 121$$
$$S_1 = 10.5, \quad S_2 = 11$$
$$n_1 = 250, \quad n_2 = 55$$

Null hypothesis H_0: Assume that there is no difference in their means.

Alternative hypothesis H_1: The difference is significant since the samples are from the same population.

$$\sigma^2 = \frac{n_1 S_1^2 + n_2 S_2^2}{n_1 + n_2} = \frac{(250)(10.5)^2 + (55)(11)^2}{250 + 55} = 112.188$$

$$\therefore \quad \sigma = 10.59$$

Test statistic:

$$Z = \frac{\bar{x}_1 - \bar{x}_2}{\sqrt{\sigma^2\left(\frac{1}{n_1} + \frac{1}{n_2}\right)}} = \frac{117.5 - 121}{10.59\sqrt{\frac{1}{250} + \frac{1}{55}}}$$

$$= \frac{-3.5}{(10.5)(0.1489)} = \frac{-3.5}{1.5772} = -2.219$$

$$\therefore \quad |Z| = 2.219 > 1.96$$

Hence the difference is significant at 5% level since $|Z| = 2.219 < 2.58$; the difference is not significant at 1% level.

For females, we have

$$\overline{x}_1 = 111, \quad \overline{x}_2 = 112$$
$$S_1 = 9.4, \quad S_2 = 9.5$$
$$n_1 = 105, \quad n_2 = 50$$

H_0: The difference is not significant.
H_1: The difference is significant.

$$\sigma^2 = \frac{n_1 S_1^2 + n_2 S_2^2}{n_1 + n_2} = \frac{(105)(9.4)^2 + (50)(9.5)^2}{105 + 50}$$

$$\sigma = \sqrt{88.9646} = 9.432$$

$$Z = \frac{\overline{x}_1 - \overline{x}_2}{\sqrt{\sigma^2\left(\frac{1}{n_1} + \frac{1}{n_2}\right)}} = \frac{111 - 112}{9.432\sqrt{\frac{1}{105} + \frac{1}{50}}}$$

$$= \frac{-1}{9.432\sqrt{\frac{155}{105 \times 50}}} = \frac{-1}{9.432(0.1778)} = -0.6172$$

$$\therefore \quad |Z| = 0.6172 < 1.96$$

Hence H_0 is accepted at 5% level of significance and at 1% level of significance.

Therefore the difference is not significant with respect to the females.

Example 8.17: An experiment is performed to determine whether the average nicotine content of one kind of cigarette exceeds that of another kind by 0.22 mg. If $n_1 = 50$ cigarettes of the first kind had an average nicotine content of $\overline{x}_1 = 2.62$ mg with a SD of $S_1 = 0.12$ mg whereas $n_2 = 40$ cigarettes of the other kind had an average nicotine content of $\overline{x}_2 = 2.40$ mg with a SD of $S_2 = 0.16$ mg. Test the null hypothesis that $\mu_1 - \mu_2 = 0.22$ against the alternative hypothesis $\mu_1 - \mu_2 \neq 0.22$

Solution: The given values are

$$\overline{x}_1 = 2.66, \quad \overline{x}_2 = 2.40$$
$$S_1 = 0.12 \text{ mg}, \quad S_2 = 0.16 \text{ mg}$$
$$n_1 = 60, \quad n_2 = 2.40$$

Null hypothesis H_0: $\mu_1 - \mu_2 = 0.22$
Alternative hypothesis H_1: $\mu_1 - \mu_2 \neq 0.22$
The test statistic is

$$Z = \frac{\overline{X}_1 - \overline{X}_1 - \delta}{\sqrt{\frac{\sigma_1^2}{n_1} + \frac{\sigma_2^2}{n_2}}}$$

Substituting S_1 for σ_1 and S_2 for σ_2 we have

$$Z = \frac{\overline{X}_1 - \overline{X}_2 - \delta}{\sqrt{\frac{S_1^2}{n_1} + \frac{S_2^2}{n_2}}} = \frac{2.66 - 2.40 - 0.22}{\sqrt{\frac{(0.12)^2}{60} + \frac{(0.16)^2}{40}}}$$

$$= \frac{0.04}{\sqrt{\frac{0.0144}{60} + \frac{0.0256}{40}}} = \frac{0.04}{\sqrt{0.00024 + 0.00064}}$$

$$= \frac{0.04}{0.0296} = 1.351$$

The corresponding probability value is 2(0.5000−0.4115) = 0.0885
Since 0.0855 exceeds 0.05, the null hypothesis cannot be rejected.

Exercise 8.1

1. Define the terms
   a. Hypothesis
   b. Null hypothesis
   c. Alternative hypothesis
2. Define
   a. Level of significance
   b. Hypothesis testing
3. Define
   a. Type I error
   b. Type II error
   c. Confidence interval
4. Explain the procedure for hypothesis testing.
5. A random sample of 50 items drawn from a particular population has a mean 30 with a SD 28. Construct a 98% confidence interval estimate of the population mean.

   Ans: 20.68, 39.32

6. For a given sample of 200 items drawn from a large population, the mean is 65 and the standard deviation (SD) is 8. Find the 95% confidence limits for the population mean?

   Ans: 65 ± 1.109

7. A sample of size 400 was drawn and the sample mean was found to be 99. Test whether this sample could have come from a normal population with mean 100 and standard deviation (SD) 8 at 5% level of significance?

   Ans: H_0 rejected

8. A stenographer claims that she can type at the rate of 120 words per minute. Can we reject her claim on the basis of 100 trials in which she demonstrates a mean of 116 words with a standard deviation (SD) of 15 words. Use 5% level of significance.

   Ans: H_0 rejected, i.e., the stenographer's claim is rejected

9. A sample of 400 items is taken from a normal population whose mean is 4 and whose variance is also 4. If the sample mean is 4.45 can the sample be regarded as truly a random sample?

   Ans: H_0 rejected, i.e., the sample cannot be regarded as having been drawn from the same population with mean 4

10. A sample of 900 members has a claim 3.4 cm and SD 2.61 cm. Can the sample be regarded as one drawn from a population with mean 3.25 cm. Using the level of significance as 0.05, test whether the claim is acceptable?

    Ans: The sample is drawn from a population with mean 3.25 cm

11. According to the norms established for a mechanical aptitude test, persons who are 18 years old have an average height of 73.2″ with a SD of 8.6. If "45 randomly selected persons of that age averaged 76.7" test the null hypothesis $\mu = 73.2$ against the alternative hypothesis $\mu \neq 73.2$ at 0.01 level of significance.

    Ans: H_0 rejected

12. In a study of an automobile insurance a random sample of 80 body repair costs had a mean of 473.36 and a standard deviation (SD) of Rs. 62.35. If $\bar{x}$ is used as a point estimate to the true average repair costs, with what confidence we can assert that the maximum error does not exceed Rs. 107?

    Ans: H_0 rejected

13. Random samples drawn from two places gave the following data relating to the heights of the children:

	Place A	Place B
Mean height (cm)	68.50	68.58
Standard deviation (cm)	2.5	3.0
No. of items in the sample	1200	1500

Test at 5% level that the mean height is the same for children at two places.

Ans: Null hypothesis H_0 is accepted

14. A random sample of 50 male employees are taken at the end of a year where the mean number of hours of absenteeism for the year is found to be 63 hours. A similar sample of 50 female employees has a mean of 66 hours, could these samples be drawn from a population with the same mean and SD 10 hours. State clearly the assumption made by you.

Ans: H_0 accepted. The two samples have been drawn from the same population

15. The standard deviation (SD) of the height of students of a college is 4.0 cm. Two samples are taken. The SD of 100 BCom (Honors) students is 3.5 cm and 50 BA (Economics) students is 4.5 cm. Test the significance of the difference of SD of the samples.

Ans: H_0: rejected

16. Intelligence test of two groups of boys and girls gave the following results: Girls: 84, Standard deviation (SD) $S_1 = 10$, $n_1 = 121$; Boys: 81, SD $S_2 = 12$, $n_2 = 81$.

Is the difference between means significant?

Ans: The difference between the mean is not significant

17. For two samples, the following data are given

$$n_1 = 1000, \quad \overline{x}_1 = 67.42 \quad S_1 = 2.58$$
$$n_2 = 1200, \quad \overline{x}_2 = 67.25 \quad S_2 = 2.50$$

Is the difference between standard deviations (SDs) significant?

Ans: The sample SD do not differ significantly

18. The following data are from an investigation:

	No. of cases	Mean wages	SD of the wages (Rs.)
Sample I	400	47	3.1
Sample II	900	50.4	3.3

Find out whether the two mean wages differ significantly?

Ans: The difference is very much significant

19. From the data given below, compute the standard error of the difference for the two sample means and find out if the two means significantly differ at 5% level of significance.

	No. of items	Mean	SD of the wages (in Rs.)
group I	50	181	3.0
group II	75	179	3.6

***Ans*:** The difference is significant.

20. In a study of the effect of chemicals on the laborers in a chemical unit, the following results were obtained on their systolic blood pressures. Examine whether differences in the blood pressures are significant?

	Males		Females	
	Exposed group	Controlled group	Exposed group	Controlled group
Number	250	55	103	50
Mean	117.5	121.6	111.7	112.5
SD	10.58	10.82	9.33	9.38

***Ans*:** The difference is significant for males. The difference is not significant for females

21. A sample of 400 male students is found to have a mean height of 67.47 units. Can it be reasonably regarded as a sample from large population with mean 67.39 units and 1.30 units?

***Ans*:** H_0 accepted

22. A random sample of 100 students gave a mean weight of 58 kg with a SD of 4 kg. Test the hypothesis that the mean weight in the population is 60 kg.

***Ans*:** H_0 rejected, mean weight cannot be 60 kg

23. A random sample of 200 measurements from a large population gave a mean value 50 and a SD of 9. Determine the 95% confidence interval for the mean of population.

***Ans*:** 48.8, 51.2

24. According to the norms established for a mechanical aptitude test, persons who are 18 year old have an average height of 73.2 with a standard deviation (SD) of 8.6. If four randomly selected persons of that age averaged 76.7, test the hypothesis $\mu = 73.2$ against the alternative hypothesis $\mu > 73.2$ at 0.01 level of significance.

***Ans*:** Null hypothesis is accepted

25. An ambulance service claims that it takes on the average less than 10 minutes to reach its destination during emergency calls. A sample of 36 calls has a mean of 11 minutes and the variance of 16 minutes. Test the significance at 0.05 level.

Ans: Null hypothesis is accepted

26. The means of two large samples of sizes 1000 and 2000 members are 67.5 in. and 68.0 in., respectively. Can the samples be rejected as drawn from the same population with SD of 2.5 in?

Ans: The samples are not drawn from the same normal population of standard deviation (SD) of 2.5 in

27. In a city of 250 men out of 750 were found to be smokers. Does this information support the conclusion that the majority of men in the city are smokers?

Ans: Null hypothesis is rejected

28. Given the following information relating to two planes A and B. Test whether there are any significant difference between their mean wages?

	A	B
Mean wages	47	49
SD	28	40
No. of workers	1000	1500

Ans: H_0 accepted

29. The number of accidents per day were studied for 144 days in town A and for 100 days in town B and the following information was obtained:

	Town A	Town B
Mean number of accidents	4.5	5.4
Standard deviation	1.2	1.5

Is the difference between the mean accidents of the two towns statistically significant?

Ans: $|Z| = 5$, significant

30. A normal population has a mean of 0.1 and standard deviation (SD) of 2.1. Find the probability that mean of a sample of 900 will be negative?

Ans: 0.0764

31. The mean produce of wheat of a sample of 100 fields comes to 200 kg/acre and another sample of 150 fields gives the mean of 220 kg. Assuming the standard deviation (SD) of the yield at 11 kg for the universe test if there is a significant difference between the means of the samples.

Ans: The difference is significant

32. An oceanographer wants to check whether the average depth of the ocean in a certain region is 574 fathoms as had previously had recorded. What can the oceanographer conclude at the level of significance $\alpha = 0.05$, if surroundings taken at 40 random locations in the given region yielded a mean of 59.1 fathoms with a standard deviation (SD) of 5.2 fathoms?

Ans: Null hypothesis rejected

33. In a random sample of 400 men the mean is found to be 40. In another sample of 50 men the mean is 45. Can these samples be drawn from the population whose SD is 5?

Ans: Significant

34. It was found that the average income of 36 workers is Rs. 12,000 per annum. Was it taken from a population whose mean is Rs. 11,800 per annum and SD 800?

Ans: Significant

35. Test whether the mean of the samples are significantly different, given the following data:

	Size	Mean	SD
Sample I	100	110	13
Sample II	100	112	16

36. A random sample of 400 items has an average length of 15 cm. Can this be regarded as a sample from a large population with mean 16 cm and SD 5 cm?

Ans: Significant

8.14 TEST OF SIGNIFICANCE FOR SINGLE PROPORTION

Consider a sample which is drawn from a population with a population proportion (percentage) P. If n denotes the sample and p is the sample proportion or percentage, then we have

$$\text{Standard error percentage} = \sqrt{\frac{P(100 - P)}{n}}$$

P = Population percentage

$E(p) = P,\ \text{Var}\ (p) = \dfrac{PQ}{n}$

$SE(p) = \sqrt{\dfrac{PQ}{n}}$

And $Z = \dfrac{p - E(p)}{SE(p)} = \dfrac{p - P}{\sqrt{\dfrac{PQ}{n}}} \sim N(0, 1)$

Z is called the test statistic and is used as a test for single proportion.

If

$$|Z| = \left| \frac{p - P}{\sqrt{\frac{PQ}{n}}} \right| < Z_\alpha \tag{8.4}$$

The difference of proportions ($p-P$) is not significant at the level of significance α.

From Eq. (8.4), we have

$$-Z_\alpha < \frac{p - P}{\sqrt{\frac{PQ}{n}}} < Z_\alpha$$

i.e.,

$$P - Z_\alpha\sqrt{\frac{PQ}{n}} < p < P + Z_\alpha\sqrt{\frac{PQ}{n}}$$

1. The probable limits for observed proportion of success, i.e., sample proportion are $P \pm Z_\alpha\sqrt{PQ/n}$.
2. If P is unknown, the confidence limits for the proportion in the population are $P \pm Z_\alpha\sqrt{PQ/n}$ where $Q = 1 - P$.
3. If is not given, the confidence limits for P are $P \pm 3\sqrt{PQ/n}$ and the confidence limits for P are $P \pm 3\sqrt{PQ/n}$.
4. If it is the level of significance under two-tailed test, then

$$\int_{-Z_\alpha}^{Z_\alpha} \varphi(t)\mathrm{d}t = 1 - \alpha$$

i.e.,

$$\int_{0}^{Z_\alpha} \varphi(t)\mathrm{d}t = 0.5 - \left(\frac{\alpha}{2}\right)$$

where α is the level of significance.

8.14.1 Solved Examples

Example 8.18: A random sample of 600 toys was taken from a large consignment and 80 were found to be defective. Find the percentage of defective toys in the consignment?

Solution: Here we have

$$n = 600, \quad \text{No. of toys} = 80$$

$$P = \text{Proportion of defective toys in the sample}$$

$$= \frac{80}{600} = \frac{8}{60} = 0.133$$

$$Q = 1 - 0.133 = 0.867$$

The limits for population proportion P are given by $P \pm 3\sqrt{PQ/n}$.

$$= 0.133 \pm 3\sqrt{\frac{0.133 \times 0.867}{600}}$$

$$= 0.133 \pm 0.041589$$

$$= 0.17458 \text{ and } 0.091411$$

Therefore the percentage of defective toys in the consignment lies between 17.458 and 9.1411.

Example 8.19: Out of a simple sample of 1000 individuals from the inhabitants of a country we find 36% of them have blue eyes the remaining have eyes of some other color. What can be inferred about the proportion of blue eyed individuals in the whole population?

Solution: Here $n = 1000$, $P =$ Population of individuals having blue eyes

$$= \frac{36}{100} = 0.36$$

$$Q = 1 - P = 1 - 0.36 = 0.64$$

Probable limits are $P \pm 3\sqrt{\frac{PQ}{n}}$

$$= 0.36 \pm 3\sqrt{\frac{0.36 \times 0.64}{1000}}$$

$$= 0.36 \pm 3(0.01517)$$

$$= 0.36 \pm 0.0455$$

$$= 0.4055 \text{ and } 0.3145$$

Hence the limits in percentage are 40.55% and 31.45%.

Example 8.20: In a locality of 20,000 families a sample of 800 families was selected. Of these 800 families, 260 families were found to have a monthly income Rs. 150 or less. It is desired to estimate how many out of 20,000 families have a monthly income Rs. 150 or less. Within what limits would you place your estimate?

Solution: Here we have $n = 800$, $P = \frac{260}{800} = \frac{26}{80} = 0.325$

$$Q = 1 - P = 0.675$$

Standard error of the proportion of families having monthly income of Rs. 150 or less

$$= \sqrt{\frac{PQ}{n}} = \sqrt{\frac{0.325 \times 0.675}{800}}$$

$$= 0.0165, \text{ i.e., } 1.65\%$$

Hence the required limits in percent are $32.5\% \pm (3 \times 1.65)\%$

$$= 32.5\% \pm 4.95\% = 37.45\% \text{ and } 27.55\%$$

Therefore between 5510 and 7490 families have a monthly income of Rs. 150 or less.

Example 8.21: If we can assert with 95% confidence that the maximum error is 0.5 and P is given as 0.2. Find the size of the sample?

Solution: We have $P = 0.2$, $Q = 1 - P = 1 - 0.2 = 0.8$

$$Z_{\alpha/2} = 1.96$$

Substituting in

$$E = Z_{\alpha/2}\sqrt{\frac{PQ}{n}} \text{ we get}$$

$$0.05 = 1.9\sqrt{\frac{0.2 \times 0.8}{n}}$$

or

$$n = \frac{(1.96)^2}{(0.05)^2}(0.16) = 245.86$$

i.e.,

$$n = 246 \text{ (approx.)}$$

Example 8.22: If 80 patients are treated with an antibiotic, 59 got cured. Find a 99% confidence limits to the true population of curve?

Solution: We have $n = 80$, $P = \sqrt{\frac{PQ}{n}} = 0.7375$

$$Q = 1 - P = 1 - 0.7375 = 0.2625$$

The required confidence interval is

$$= \left(0.7375 - 3\sqrt{\frac{PQ}{n}},\ 0.7375 + 3\sqrt{\frac{PQ}{n}}\right)$$

$$= \left(0.7375 - 3\sqrt{\frac{0.7375 \times 0.2625}{80}},\right.$$

$$\left. 0.7375 + 3\sqrt{\frac{0.7375 \times 0.2625}{80}}\right)$$

$$= (0.7375 - 3(0.049),\ 0.7375 + 3(0.049))$$

$$= (0.59,\ 0.88)$$

Example 8.23: In a random sample of 400 industrial accidents, it was found trial 231 were due at least partially to unsafe working conditions. Construct a 99% confidence interval for the corresponding true proportion.

Solution: We have

$$n = 400,\quad x = 231$$

$$P = \frac{x}{n} = \frac{231}{400} = 0.5775$$

$$Q = 1 - P = 1 - 0.5775 = 0.4225$$

The required confidence interval is $\left(P - 3\sqrt{\frac{PQ}{n}},\ P + 3\sqrt{\frac{PQ}{n}}\right)$

$$= \left(0.5775 - 3\sqrt{\frac{0.5775 \times 0.4225}{400}},\ 0.5775 + 3\sqrt{\frac{0.5775 \times 0.4225}{400}}\right)$$

$$= (0.5775 - 3(0.0247),\ 0.5775 + 3(0.0247))$$

$$= (0.5034,\ 0.6516)$$

Example 8.24: Among 900 people in a state, 90 were found to be chapathi eaters. Construct a 99% confidence interval for the true proportion.

Solution: Here we have $x = 90$, $n = 900$

$$P = \frac{x}{n} = \frac{90}{900} = \frac{1}{10}$$

$$Q = 1 - P = 1 - \frac{1}{10} = \frac{9}{10}$$

$$\sqrt{\frac{PQ}{n}} = \sqrt{\frac{\left(\frac{1}{10}\right)\left(\frac{9}{10}\right)}{900}} = \sqrt{\frac{1}{(100)(100)}}$$

$$= \frac{1}{100} = 0.01$$

The confidence interval is $\left(P - 3\sqrt{\frac{PQ}{n}},\ P + 3\sqrt{\frac{PQ}{n}}\right)$

$$= (0.1 - 3 \times 0.01,\ 0.1 + 3 \times 0.01)$$
$$= (0.07,\ 0.13)$$

Remarks:

The confidence limits can also be computed by using

$$\left(P - Z_{\alpha/2}\sqrt{\frac{PQ}{n}},\ P + Z_{\alpha/2}\sqrt{\frac{PQ}{n}}\right)$$

Example 8.25: A sample size of two persons selected at random from a large city show that the percentage of males in the sample is 52%. It is believed that male to the total population ratio is Test whether this belief is confirmed by the observation.

Solution: It is given that $P = 50$, $Q = 100 - 50 = 50$

$$p = 52$$

Null hypothesis: H_0: $P = 50\%$

Alternative hypothesis: H_1: $P \neq 50\%$

Level of significance $\alpha = 0.05$, $Z_\alpha = 1.96$

$$SE(p) = \sqrt{\frac{PQ}{n}} = \sqrt{\frac{P(100 - P)}{n}} = \sqrt{\frac{50(100 - 50)}{600}} = 2.04$$

Test statistic:

$$Z = \frac{p - P}{SE(p)} = \frac{52 - 50}{2.04} = 0.9803$$

Since calculated value of Z is less than the table value of Z at 5% level of significance, null hypothesis H_0 is accepted, i.e., male population is 50%.

Example 8.26: If 20 people were attacked by a disease and only 18 survived. Will you reject the hypothesis that the survival rate if attacked by this disease is 85% in favor of the hypothesis that is more at 5% level?

Solution: Number of people survived = $x = 18$

Size of the sample = $n = 20$

p = Population of the people survived = $\frac{x}{n} = \frac{18}{20} = \frac{9}{10} = 0.9$

It is given that $P = 85\% = 0.85$

$Q = 1 - P = 1 - 0.85 = 0.15$

Null hypothesis H_0: $P = 0.85$

Alternative hypothesis $H_1 = P > 0.85$ (one-tailed test)

Level of significance $\alpha = 0.05$

Table value of $Z = 1.645$

Test statistic:

$$Z = \frac{p - P}{SE(p)} = \frac{0.9 - 0.85}{\sqrt{\frac{(0.85)(0.15)}{20}}} = \frac{0.05}{0.0798} = 0.6265$$

Calculated value of Z is less than the table value of Z at 5% level of significance, null hypothesis H_0 is accepted.

Hence the proportion of the survived people is 0.85.

Example 8.27: In a random sample of 125 cola drinks, will you accept null hypothesis $P = 0.5$ against the alternative hypothesis $P > 0.5$. At 5% level of significance given $Z\alpha = 1.645$

Solution: Here we have

$$n = 125, \quad x = 68, \quad p = \frac{x}{n} = \frac{68}{125} = 0.544$$

Null hypothesis H_0: $P = 0.5$

Alternative hypothesis H_1: $P > 0.5$

Assume that, level of significance $\alpha = 0.05$, $Z_\alpha = 1.645$

The test statistic $= \frac{p - P}{SE(p)} = \frac{0.544 - 0.5}{\sqrt{\frac{(0.5)(0.5)}{125}}} = \frac{0.044}{0.045} = 0.9839$

Calculated value of Z is less than the table value of Z at 5% level of significance, null hypothesis H_0 is accepted.

Exercise 8.2

1. A dice was thrown 400 times and "6" resulted 80 times. Do the data justify the hypothesis of an unbiased dice?
   Ans: The die is unbiased
2. A dice was thrown 9000 times. Of these 3220 yielding 3 or 4. Is this consistent with the hypothesis that the die was unbiased?
   Ans: The die is unbiased
3. In a sample of 400 parts manufactured by a factory, the number of defective parts was found to be 30. The company, however, claimed that only 5% of their products are defective. Is the claim tenable?
   Ans: The claim is rejected
4. Experience has shown that 20% of a manufactured product is of the top quality. In one day's production of 400 articles only 50 are of the top quality. Test the hypothesis at 0.05 level.
   Ans: Null hypothesis is rejected
5. A manufacturer claims that only 4% of his products are defective. A random sample of 500 were taken among which 100 were defective. Test the hypothesis at 5% level.
   Ans: Null hypothesis is rejected
6. In a random sample of 160 workers exposed to a certain amount of radiation, 24 experienced some ill effects. Construct 99% confidence interval for the corresponding true percentage.
   Ans: 0.065, 0.234
7. Experiences have shown that 10% of a manufactured product is of quality what can you say about the maximum error with 95% confidence for 100 items.
   Ans: 0.0588
8. Experience has shown that 20% of a manufactured product is of the top quality. In one day's production of 400 articles only 50 are of the top quality. Show that whether the production of the day taken was not a representative sample or the hypothesis of 20% was wrong.
   Ans: Null hypothesis is rejected (20%)
9. 500 apples are taken from a large consignment and 50 are found to be bad. Estimate the percentage of bad apples in the consignment and assign the limits within which the percentage probability lies.
   Ans: 0.075, 0.125
10. In a city, 250 men out of 750 men were found to be smokers. Does this information support the conclusion that the majority of men in this city are smokers?
    Ans: Majority are smokers

11. In a sample of 500 people from a village in Rajasthan, 280 are found to be rice eaters and the rest are wheat eaters. Can we assume that both food practices are equally popular?

Ans: Both the food eaters are not equally popular

12. A coin is tossed 400 times and it turns up head 216 times. Discuss whether the coin may be unbiased one, and explain briefly the theoretical principles you would use for this purpose.

Ans: Coin is unbiased

13. A cultivator of bananas claims that only 3 out of 100 supplied by him are defective. A random sample of 700 bananas contained 45 defective bananas. Test whether the claim of cultivator is correct?

Ans: Cultivator's claim cannot be accepted

14. A wholesaler in apples claims that only 4% of the apples supplied by him are defective. A random sample of 600 apples contained 36 defective apples. Test the claim of the wholesaler.

Ans: Claim cannot be accepted

8.15 TESTING OF SIGNIFICANCE FOR DIFFERENCE OF PROPORTIONS

Suppose two samples of sizes n_1 and n_2 are drawn from two different populations. To test the significance of difference between two proportions we consider the following cases:

Case I: When the population proportions P_1 and P_2 are known:

In this case $Q_1 = 1 - P_1$ and $Q_2 = 1 - P_2$

The test statistic is

$$Z = \frac{P_1 - P_2}{\sqrt{\frac{P_1 Q_1}{n_1} + \frac{P_2 Q_2}{n_2}}}$$

Case II: When the population proportions P_1 and P_2 are not known but sample proportions p_1 and p_2 are known:

The test statistic is

$$Z = \frac{p_1 - p_2}{\sqrt{\frac{p_1 q_1}{n_1} + \frac{p_2 q_2}{n_2}}}$$

Case III: Method of pooling:

In this method, the sample proportions p_1 and p_2 are pooled into a single proportion P, by using the formula

$$P = \frac{n_1 p_1 + n_2 p_2}{n_1 + n_2}$$

So that $Q = 1 - P$

The test statistic in this case is

$$Z = \frac{p_1 - p_2}{\sqrt{PQ\left(\frac{1}{n_1} + \frac{1}{n_2}\right)}}$$

8.15.1 Solved Examples

Example 8.28: A machine puts out 16 imperfect articles in a sample of 500. After the machine is overhauled, it puts 3 imperfect articles in a batch of 100. Has the machine improved?

Solution: Here we have

$$p_1 = \frac{16}{500} = 0.032$$

$$p_2 = \frac{3}{100} = 0.03$$

$$q_1 = 1 - p_1 = 1 - 0.032 = 0.968$$

$$q_2 = 1 - p_2 = 1 - 0.03 = 0.970$$

Null hypothesis H_0: $p_1 = p_2$

Alternative hypothesis H_1: $p_1 < p_2$ (one-tailed test)

Level of significance = 0.05, value of $Z = 1.645$

$$\text{Standard error of difference} = \sqrt{\frac{p_1 q_1}{n_1} + \frac{p_2 q_2}{n_2}}$$

$$= \sqrt{\frac{(0.032)(0.968)}{500} + \frac{(0.03)(0.97)}{100}}$$

$$= 0.0187$$

The test statistic is

$$Z = \frac{p_1 - p_2}{\sqrt{\frac{p_1 q_1}{n_1} + \frac{p_2 q_2}{n_2}}} = \frac{0.032 - 0.030}{0.0187}$$

$$= 0.107$$

Since the calculated value of Z at 5% level of significance is less than the table value of Z, we accept null hypothesis, i.e., the machine has not improved after overhauling.

Example 8.29: During a country wide investigation, the incidence of TB was found to be 1%. In a college of 400 strong 5 were affected, whereas in another 1200 strong, 10 were affected. Does this indicate any significance difference?

Solution: Here we have

$$p_1 = \frac{5}{400} = 0.0125, \quad p_2 = \frac{10}{1200} = 0.008333$$

$$P = 1\% = \frac{1}{100} = 0.01, \quad Q = 0.99$$

Null hypothesis H_0: $p_1 = p_2$, i.e., there is no significance difference
Alternative hypothesis H_1: $p_1 \neq p_2$ (two-tailed test)
Level of significance = 0.05, Table value of $Z = 1.645$
The test statistic is

$$Z = \frac{p_1 - p_2}{\sqrt{PQ\left(\frac{1}{n_1} + \frac{1}{n_2}\right)}} = \frac{0.0125 - 0.008333}{\sqrt{(0.01)(0.99)\left(\frac{1}{400} + \frac{1}{1200}\right)}}$$

$$= \frac{0.0041667}{\sqrt{\frac{1}{100} \times \frac{99}{100}\left(\frac{3+1}{1200}\right)}} = \frac{0.0041667}{\sqrt{\frac{99}{10^2}}\sqrt{\frac{1}{300}}}$$

$$= \frac{0.0041667 \times 10^3}{\sqrt{99}\sqrt{\frac{1}{3}}}$$

$$= \frac{4.1667}{\sqrt{33}} = = \frac{4.1667}{5.744} = 0.7254$$

Since the calculated value of Z at 5% level of significance is less than the table value of Z, we accept null hypothesis, i.e., the difference is not significant.

Example 8.30: In a random sample of 100 persons from town A, 400 are found to be consumers of wheat. In a sample of 800 from town B, 400 are found to be consumers of wheat. Do these data reveal a significant difference between town A and town B so far as the proportion of wheat consumers is concerned?

Solution: We have

$$p_1 = \frac{400}{1000} = 0.4, \quad n_1 = 1000$$

$$p_2 = \frac{400}{800} = 0.5, \quad n_2 = 800$$

$$P = \frac{n_1 p_1 + n_2 p_2}{n_1 + n_2} = \frac{(1000)(0.4) + (800)(0.5)}{1000 + 800} = \frac{4}{9}$$

$$Q = 1 - P = 1 - \frac{4}{9} = \frac{5}{9}$$

Null hypothesis H_0: $p_1 = p_2$
Alternative hypothesis H_1: $p_1 \neq p_2$ (two-tailed test)
Level of significance = 1%
Table value of $Z = 2.58$
The test statistic is

$$Z = \frac{p_1 - p_2}{\sqrt{PQ\left(\frac{1}{n_1} + \frac{1}{n_2}\right)}} = \frac{0.4 - 0.5}{\sqrt{\frac{4}{9} \times \frac{5}{9}\left(\frac{1}{100} + \frac{1}{800}\right)}}$$

$$= \frac{-0.1}{0.024} = -4.17$$

$$\therefore \quad |Z| = 4.17 > 2.58$$

Since the difference is less than 2.58 at 1% level of significance, we reject null hypothesis.

Therefore the data reveal a significant difference between town A and town B so far as the proportion of wheat consumers is concerned.

Example 8.31: Random samples of 400 men and 600 women were asked whether they would like to have a school near their residence. 200 men and 350 women were in favor of the proposal. Test the hypothesis that the proportions of men and women in favor of the proposal are the same at 5% level of significance?

Solution: Size of the sample of men = $n_1 = 400$
Size of the sample of women = $n_2 = 600$
Proportion of men = $p_1 = \frac{200}{400} = 0.5$

Proportion of women $= p_2 = \dfrac{350}{600} = 0.5833$

Null hypothesis H_0: $p_1 = p_2$

That is, there is no significant difference between the attitude of men and women as far as the proposal is concerned.

Alternative hypothesis H_1: $p_1 \neq p_2$ (two-tailed test)

Level of significance = 0.05 (5%)

Table value of $Z = 1.96$

The test statistic is

$$Z = \frac{p_1 - p_2}{\sqrt{PQ\left(\frac{1}{n_1} + \frac{1}{n_2}\right)}}$$

We have

$$P = \frac{n_1 p_1 + n_2 p_2}{n_1 + n_2} = \frac{(400)(0.5) + (600)(0.5833)}{400 + 600}$$

$$= 0.5499$$

$$Q = 1 - P = 1 - 0.5499 = 0.4501$$

$$\sqrt{PQ\left(\frac{1}{n_1} + \frac{1}{n_2}\right)} = \sqrt{(0.5499)(0.4501)\left(\frac{1}{400} + \frac{1}{600}\right)}$$

$$= 0.03207$$

$$|Z| = \left|\frac{p_1 - p_2}{\sqrt{PQ\left(\frac{1}{n_1} + \frac{1}{n_2}\right)}}\right| = \left|\frac{0.5 - 0.5833}{0.03207}\right| = 2.597$$

Since $|Z| = 2597 > 1.96$, the null hypothesis is rejected significantly in their attitude. Men and women differ.

Example 8.32: In two large populations there are 30% and 25% respectively of fair haired people. Is this difference likely to be hidden in samples of 1200 and 900 respectively from the two populations?

Solution: Here we have

P_1: Population of fair haired people in the first population = 30% = 0.3

P_2: Population of fair haired people in another population = 25% = 0.25

$Q_1 = 1 - P_1 = 1 - 0.3 = 0.7$

$Q_2 = 1 - P_2 = 1 - 0.25 = 0.75$

Null hypothesis H_0: $P_1 = P_2$

That is, the difference in population proportions is likely to be hidden in sampling.

H_1: $P_1 \neq P_2$

Level of significance = 0.05

$$Z = \frac{P_1 - P_2}{\sqrt{\frac{P_1Q_1}{n_1} + \frac{P_2Q_2}{n_2}}} = \frac{0.30 - 0.25}{\sqrt{\frac{(0.3)(0.7)}{1200} + \frac{(0.25)(0.75)}{900}}}$$

$$= \frac{0.05}{\frac{1}{10}\sqrt{\frac{0.21}{12} + \frac{0.1875}{9}}}$$

$$= \frac{0.05}{\sqrt{0.0175 + 0.0208}}$$

$$= 2.5537 > 1.58$$

At 5% level of significance calculated value of Z is greater than the table value. Hence H_0 is rejected. Since $|Z| < 2.58$, H_0 is accepted at 1% level.

Exercise 8.3

1. One thousand articles from a factory are examined and found to be 3% defective. Fifteen hundred similar articles from a second factory are found to be only 2% defective. Can it be reasonably be concluded that the product of the first factory is inferior to the second?

   Ans: The difference is not significant

2. Out of a sample of 1000 persons, 800 persons were found to be coffee drinkers. Subsequently, the excise duty on coffee was increased. After the increase in excise duty of coffee seeds, 800 people were found to take coffee out of a sample of 1200. Test whether there is any significant increase in the consumption of coffee after the increase in excise duty?

   Ans: H_0 is rejected. There is significant difference in the consumption of coffee duty increase in excise duty

3. A machine produced 20 defective units in a sample of 400. After over-hauling the machine it produced 10 defective items in a batch of 300. Has the machine improved due to over hauling?

   Ans: H_0 is accepted

4. In a random sample of 400 students in the university teaching departments, it was found that 300 students failed in the examination. In another random sample of 500 students of the affiliated college, the number of students failed in the examination was found to be 300. Find out whether the population of failures in the university, teaching department is significantly greater than the proportion of failures in the university teaching department and affiliated college taken together?

   Ans: H_0 rejected

5. In village A, out of random sample of 100 persons 100 were found to be vegetarians, while in village B out of 1500 persons 180 were found to be vegetarians. Do you find a significant difference in the food habits of the people of the two villages?

   Ans: There is no significant difference in the food habits of the people of the two villages

6. In a simple random sample of 600 men taken from a big city, 400 are found to be smokers. In another simple random sample of 900 men taken from another city are smokers. Do the data indicate that there is a significant difference in the habit of smoking in the cities?

   Ans: There is a significant difference in the habit of smoking in men of the two cities

7. In a sample of 600 students, 400 were found to use pens. In another college from a sample of 900 students, 450 were found to use pens. Test whether the two colleges are significantly different with regard to using pens?

   Ans: H_0 is rejected

8. In town A, there were 956 births of which 52.5% were male births, while in towns A and B combined. This population in a total of 1406 births was 0.496. Is there any significant difference in the proportion of male births in the two towns.

   Ans: There is a significant difference in male births in the two towns A and B

9. A machine puts out 6 defective articles in a sample of 500. After the machine is overhauled, it puts out 3 defective articles in a sample of 100. Has the machine been improved

***Ans*:** Improved

$$\text{Hint: } P_1 = \frac{16}{500} = 0.032, \quad P_2 = \frac{3}{100} = 0.03, \quad \mu_{P_1 - P_2} = 0$$

$$\sigma_{P_1 - P_2} = \sqrt{\frac{(0.032)(0.968)}{500} + \frac{(0.03)(0.97)}{100}} = 0.0187$$

And

$$\sigma_{P_1 - P_2} = 0.0187, \quad Z = \frac{P_1 - P_2 - \mu_{P_1 - P_2}}{\sigma_{P_1 - P_2}} = \frac{0.032 - 0.03}{0.0187} = 0.107 < 1.96$$

Calculated value of Z is less than the table value of Z at 5% level of significance. It is likely that the machine has not been improved.

CHAPTER 9

Chi-Square Distribution

9.1 INTRODUCTION

In this chapter we introduce chi-square distribution, the measure of which enables us to find out the degree of discrepancy between the observer and expected frequencies and then to determine whether the discrepancy between the observed and expected frequencies is due to error of sampling or due to chance.

The chi-square is denoted by the symbol χ^2. It was discovered by Helmert in 1875 and was rediscovered by Karl Pearson in 1900. Chi-square is always positive. The value of χ^2 lies between 0 and ∞.

Since χ^2 is not derived from the observations in a population, it is not a parameter.

Chi-square test is not a parametric test.

The χ^2 distribution is used for testing the goodness of fit. It is used for finding association and relation between attributes. It is also used to test the homogeneity of independent estimates of population. Chi-square is computed on the basis of frequencies in a sample and the value of χ^2 so obtained is a statistic.

Chi-Square Test (χ^2):

χ^2 test is defined as

$$\chi^2 = \sum \frac{(O_i - E_i)^2}{E_i}$$

where O_i = Observed frequency; E_i = Expected frequency.

9.2 CONTINGENCY TABLE

A classification table containing r rows and c columns figuring observed frequencies is called contingency table. A 2×2 contingency table is of the form:

a	b
c	d

Statistical Techniques for Transportation Engineering.
DOI: http://dx.doi.org/10.1016/B978-0-12-811555-8.00009-X

If the data is divided into m classes $A_1, A_2, \ldots, A_m$ according to an attribute A and n classes $B_1, B_2, \ldots, B_n$ according to another attribute B, the $m \times n$ contingency table can be formed as follows:

Attributes	B_1	B_2	$\ldots$	B_j	$\ldots$	B_n
A_1	A_{11}	A_{12}	$\ldots$	A_{1j}	$\ldots$	O_{1n}
A_2	A_{21}	A_{22}	$\ldots$	A_{2j}	$\ldots$	O_{2n}
$\vdots$	$\vdots$	$\vdots$		$\vdots$		$\vdots$
A_i	A_{i1}	A_{i2}	$\ldots$	A_{ij}	$\ldots$	O_{in}
$\vdots$	$\vdots$	$\vdots$		$\vdots$		$\vdots$
A_m	A_{m1}	A_{m2}	$\ldots$	A_{mj}	$\ldots$	O_{mn}

where O_{ij} denote the ith row jth column frequency of the cell belonging to both the classes A_i and B_j.

9.3 CALCULATION OF EXPECTED FREQUENCIES

Consider the 2×2 contingency table (Table 9.1).

Table 9.1 2×2 Contingency table

		Total
a	b	$a+b$
c	d	$c+d$
Total $a+c$	$b+d$	$N=a+b+c+d$

where a, b, c, and d are observed frequencies.

The expected frequencies corresponding the cell frequencies a, b, c, and d can be calculated and expressed in the form of the given Table 9.2.

This method can be extended to compute the expected frequencies of a $m \times n$ contingency table.

Table 9.2 Expected frequency table for 2×2 contingency table

$\frac{(a+c)(a+b)}{N}$	$\frac{(b+d)(a+b)}{N}$
$\frac{(a+c)(c+d)}{N}$	$\frac{(b+d)(c+d)}{N}$

9.4 CHI-SQUARE DISTRIBUTION

Let samples of size n be drawn from a normal population with standard deviation σ and if for each sample we calculate: a sampling distribution of χ^2 can be obtained. It is given by

$$y = f(\chi^2) = y_0 e^{-\frac{1}{2}\chi^2} \cdot (\chi^2)^{\frac{1}{2}(\nu-2)}$$
$$= y_0 e^{-\frac{1}{2}\chi^2} \cdot (\chi)^{(\nu-2)}$$

where $\nu = n - 1$ is the number of degrees of freedom and y_0 is a constant depending on such that the total area under the curve one. The χ^2 distribution corresponding to various values of are shown in the Fig. 9.1.

Maximum value of y:

We have

$$y = y_0 e^{-1/2\chi^2} (\chi^2)^{1/2(\nu-2)} \tag{9.1}$$

Differentiating Eq. (9.1) with respect to χ^2 we get

$$\frac{dy}{d\chi^2} = y_0 \left[\frac{1}{2}(\nu-2)(\chi^2)^{1/2(\nu-4)} e^{-1/2\chi^2} - \frac{1}{2} e^{-\chi^2/2} (\chi^2)^{1/2(\nu-2)} \right]$$

Taking $\frac{dy}{dx} = 0$ for maximum value of y we get

$\chi^2 = \nu - 2$ for $\nu \geq 2$ (since χ^2 cannot be negative)

The constant y_0 is related in such a way that the area $\chi^2 = 0$ to ∞ is unity.

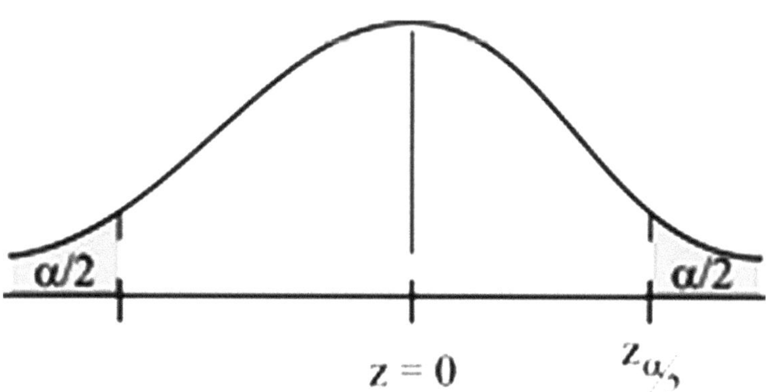

Figure 9.1 Level of significance

In this case

$$y_0 = \frac{1}{2^{\nu/2}\Gamma(\nu/2)}$$

9.4.1 Characteristic Function of χ^2 Distribution

We have

$$\begin{aligned}
\varphi(\chi^2(t)) &= E\left(e^{it\chi^2}\right) \\
&= \int_0^\infty e^{it\chi^2} f(\chi^2) d\chi^2 \\
&= \frac{1}{2^{\nu/2}\Gamma(\nu/2)} \int_0^\infty e^{-1/2(1-2it)\chi^2} (\chi^2)^{1/2(n-1)} d\chi^2 \\
&= (1-2it)^{-n/2}
\end{aligned}$$

9.5 MEAN AND VARIANCE OF CHI-SQUARE

The moment generating function of χ^2 with respect to the origin is

$$\begin{aligned}
M_0(t) &= \int_0^\infty e^{t\chi^2} \frac{1}{2^{\nu/2}\Gamma(\nu/2)} e^{-\frac{1}{2}(\chi^2)} \left(\frac{1}{2}\chi^2\right)^{\frac{1}{2}(\nu-1)} d(\chi^2) \\
&= \frac{1}{2^{\frac{\nu}{2}}\Gamma(\nu/2)} \int_0^\infty \exp\left\{-\left(\frac{1-2t}{2}\right)\chi^2\right\}(\chi^2)^{\frac{\nu}{2}-1} d(\chi^2) \\
&= \frac{1}{2^{\frac{\nu}{2}}\Gamma(\nu/2)} \cdot \frac{\Gamma\left(\frac{1}{2}\nu\right)}{\left\{\frac{1}{2}(1-2t)\right\}^{\frac{\nu}{2}}} \\
&= (1-2t)^{-\frac{\nu}{2}}, (|2t| < 1) \\
&= 1 + \frac{\nu}{2}(2t) + \frac{1}{2}\frac{\nu\left(\frac{1}{2}\nu+1\right)}{2!}(2t)^2 + \cdots \\
&\quad + \frac{\frac{1}{2}\nu\left(\frac{1}{2}\nu+1\right)\cdots\left(\frac{1}{2}\nu+r-1\right)}{r!}(2t)^r + \cdots
\end{aligned}$$

$$\mu'_r = \text{coefficient}\,\frac{t^r}{r!}$$

$$= 2^r \frac{1}{2}\nu\left(\frac{1}{2}\nu + 1\right)\left(\frac{1}{2}\nu + 2\right)\cdots\left(\frac{1}{2}\nu + r - 1\right)$$

$$= \nu(\nu + 2)(\nu + 4)\cdots(\nu + 2r - 2)$$

Hence the mean value of χ^2 is ν and variance is $\nu\ (\nu + 2) - \nu^2 = 2\nu$. Thus $\chi^2 - n/\sqrt{2n}$ is a standard variate.

9.6 ADDITIVE PROPERTY OF INDEPENDENT CHI-SQUARE VARIATE

Theorem 1:
If χ_1^2 and χ_2^2 are independent χ^2 variates with n_1 and n_2 degrees of freedom, then $\chi_1^2 + \chi_2^2$ is a χ^2 variate with $n_1 + n_2$ degrees of freedom.

Proof:
The moment generating function of

$$(\chi_1^2 + \chi_2^2) = (\text{m.g.f of } \chi_1^2)\ (\text{m.g.f of } \chi_2^2)$$

$$= (1-2t)^{-\frac{n_1}{2}}(1-2t)^{-\frac{n_2}{2}}$$

$$= (1-2t)^{-\frac{(n_1+n_2)}{2}}$$

which is the moment generating function of χ^2 variate with $n_1 + n_2$ degrees of freedom.

Hence proved.

Theorem 2:
The chi-square distribution tends to normal distribution as n tends to infinity.

Proof:
We have

$$M(t) = e^{-\frac{nt}{\sqrt{2n}}} M_o\left(\frac{t}{\sqrt{2n}}\right)$$

$$= e^{-\frac{nt}{\sqrt{2n}}}\left(1 - \frac{2t}{\sqrt{2n}}\right)^{\frac{-n}{2}}$$

Therefore

$$\log_e M(t) = \frac{-nt}{\sqrt{2n}} - \frac{n}{2}\log_e\left(1 - \frac{2t}{\sqrt{2n}}\right)$$

or

$$\log_e M(t) = \frac{-nt}{\sqrt{2n}} + \frac{n}{2}\left(\frac{2t}{\sqrt{2n}} + \frac{1}{2}\left(\frac{2t}{\sqrt{2n}}\right)^2 + \cdots\right)$$

or

$$\log_e M(t) = \frac{1}{2}t^2 + O\left(\frac{1}{n}\right)$$

Further, as

$$n \to \infty \ \log_e M(t) \to \frac{1}{2}t^2$$

or

$$M(t) \to e^{\frac{1}{2}t^2}$$

Hence the result.

Example 9.1: Show that the value of χ^2 for the contingency table.

Classes	A	A[1]	Total
B	a	b	$a+b$
B[1]	c	d	$c+d$
Total	$a+c$	$b+d$	$N=a+b+c+d$

a, b, c, and d are cell frequencies

Calculated from the independent frequencies is

$$\chi^2 = \frac{N(ad-bc)^2}{(a+b)(c+d)(b+d)(a+c)}, \quad (N = a+b+c+d)$$

Solution: Since the marginal total is fixed the probability for a member belonging to class A is $\frac{a+c}{N}$ and is a constant.

Further the attributes being independent, the probability for it to belong to both classes A and B is $\left(\frac{a+b}{N}\right)\left(\frac{a+c}{N}\right)$.

The expected frequency of the class A, denoted by $E(A)$ is given by

$$E(A) = N\cdot\left(\frac{a+b}{N}\right)\left(\frac{a+c}{N}\right) = \frac{(a+b)(a+c)}{N}$$

The expected frequency in each cell = product of column total and row total/whole total

Similarly we get

$$E(B) = \frac{(a+b)(b+d)}{N}$$

$$E(C) = \frac{(c+d)(a+c)}{N}$$

$$E(D) = \frac{(c+d)(b+d)}{N}$$

By the definition of we have

$$\chi^2 = \frac{[a-E(A)]^2}{E(A)} + \frac{[b-E(B)]^2}{E(B)} + \frac{[c-E(C)]^2}{E(C)} + \frac{[d-E(D)]^2}{E(D)}$$

$$-\frac{\left[a-\frac{(a+b)(a+c)}{a+b+c+d}\right]^2}{\frac{(a+b)(a+c)}{a+b+c+d}} + \frac{\left[b-\frac{(a+b)(b+d)}{a+b+c+d}\right]^2}{\frac{(a+b)(b+d)}{a+b+c+d}}$$

$$+\frac{\left[c-\frac{(c+d)(a+c)}{a+b+c+d}\right]^2}{\frac{(c+d)(a+c)}{a+b+c+d}} + \frac{\left[d-\frac{(c+d)(b+d)}{a+b+c+d}\right]^2}{\frac{(c+d)(b+d)}{a+b+c+d}}$$

$$= \frac{(ad-bc)^2}{a+b+c+d}\left[\frac{1}{(a+b)(a+c)} + \frac{1}{(a+b)(b+d)} + \frac{1}{(c+d)(a+c)} + \frac{1}{(c+d)(b+d)}\right]$$

$$= \frac{(ad-bc)^2}{a+b+c+d}\left[\frac{a+b+c+d}{(a+b)(a+c)(b+d)} + \frac{a+b+c+d}{(a+c)(c+d)(b+d)}\right]$$

$$= (ad-bc)^2\left[\frac{1}{(a+b)(a+c)(b+d)} + \frac{1}{(a+c)(c+d)(b+d)}\right]$$

$$= (ad-bc)^2\left[\frac{c+d+a+b}{(a+b)(a+c)(b+d)(c+d)}\right]$$

$$= \left[\frac{(c+d+a+b)(ad-bc)^2}{(a+b)(a+c)(b+d)(c+d)}\right]$$

$$= \frac{N(ad-bc)^2}{(a+b)(c+d)(b+d)(a+c)}$$

Example 9.2: Show that for 2 degrees of freedom the probability p of a value of χ^2 greater than χ_0^2 is $e^{-\frac{1}{2}\chi_0^2}$ and hence that $\chi_0^2 = 2\log_e\left(\frac{1}{p}\right)$. Deduce the value of when $p = 0.05$.

Solution: We know that

$$p(\chi^2) = \frac{1}{2^{\nu-2/2}\Gamma(\nu/2)}\int_{\chi_0}^{\infty} e^{-\frac{\chi^2}{2}}\chi^{\nu-1}d(\chi^2)$$

when $\nu = 2$

$$p(\chi^2) = \frac{1}{2^0\Gamma(2/2)}\int_{\chi_0}^{\infty} e^{-\frac{\chi^2}{2}}\chi d(\chi^2)$$

$$= \int_{\chi_0}^{\infty} e^{-\frac{\chi^2}{2}}\chi d(\chi^2)$$

$$= \left[-e^{-\frac{\chi^2}{2}}\right]_{\chi 0}^{\infty} = e^{-\frac{1}{2}\chi_0^2}$$

$$\therefore \quad e^{-\frac{1}{2}\chi_0^2} = p \text{ or } e^{\frac{1}{2}\chi_0^2} = \left(\frac{1}{p}\right)$$

or

$$\frac{\chi_0^2}{2} = \log_e\left(\frac{1}{p}\right)$$

or

$$\chi_0^2 = 2\log_e\left(\frac{1}{p}\right).$$

When $p = 0.05$ we have

$$\chi_0^2 = 2\log_e\left(\frac{1}{0.05}\right)$$

$$\chi_0^2 = 2\log_e\left(\frac{1}{\frac{1}{20}}\right)$$

$$\chi_0^2 = 2\log_e(20) = 3.012$$

Example 9.3: Prove that for a χ^2 distribution with n degrees of freedom

$$\mu_{r+1} = 2r(\mu_r + n\mu_{r-1}), \quad (\nu > 0)$$

Solution: The moment generating function of χ^2 distribution with n degrees of freedom about the mean is

$$\mu(t) = e^{-nt}(1-2t)^{n/2}$$

Applying logarithms on both sides and differentiating we get

$$\frac{\mu'(t)}{\mu(t)} = -n + \frac{n}{2}\left(\frac{2}{1-2t}\right)$$

i.e.,

$$(1-2t)\mu'(t) = 2nt\,\mu(t)$$

Using Leibnitz theorem and differentiating with respect to t we get

$$(1-2t)\,\mu^{r+1}(t) + 2(-2)\mu^r(t) = 2nt\,\mu^r(t) + 2nr\,\mu^{r-1}(t)$$

Pitting $t = 0$ and using the relation

$$\mu_r = \left[\frac{\mathrm{d}}{\mathrm{d}t}\mu^r(t)\right]_{t=0} = \mu^r(\nu)$$

We get

$$\mu_{r+1} = 2r(\mu_r + n\mu_{r-1})$$

9.7 DEGREES OF FREEDOM

It is the number of values in a set, which may be arbitrarily assigned. The number of independent variables usually called the degrees of freedom. It is denoted by the symbol v. If these are normalized variables subject to k linear constraints then the degrees of freedom is $v = n - k$.

If the data is given in the form of a row containing n observations, then the degrees of freedom is $n - 1$. Similarly if the data is given in the form of a column, the degrees of freedom is $n - 1$. Where n is the number of observation in the column. If there are r rows and c columns. The degree of freedom is given by $(r - 1)\ (c - 1)$.

9.8 CONDITIONS FOR USING CHI-SQUARE TEST

1. The total number observations used in this test must be large (i.e., $n \geq 50$).
2. Each of the observations making up sample for χ^2 test should be independent of each other.
3. The test is wholly dependent on the degree of freedom.
4. The frequencies used in χ^2 test should be absolute and not relative in terms.
5. The expected frequency of any item or cell should not be less than 5. If it is less than 5, then the frequencies from the adjacent items or cells should be pooled together in order to make it 5 or more than 5 (preferably not less than 10).
6. The observations collected for χ^2 should be based on the method of random sampling.

   The constraints on the cell frequencies if any should be linear.

9.9 USES OF CHI-SQUARE TEST

Chi-square test an important test. If we require only the degrees of freedom for using this test. It is a powerful test. It is used

1. as a test of goodness of fit.
2. as a test of independence of attributes, and
3. as a test of homogeneity.

9.9.1 Chi-Square Test as a Test of Goodness of Fit

Chi-square test is applied as a test of goodness of fit to determine whether the actual (i.e., observed) the expected (i.e., theoretical) frequencies. The degrees of freedom in this case are $\nu = n - 1$ where n is the number of observations.

Example 9.4: In 90 throws of a die, face 1 turned 9 times, face 2 or 3 turned 24 times, face 4 or 5 turned 36 times, and face 6 turned 18 times. Test at 10% level, if the die is honest, it being given that χ^2 for 3 df = 6.25 at 10% level of significance.

Solution:

H_0: The die is honest.

H_1: The die is not honest.

Expected frequencies for each face $= 90 \times \frac{1}{6} = 15$

Level of significance = 10% (i.e., 0.01)

Degrees of freedom $= 4 - 1 = 3$

Chi-square value for 3 df at 10% level of significance = 6.25

We have the following table:

Face turned	Observed O_i	Expected E_i	$(O_i - E_i)$	$\frac{(O_i - E_i)^2}{E_i}$
1	9	15	−6	2.4
2 or 3	27	30	−3	0.3
4 or 5	36	30	6	1.2
6	18	15	3	0.6
Total	90	90		4.5

$$\chi^2 = \sum \frac{(O_i - E_i)^2}{E_i} = 4.5$$

Since the calculated value of χ^2 is less than the table value at 10% level of significance and for 3 df, we accept the null hypothesis and conclude that the die is honest.

Example 9.5: Find whether or not the following observed distribution of phenotypes in a sample of 384 *Drosophila* flies have a significance goodness of fit with proposed median 9:3:3:1 distribution (test at 5% level of significance)?

Phenotypes	AB	Ab	aB	ab	Total
Number of files	232	76	58	18	384

Solution: Degrees of freedom = 4 − 1 = 3
Null hypothesis H_0: 9:3:3:1
i.e., Variation is not significant.
Alternative hypothesis H_1**:** 1:1:1:1
Level of significance = $\alpha = 0.05$
Table value $\chi^2 = 7.82$ for 3 df

Observed O_i	Expected E_i	$(O_i - E_i)$	$(O_i - E_i)^2$	$\frac{(O_i - E_i)^2}{E_i}$
232	$\frac{384}{16} \times 9 = 216$	16	256	1.185
76	$\frac{384}{16} \times 3 = 72$	4	16	0.223
58	$\frac{384}{16} \times 3 = 72$	−14	196	2.723
18	$\frac{384}{16} \times 124$	−6	36	2.25
Total = 384				6.381

The calculated value of χ^2 for 3 df at 5% level of significance is less than the table value. Hence we accept H_0, i.e., variation is not significant.

Example 9.6: Among 64 offspring's of a certain cross between guinea pigs 32 were red, 10 were black, and 22 were white. According to the genetic model these numbers should be in the ratio 9:3:4. Are the data consistent with the model at 5% level?

Solution:

H_0: Data are consistent with the model.

H_1: Data are not consistent with the model.

Level of significance = 5%

Degrees of freedom = $n - 1 = 3 - 1 = 2$

Table value of χ^2 for 2 df at 5% level = 5.991

Expected frequencies are in the ratio 9:3:4

i.e.,

$$64\left(\frac{9}{16}\right) = 36, \quad 64\left(\frac{3}{16}\right) = 12, \quad 64\left(\frac{4}{16}\right) = 16$$

Observed O_i	**Expected E_i**	**$(O_i - E_i)^2$**	**$\frac{(O_i - E_i)^2}{E_i}$**
32	36	16	0.4444
10	12	4	0.3333
22	16	36	2.25
Total	64	56	3.0277

$$\chi^2 = \sum \frac{(O_i - E_i)^2}{E_i} = 3.0277$$

The calculated value of χ^2 for 2 df at 5% level of significance is less than the table value. Hence we accept H_0, i.e., data are consistent with the model.

Example 9.7: A sample analysis of examination results of 500 students was made. It was found that 230 students had failed. one hundred sixty had secured third class, 80 were placed in second class, and 30 got first class. Do these figures commensurate with the general examination result, which is in the ratio 4:3:2:1 for various categories respectively?

Solution: Null Hypothesis H_0: The observed results commensurate with general examination result

Alternative Hypothesis H_1: It is not true that the observed results commensurate with general examination result.

Level of significance = 5%

Degrees of freedom = $n - 1 = 4 - 1 = 3$

The total frequency = $N = 500$

Table value χ^2 for 3 df at 5% level of significance = 7.81 for 3 df

Dividing 500 in the ratio 4:3:2:1

We get 200, 150, 100, and 50

Therefore the expected frequencies are 200, 150, 100, and 50 corresponding to the observed frequencies 230, 160, 80, and 30.

Class/division	**Observed O_i**	**Expected E_i**	**$(O_i - E_i)$**	**$\frac{(O_i - E_i)^2}{E_i}$**
Failed	230	200	30	4.500
Third	160	150	10	0.6666
Second	80	100	−20	4.0000
First	30	50	−20	8.00

$$\chi^2 = \sum \frac{(O_i - E_i)^2}{E_i} = 17.1666$$

Since the calculated value χ^2 is greater than the table value of χ^2—the null hypothesis is rejected.

Example 9.8: The table below gives the number of aircraft accidents that occurred during the various days of the week. Test whether the accidents are uniformly distributed over the week?

Days	Mon	Tue	Wed	Thu	Fri	Sat
No. of accidents	14	18	12	11	15	14

Solution: Null Hypothesis H_0: The accidents are uniformly distributed over the week.

Alternative Hypothesis H_1: The accidents are not uniformly distributed over the week.

Total number of accidents = $14 + 18 + 12 + 11 + 15 + 14 = 84$

Level of significance = 5%

Degrees of freedom = $6 - 1 = 5$

Table value of $\chi^2 = 11.07$

The expected frequency of each day accidents is = $\frac{84}{6} = 14$

Day	Observed O_i	Expected E_i	$(O_i - E_i)$	$(O_i - E_i)^2$	$\frac{(O_i - E_i)^2}{E_i}$
Mon	14	14	0	0	0
Tue	18	14	4	16	1.1428
Wed	12	14	−2	4	0.2857
Thu	11	14	−3	9	0.6428
Fri	15	14	1	1	0.0714
Sat	14	14	0	0	0.000
Total	84	84			2.1427

Since $2.1427 < 11.07$, the calculated value of χ^2 at 5% level, for 5 df is less than the table value of χ^2.

The null hypothesis is accepted. That is the accidents are uniformly distributed over the week.

Example 9.9: Four coins are tossed 160 times and the following results were obtained:

Number of heads	0	1	2	3	4
Observed frequencies	17	52	54	31	6

Under the assumption that coins are balanced, find the expected frequencies of getting 0, 1, 2, 3, or 4 heads and test the goodness of fit.

Solution:

Null Hypothesis H_0: The coins are balanced.

Alternative hypothesis H_1: The coins are not balanced.

Level of significance = 5%

Degrees of freedom = 5 − 1 = 4

Table value of $\chi^2 = 9.488$

We have $N = 160$, $p = \frac{1}{2}$, $q = \frac{1}{2}$

The expected frequencies of 0, 1, 2, 3, or 4 successes are

$$160 \times {}^4C_0\left(\frac{1}{2}\right)^0\left(\frac{1}{2}\right)^4 = 10 \quad 160 \times {}^4C_1\left(\frac{1}{2}\right)^1\left(\frac{1}{2}\right)^3 = 40$$

$$160 \times {}^4C_2\left(\frac{1}{2}\right)^2\left(\frac{1}{2}\right)^2 = 60 \quad 160 \times {}^4C_3\left(\frac{1}{2}\right)^3\left(\frac{1}{2}\right)^1 = 40$$

$$160 \times {}^4C_4\left(\frac{1}{2}\right)^4\left(\frac{1}{2}\right)^0 = 10$$

No. of heads	Observed O_i	Expected E_i	$(O_i - E_i)$	$(O_i - E_i)^2$	$\frac{(O_i - E_i)^2}{E_i}$
0	17	10	7	49	4.900
1	52	40	12	144	3.600
2	54	60	−6	36	0.600
3	31	40	−9	81	2.025
4	6	10	−4	16	1.600
Total	160	160			12.725

$$\chi^2 = \frac{\sum (O_i - E_i)^2}{E_i} = 12.725$$

The calculated value of χ^2 is greater than the table value of χ^2 at 5% level of significance and 4 df. Therefore null hypothesis is rejected. The coins are not balanced. Hence the fit is poor.

Example 9.10: Fit a Poisson distribution to the following data and test the goodness of fit:

x	0	1	2	3	4	5	6
F	275	72	30	7	5	2	1

Solution: Null Hypothesis H_0: The Poisson fit is good to the given data.

Alternative Hypothesis H_1: The Poisson fit is not a good fit to the given data.

Level of significance = 5%

$$\text{Mean of the distribution} = \frac{\sum f_i x_i}{\sum f_i}$$

$$= \frac{0 + 72 + 60 + 21 + 20 + 10 + 6}{275 + 72 + 30 + 7 + 5 + 2 + 1}$$

$$= \frac{189}{392} = 0.482$$

The frequencies of 0, 1, 2, 3, 4, 5, and 6 successes by using recurrence formula for Poisson distribution are the expected frequencies.

Therefore the expected frequencies are

$$N(0) = 392\, e^{-0.482} = 242.1$$

$$N(1) = 392\, e^{-0.482}(0.482) = 116.7$$

$$N(2) = 116.7\left(\frac{0.482}{2}\right) = 28.12$$

$$N(3) = 28.12\left(\frac{0.482}{3}\right) = 4.52$$

$$N(4) = 4.52\left(\frac{0.482}{4}\right) = 0.54$$

$$N(5) = 0.54\left(\frac{0.482}{5}\right) = 0.052$$

$$N(6) = 0.052\left(\frac{0.482}{6}\right) = 0.004$$

The frequency table is:

x	0	1	2	3	4	5	6
Observed frequency	275	72	30	7	5	2	1
Expected frequency	242.1	116.7	28.1	4.5	0.5	0.1	0

Since the last four frequencies are small, we regroup the last four frequencies and obtain the following table:

Calculation of χ^2

Observed O_i	**Expected E_i**	**$(O_i - E_i)$**	**$(O_i - E_i)^2$**	**$\frac{(O_i - E_i)^2}{E_i}$**
275	242.1	32.9	1082.41	4.4709
72	116.7	44.7	1998.09	17.1216
30	28.1	1.9	3.61	0.1285
15	5.1	9.9	98.01	19.217
Total				40.938

$$\chi^2 = \sum \frac{(O_i - E_i)^2}{E_i} = 40.938$$

Degrees of freedom for the Poisson fit $= n - 2 = 4 - 2 = 2$

Table value of χ^2 at 5% for 2 df $= 5.991$

The calculated value of χ^2 is 40.938 which is greater than the table value.

We reject null hypothesis and conclude that the Poisson fit is not a good fit to the given data.

9.9.2 Test for Independence of Attributes

The χ^2 test can also be applied to test the association between the attributes such as honesty, smoking, drinking, etc. when the sample data is presented in the form of contingency table with any number of rows and columns.

Example 9.11: The following table gives classification of 150 workers according to sex and nature of work. Test whether the nature of work is independent of the sex of the work?

	Stable	Unstable	Total
Males	60	30	90
Females	15	45	60
Total	75	75	150

Solution: Null Hypothesis H_0: The nature of the work is independent of the sex of the worker.

Alternative Hypothesis H_1: The nature of the work is not independent of the sex of the worker.

Degrees of freedom $= (r-1)\ (c-1) = (2-1)\ (2-1) = 1$

Level of significance = 5%

Table value of $\chi^2 = 3.84$

Expected frequencies are given in the following table:

	Stable	Unstable
Males	$\frac{75 \times 90}{150} = 45$	$\frac{75 \times 90}{150} = 45$
Females	$\frac{75 \times 90}{150} = 30$	$\frac{75 \times 90}{150} = 30$

Calculation of χ^2

Observed O_i	Expected E_i	$(O_i - E_i)$	$(O_i - E_i)^2$	$\frac{(O_i - E_i)^2}{E_i}$
60	45	15	225	5.00
15	30	−15	225	7.50
30	45	−15	225	5.00
45	30	15	225	7.50
Total				25

$$\chi^2 = \sum \frac{(O_i - E_i)^2}{E_i} = 25$$

The calculated value of χ^2 is greater than the table value at 5% and df. Hence we reject the null hypothesis.

We conclude that the nature of work is independent of the sex of the worker.

Example 9.12: In a certain sample of 2000 families, 1400 families are consumers of tea, out of 1800 Hindu families 1236 families consume tea. Use χ^2 test and state whether there is any significant difference between consumption of tea among Hindu and nonhindu families?

Solution: From the given data, the 2 × 2 contingency table that can be formed is given below.

	Hindu	Nonhindu	Total
Consuming tea	1236	164	1400
Not consuming tea	564	36	600
Total	1800	200	200

Null Hypothesis H_0: The attributes are independent, i.e., there is no significant difference between the communities as far as consuming of tea is concerned.

Alternative Hypothesis H_1: The attributes are not independent.

Level of significance: 5%

df = 1, Table value of χ^2 at 5% for 1 df = 3.841

The expected frequencies corresponding to the given observed frequencies are given in the table below:

$$\frac{1800 \times 1400}{2000} = 1264 \qquad \frac{200 \times 1400}{2000} = 140$$

$$\frac{1800 \times 600}{2000} = 540 \qquad \frac{200 \times 600}{2000} = 60$$

Computation of χ^2

Observed O_i	Expected E_i	$(O_i - E_i)$	$(O_i - E_i)^2$	$\frac{(O_i - E_i)^2}{E_i}$
1236	1260	−24	576	0.457
564	540	24	576	1.068
164	140	24	576	4.114
36	60	−24	576	9.600
Total				15.239

$$\chi^2 = \sum \frac{(O_i - E_i)^2}{E_i} = 15.239$$

The calculated value of χ^2 is higher than the table value of χ^2 at 5% and 1 df. Therefore the null hypothesis is rejected. The two communities differ significantly as far as consumption of tea is concerned.

Example 9.13: A tobacco company claims that there is no relationship between smoking and lung ailments. To investigate the claims, a random sample of 300 males in the age group of 40 and 50 years is given a medical test. The observed sample results are tabulated below:

	Lung ailment	Nonlung ailment	Total
Smokers	75	105	180
Nonsmokers	25	95	120
Total	100	200	300

On the basis of this information, can it be concluded that smoking and long ailments are independent (given $\chi^2_{0.05} = 3.841$ for 1 df).

Solution: Null hypothesis H_0: The smoking and lung ailments are not associated.

Alternative Hypothesis H_1: The smoking and lung ailments are associated.

Level of significance = 5%

Table value of χ^2 for 1 df = 3.841

The expected frequencies are:

$$\frac{100 \times 180}{300} = 60 \qquad \frac{200 \times 180}{300} = 120$$

$$\frac{100 \times 120}{300} = 40 \qquad \frac{200 \times 120}{300} = 80$$

Computation of χ^2

Observed O_i	Expected E_i	$(O_i - E_i)$	$(O_i - E_i)^2$	$\frac{(O_i - E_i)^2}{E_i}$
75	60	15	225	3.750
25	40	−15	225	5.625
105	120	−15	225	1.875
95	80	15	225	2.813
Total				14.063

$$\chi^2 = \sum \frac{(O_i - E_i)^2}{E_i} = 14.063$$

For 1 df at 5% level of significance the calculated value of χ^2 is more than the table value. Hence we reject the null hypothesis.

Therefore smoking and lung ailments are not independent.

Example 9.14: Given the following contingency table for hair color and eye color, find the value of χ^2? Is there good association between the two?

		Hair color			Total
		Fair	Brown	Black	
Eye color	Blue	15	5	20	40
	Gray	20	10	20	50
	Brown	25	15	20	60
Total		60	30	60	150

Solution: Null Hypothesis H_0: The two attributes hair color and eye color are independent.

Alternative Hypothesis H_1: The two attributes hair color and eye color are not independent.

Level of significance = 9.488 for 4 df at 5% level of significance.

Table of expected frequencies are:

$$\frac{60 \times 40}{150} = 16 \quad \frac{30 \times 40}{150} = 8 \quad \frac{60 \times 40}{150} = 16$$

$$\frac{60 \times 50}{150} = 20 \quad \frac{30 \times 50}{150} = 10 \quad \frac{60 \times 50}{150} = 20$$

$$\frac{60 \times 60}{150} = 24 \quad \frac{30 \times 60}{150} = 12 \quad \frac{60 \times 60}{150} = 24$$

Computation of χ^2

Observed O_i	Expected E_i	$(O_i - E_i)$	$(O_i - E_i)^2$	$\frac{(O_i - E_i)^2}{E_i}$
15	16	−1	1	0.0625
5	8	−3	9	1.125
20	16	4	16	1
20	20	0	0	0
10	10	0	0	0
20	20	0	0	0

(Continued)

Observed O_i	Expected E_i	$(O_i - E_i)$	$(O_i - E_i)^2$	$\frac{(O_i - E_i)^2}{E_i}$
25	24	1	1	0.042
15	12	3	9	0.75
20	24	−4	16	0.666
Total				3.6458

$$\chi^2 = \frac{\sum (O_i - E_i)^2}{E_i} = 14.063$$

Since the calculated value of χ^2 for $(3-1)(3-1) = 2 \times 2 = 4$ df at 5% level, is less than the table value 9.488. Hence we accept null hypothesis, i.e., the hair color and eye color are independent.

Example 9.15: The following table shows the result of an experiment to investigate the effect of vaccination induced on the animals against a particular disease. Use the χ^2 test to test the hypothesis that the vaccinated and unvaccinated groups, i.e., vaccination and the disease are independent.

	Got disease	Did not get disease
Vaccinated	9	42
Not vaccinated	17	28

(Value of χ^2 for 1 df at 5% level is equal to 3.841).

Solution: We have $a = 9$, $b = 42$, $c = 17$, $d = 28$

$N = a + b + c + d = 9 + 42 + 17 + 28 = 96$

Null Hypothesis H_0: The vaccination and disease are independent.

Alternative Hypothesis H_1: The vaccination and disease are not independent.

Table value of χ^2 for 1 df at 5% level = 3.841

$$\chi^2 = \frac{N(ad - bc)^2}{(a+b)(c+d)(b+d)(a+c)} = \frac{96(9 \times 28 - 17 \times 42)^2}{(51)(26)(70)(45)}$$

$$= \frac{96(-462)^2}{4176900} = 4.906$$

The calculated value of χ^2 is more than the table value of χ^2 at 5% level and 1 df. Therefore we reject H_0 and conclude that the disease and vaccination are not independent.

9.9.2.1 Yate's Correction

In a 2 × 2 contingency table, if the cell frequency is small, Yate's correction is necessary. If the expected value or frequency in any observation less than 5 in 1 df we apply Yate's correction. We subtract 0.5 from the absolute difference between observed and expected frequencies. By making Yate's correction becomes continuous and the formula after making Yate's correction is

$$\chi^2 = \sum \frac{(|O_i - E_i| - 0.5)^2}{E_i} \quad \text{(0.5 is Yate's correction)}$$

Example 9.16: Two batches each of 12 animals are taken for test of inoculation. One batch was inoculated and the others batch was not inoculated. The frequencies of the dead and surviving animals are given below in both cases. Can the inoculation be regarded as effective against the disease?

	Dead	Survived	Total
Inoculated	2	10	12
Not inoculated	8	4	12
Total	10	14	

($\chi^2_{0.05}$ for 1 df = 3.841)

Solution: Null Hypothesis H_0: Inoculation and the disease are independent.

Alternative Hypothesis H_1: Inoculation and the disease are not independent.

$$Df = (r - 1)(c - 1) = (2 - 1)(2 - 1) = 1$$

Level of significance = 5%

Table value of χ^2 for 1 df at 5% level = 3.841

Frequencies in the cells are small (a and d are small) so, we make Yate's correction. The expected frequencies are given in the table below:

$$\frac{10 \times 12}{24} = 5 \qquad \frac{14 \times 12}{24} = 7$$

$$\frac{10 \times 12}{24} = 5 \qquad \frac{14 \times 12}{24} = 7$$

Computation of χ^2

Observed O_i	Expected E_i	$\lvert O_i - E_i \rvert$	$\lvert O_i - E_i \rvert - 0.5$	$(\lvert O_i - E_i \rvert - 0.5)^2$	$\frac{(\lvert O_i - E_i \rvert - 0.5)^2}{E_i}$
2	5	−3	2.5	6.25	1.25
10	7	3	2.5	6.25	0.89285
8	5	3	2.5	6.25	1.25
4	7	−3	2.5	6.25	0.89285
Total					4.2857

$$\chi^2 = \sum \frac{(|O_i - E_i| - 0.5)^2}{E_i} = 4.2857$$

For 1 df at 5% level of significance, the calculated value of χ^2 is greater than the table value. Hence H_0 is rejected. The inoculation and disease and independent we conclude that inoculation is effective against the disease.

9.9.3 Homogeneity Chi-Square

Chi-square test may be used to test the homogeneity of the attributes in respect of particular characteristics. It is performed to decide whether separate samples are sufficiently uniform to be added together. Chi-square test may also be used to test the population variance.

9.9.4 Chi-Square Distribution of Sample Variance

Let σ^2 denote the population variance and S^2 denote the sample variance. The sampling distribution $(n-1)\frac{S^2}{\sigma^2}$ has χ^2 distribution with $n-1$ degrees of freedom. It is very useful in making inference about the population variance σ^2 by using sample variance S^2. Also I is used in making interval estimate of the population variance which is given by

$$\frac{(n-1)S^2}{\chi^2_{\alpha/2}} \leq \alpha \leq \frac{(n-1)S}{\chi^2_{1-\alpha/2}}$$

where α is the level of significance, and χ^2_α is the value of χ^2 distribution giving an area to the right of χ^2_α.

The test may also be used as hypothesis test for the value of the population variance.

9.9.5 Testing a Hypothesis About the Variance of Normally Distributed Population—Decision Rule

Let σ_0^2 denote the hypothesized value of the population variance. Then the decision rule for accepting or rejecting H_0 is as follows:

1. Two-tailed test	**Decision rule**
$H_0{:}\sigma^2 = \sigma_0^2$	Accept H_0 if computed value of $\chi^2 > \chi^2_{\alpha/2}$ (table value)
$H_1{:}\sigma^2 \neq \sigma_0^2$	Reject H_0 if computed value of $\chi^2 > \chi^2_{\alpha/2}$ (table value)
2. One-tailed test	
$H_0{:}\sigma^2 \geq \sigma_0^2$	Accept H_0 if computed value of $\chi^2 <$ table value of χ^2_α
$H_1{:}\sigma^2 < \sigma_0^2$	Reject H_0 if computed value of $\chi^2 >$ table value of χ^2_α

9.9.5.1 Solved Examples

Example 9.17: According to the census report of 2004, the numbers of women to every 1000 men in the following 5 states/union territories are as follows:

In Andaman 846, in Delhi 821, in Chandigarh 777, in Daman and Diu 710, in Dadar and Nagar Haveli 812. By an appropriate statistical method determine whether 1:1 sex ratio among human population can be attributed to 5 regions ($\chi^2_{0.05}$ for 4 df = 7.88).

Solution: Null hypothesis H_0: Sex ratio is 1:1 among human population of 5 states, Alternative hypothesis H_1: Sex ratio is not 1:1 among the human population of 5 states. Table value of χ^2 for 4 df at 5% level = 7.88

Computation of χ^2 for 5 states

	Observed O_i	Expected E_i	$\lvert O_i - E_i \rvert$	$\frac{(\lvert O_i - E_i \rvert)^2}{E_i}$	χ^2	df
Andaman	Men = 1000	923	77	6.42	12.84	1
	Women = 846	923	77	6.42		
Delhi	Men = 1000	910.5	89.5	8.79	17.58	1
	Women = 821	910.5	89.5	8.79		
Chandigarh	Men = 1000	888.5	111.5	13.99	27.98	1
	Women = 777	888.5	111.5	19.99		
Damn and Diu	Men = 1000	855	145	24.59	49.18	1
	Women = 710	855	145	24.59		
Dadar and Nagar Haveli	Men = 1000	906	94	9.75	19.5	1
	Women = 812	906	94	9.75		
Men	5000	Total			127.08	5
Women	3966					

Chi-square for summed data

Observed O_i	Expected E_i	$\lvert O_i - E_i \rvert$	$\frac{(\lvert O_i - E_i \rvert)^2}{E_i}$	df
Men = 500	4483	517	59.62	$2 - 1 = 1$
Women = 3966	4483	517	59.62	

	Chi-square	df
Total	127.08	5
Summed	119.2	1
Homogeneity	7.88	4

The calculated value of χ^2 at 5% level of significance for 4 df is less than the table value of χ^2. We accept H_0.

Example 9.18: Heights in cm of 10 students are given below:

61, 65, 67, 66, 68, 70, 64, 65, 68, 71

Can we say that variance of the distribution heights of all students from which the above sample of 10 students was drawn is equal to 50?

Solution: We have

$$\text{Mean of the sample} = \frac{61 + 65 + 62 + 66 + 68 + 70 + 64 + 65 + 68 + 71}{10}$$

$$= \frac{660}{10} = 66$$

Null Hypothesis H_0: $\sigma^2 = \sigma_0^2 = 50$

Alternative Hypothesis H_1: $\sigma^2 \neq \sigma_0^2$, i.e., $\sigma^2 \neq 50$ (Two-tailed test)

Level of significance: 5% ($\alpha = 0.05$) $\chi^2_{1-\alpha/2} = \chi^2_{0.975} = 2.70$

Computation of χ^2

x_i	$x_i - \bar{x}$	$(x_i - \bar{x})^2$
61	−5	25
65	−1	1
62	−4	16
66	0	0
68	2	4
70	4	16
64	−2	4
65	−1	1
68	2	4
71	5	25

Test statistic is

$$\chi^2 = (n-1)\frac{S^2}{\sigma_0^2}$$

$$= \frac{(n-1)}{\sigma_0^2} \cdot \frac{\sum (x_i - \overline{x})^2}{n-1}$$

$$= \frac{\sum (x_i - \overline{x})^2}{\sigma_0^2} = \frac{96}{50} = 1.92$$

Since calculated value of χ^2 for 9 df at 5% level of significance is less than the value of $\chi^2_{1-\alpha/2}$ at 5% level, we reject null hypothesis and conclude that population variance is not 50.

Exercise 9.1

1. The following figures show the distribution of digits in numbers chosen at random from a telephone directory:

Digits	0	1	2	3	4	5	6	7	8	9
Frequency	1026	1107	997	996	1075	993	1107	972	964	853

Test whether the digits may be taken to occur equally frequently in directory ($\chi^2_{0.05} = 16.919$ for 9 df)?

(Hint: Expected frequency of each observation $= 10,000/10 = 1000$)

Ans: Null hypothesis is rejected

2. A die is thrown 264 times with the following results show that the die is biased:

No. appeared on the die	1	2	3	4	5	6
Frequency	40	32	28	58	54	60

(Given χ^2 for 5 df at 5% level = 11.07)

3. On the basis of information given below about the treatment of 200 patients suffering from a disease, state whether the new treatment is comparatively superior to the conventional treatment?

	Favorable	Not favorable	Total
New treatment	60	30	90
Conventional treatment	40	70	110

($\chi^2_{0.05} = 3.841$ for 1 df)

***Ans*:** Null hypothesis is rejected. The new treatment is superior to conventional statement

4. Two hundred digit were chosen at random from a set of tables the frequencies of the digits were

Digit	0	1	2	3	4	5	6	7	8	9
Frequency	18	19	23	21	16	25	22	20	21	15

Use χ^2 test to assess the correctness of the hypothesis that the digits were distributed in equal numbers in the tables from which these were chosen (For df at 5% level of significance χ^2 value is 16.919).

5. In a sample of owls, it is found that red male are 35, red female are 70, gray male are 50, gray female are 45. Coloration is due to the plumage. Is the coloration independent of sex of the sample?

***Ans*:** The coloration of the individuals are not independent of sex

6. The following table given the frequencies of occurrence of the digits 0, 1, 2, ..., 9, in the last place in the four-figure logarithm of numbers. Examine if there is any peculiarity?

Digit	0	1	2	3	4	5	6	7	8	9
Frequency	6	16	15	10	12	12	3	2	9	5

(χ^2 Value for 5 df at 5% level of significance is 11.07)

***Ans*:** H_0 accepted. There is peculiarity

7. A die was thrown 498 times. Denoting x to be the number appearing on the top face of it, the observed frequency of x is given below:

X	1	2	3	4	5	6
t	69	78	85	82	86	98

What opinion you would form for the accuracy of the die (χ^2 value for 5 df at 5% level of significance is 11.07)?

***Ans*:** Die is unbiased

8. Among 64 offsprings of a certain cross between guinea pigs 34 were red, 10 were black, and 30 were white. According to the genetic model these numbers should be in the ratio 9:3:4. Are the data consistent with the model at 5% level?

***Ans*:** The data are consistent with the model

9. In an investigation into the health and nutrition of the two groups of children of different social states, the following results are obtained:

Health	Social status		Total
	Poor	Rich	
Below normal	130	20	150
Normal	102	108	210
Above normal	24	96	120

Discuss the relation between the health and their social status.

Ans: Health and social status are associated, i.e., H_0 is rejected

10. The following table gives a classification of a sample of 160 plants of their leaf color and flatness:

	Flat leaves	Curled leaves	Total
White flower	99	36	135
Red flower	20	5	25
Total	119	41	160

Test whether the flower color is independent of the flatness of leaf?

Ans: H_0 is rejected, flower color is independent of the flatness of leaf

11. From the following information, state whether the two attributes viz., condition of house and condition of child are independent:

Condition of the child	Condition of the house		Total
	Clean	Dirty	
Clean	69	51	120
Fairly clean	81	20	101
Dirty	35	44	79
Total	185	115	300

($\chi 2$ at 5% level for 2 df = 5.991)

Ans: H_0 rejected. There is an association between the condition of the child and condition of the house

12. A certain drug was administered to 500 people out of a total of 800 included in the sample to test its efficiency against typhoid. The results are given below:

	Typhoid	No typhoid	Total
Drug	200	300	500
No drug	280	20	300
Total	480	320	800

On the basis of the data can we say that the drug is effective in preventing typhoid?

Ans: The drug is effective

13. In an experiment on the immunization of goats from anthrax the following results were obtained. Derive your inference on the vaccine.

	Died of anthrax	Survived	Total
Inoculated with vaccine	2	10	12
Not inoculated	6	6	12
Total	8	16	

Ans: H_0 is accepted. The vaccine is ineffective in controlling the disease

14. Fifty students were selected at random from 500 students enrolled in a computer program were classified according to age and grade points giving the following data:

	20 and under	Age in years 21–30	Age above 30
	3	5	2
5.1 to 7.5	8	7	5
7.6 to 10	4	Grade parts up to 5.0	8

Test at 5% level of significance the hypothesis that the age and grade points are independent (χ^2 at 5% level for 2 df = 5.991).

Ans: H_0 accepted. Grade points and age are independent of each other

15. The following data related to the sales in a time of a trade depression of a certain commodity demand:

District where sales	District not hit by depression	District hit by depression	Total
Satisfactory	350	80	330
Not satisfactory	140	30	170
Total	390	110	500

Do these data suggest that the sales are significantly affected by depression ($\chi^2_{0.05} = 3.841$ for 1 df)?

Ans: H_0 is accepted. The sales are not significantly affected by depression

16. A survey of 200 families having three children selected at random gave the following results:

Male births	0	1	2	3
No. of families	40	58	62	40

Test the hypothesis that male and female births are equally likely at 5 % level of significance ($\chi^2_{0.05} = 7.82$ for 3 df)?

17. Two researchers adopted different sampling techniques. While investigating same group of students to find the number of students falling into different intelligence level. The results are as follows:

Researchers	Below average	Average	Above average	Genius	Total
X	86	60	44	10	200
Y	40	33	25	2	100
Total	126	93	69	12	300

Would you say that the sampling techniques adopted by the two researchers are significantly different ($\chi^2_{0.05}$ for 2 df and 3 df are 5.991 and 7.82, respectively)?

18. In the following data find whether there is any significant liking in the habit of soft designs among categories of employees?

Soft drinks	Clerks	Teachers	Officers
Pepsi	10	25	65
Thumbs-up	15	30	65
Fanta	50	60	30

(χ^2 for 4 df at 5% level of significance = 9.4888)

Ans: H_0 rejected. Habit of drinking soft drinks depends on the category

19. A firm manufacturing rivets wants to limit variations to their length as much as possible. The lengths (in cm) of 10 rivets manufactured by a new process are

21.5	1.99	2.05	2.12	2.17
2.01	1.98	2.03	2.25	1.92

Examine whether the new process can be considered superior to the old if the population has standard deviation 0.145 cm ($\chi^2_{0.05} = 16.919$ for 9 df)?

Ans: H_0 accepted. The new process cannot be considered superior to the old process

20. Four coins were tossed 160 times and the following results were obtained:

No. of heads	0	1	2	3	4
Observed frequencies	17	52	54	31	6

Under the assumption that wins are balanced, find the expected frequencies of getting 0, 1, 2, 3, or 4 heads and test the goodness of fit ($\chi^2_{0.05}$ value is 9.488)?

Ans: The fit is poor

21. One of 8000 graduates in a city, 800 are males; out of 1600 graduate employees 120 are females. Use χ^2 test to determine if any destruction is made in appointment on the basis of sex ($\chi^2_{0.05}$ for 1 df = 3.841).

Ans: H_0 rejected. There is distribution

22. A survey of 200 families having three children selected at random gave the following results:

Male birth	0	1	2	3
No. of families	40	58	62	40

Test the hypothesis that male and female births are equally likely at 5% level of significance ($\chi^2_{0.05}$ value is 7.82).

Ans: H_0 rejected

23. In the accounting department of bank 100 accounts are selected at random and examined for errors. The following results have been obtained:

No. of errors	0	1	2	3	4	5	6
No. of accounts	36	40	19	2	0	2	1

Does this information verify that errors are distributed according poison probability ($\chi^2_{0.05} = 7.815$ for 3 df)?

Ans: H_0 accepted

CHAPTER 10

Test of Significance—Small Samples

10.1 INTRODUCTION

In this chapter we discuss tests of significance for small samples. The important tests for small samples are

1. Chi-square test
2. t-test
3. F-test.

χ^2 Distribution was already introduced in the previous chapter. We now introduce t and F distributions in this chapter.

t-Distribution

When population standard deviation (SD) is not known and size of the sample is less than or equal to 30, we use t-test. The parameter t was introduced by W.S. Gosset in the year 1908. But t-distribution was established by Fisher in the year 1920. t-Distribution is also known as students t-distribution. Let $x_1, x_2, \ldots, x_n$ be the members of a random sample drawn from a normal population with mean μ and variance σ^2. We define t-test statistic as

$$t = \frac{\overline{x} - \mu}{\frac{S}{\sqrt{n}}}$$

where $\overline{x}$ is mean of the sample given by $\sum x_i/n$.

If we take $(n-1)S^2 = \sum_{i=1}^{n}(x_i - \overline{x})^2 = nS^2$, then we get

$$\frac{t^2}{\nu} = \frac{t^2}{n-1} = \frac{(\overline{x}-\mu)^2 n}{(n-1)S^2} = \frac{(\overline{x}-\mu)^2/(\sigma^2/n)}{nS^2/\sigma^2}(\nu = n-1 \text{ is degrees of freedom})$$

Since x_i, $i = 1, 2, 3, \ldots, n$ is a random sample from the normal population with mean μ and variance σ^2, $\sigma^2 \dfrac{\overline{x}-\mu}{\sigma/\sqrt{n}} \sim N(0, 1)$.

Statistical Techniques for Transportation Engineering.
DOI: http://dx.doi.org/10.1016/B978-0-12-811555-8.00010-6

$(\overline{x}-\mu)^2 n/\sigma^2$, is the square of a standard normal variate and it is distributed as chi-square variate, with one degrees of freedom. nS^2/σ^2 is also distributed as a χ^2-variate with $n-1$ degrees of freedom. Hence t^2/ν is the ratio of two independent χ^2-variates distributed with 1 and ν degrees of freedom, respectively. Therefore t^2/ν is a $\beta_2\left(\frac{1}{2},\frac{\nu}{2}\right)$ variate. For different samples, it was shown by Fisher that its distribution is

$$y=\frac{y_0}{\left(1+\frac{t^2}{\nu}\right)^{\frac{\nu+1}{2}}},-\infty<x<\infty \tag{10.1}$$

The constant y_0 is chosen in such a way that

$$\int_{-\infty}^{\infty} y\mathrm{d}t=1$$

i.e., area of the curve given by Eq. (10.1) is unity
i.e.,

$$\int_{-\infty}^{\infty}\frac{y_0}{\left(1+\frac{t^2}{\nu}\right)^{\frac{\nu+1}{2}}}\mathrm{d}t=1$$

i.e.,

$$y_0\int_0^{\infty}\frac{\left(\frac{t^2}{\nu}\right)^{-1/2}}{\left(1+\frac{t^2}{\nu}\right)^{\frac{1}{2}\left(\frac{\nu+1}{2}\right)}}\mathrm{d}\left(\frac{t^2}{\nu}\right)=1$$

or

$$y_0\nu^{1/2}\beta\left(\frac{1}{2},\frac{\nu}{2}\right)=1$$

or

$$y_0=\frac{1}{\sqrt{\nu}\beta\left(\frac{1}{2},\frac{\nu}{2}\right)}$$

$$y=f(t)=\frac{1}{\sqrt{\nu}\beta\left(\frac{1}{2},\frac{\nu}{2}\right)\left(1+\frac{t^2}{\nu}\right)^{\frac{\nu+1}{2}}} \tag{10.2}$$

Eq. (10.2) is known as t-distribution.

It is called the equation of t-probability curve.

The probability that we get a value of t between the limits t_1 and t_2 is given by

$$P = \int_{t_1}^{t_2} \frac{y_0}{\left(1+\frac{t^2}{\nu}\right)^{\frac{\nu+1}{2}}} \mathrm{d}t$$

where

$$y_0 = \frac{1}{\sqrt{\nu}\beta\left(\frac{1}{2},\frac{\nu}{2}\right)}$$

10.2 MOMENTS ABOUT MEAN

The origin is mean for a t-distribution. All the moment of odd order about the origin vanish. The moment of even order about the origin is

$$\mu'_{2r} = 2\int_0^{\infty} \frac{t^{2r}}{\sqrt{\nu}\beta\left(\frac{\nu}{2},\frac{1}{2}\right)} \cdot \frac{\mathrm{d}t}{\left(1+\frac{t^2}{\nu}\right)^{\frac{\nu+1}{2}}}$$

$$= 2\int_0^{\infty} \frac{t^{2r}\left(\frac{t^2}{\nu}\right)^{\frac{-1}{2}}}{\beta\left(\frac{\nu}{2},\frac{1}{2}\right)} \cdot \frac{\mathrm{d}\left(\frac{t^2}{\nu}\right)}{\left(1+\frac{t^2}{\nu}\right)^{\frac{\nu+1}{2}}}$$

$$= \frac{\nu^2}{\beta\left(\frac{\nu}{2},\frac{1}{2}\right)} \int_0^{\infty} \frac{\left(\frac{t^2}{\nu}\right)^{r+\frac{1}{2}-1}}{\left(1+\frac{t^2}{\nu}\right)^{\frac{\nu+1}{2}}} \mathrm{d}\left(\frac{t^2}{\nu}\right)$$

Substituting,

$$1+\frac{t^2}{\nu} = y,$$

we get

$$\frac{t^2}{\nu} = \frac{1-y}{y}$$

and

$$\mu'_{2r} = \frac{\nu}{\beta\left(\frac{\nu}{2}\cdot\frac{1}{2}\right)}\beta\left(r+\frac{1}{2},\frac{\nu}{2}-r\right), \quad r<\frac{\nu}{2}$$

Therefore

$$\mu_{2r} = \frac{(2r-1)(2r-3)\ldots3.1}{(\nu-2)(\nu-4)\ldots(\nu-2r)}\cdot\nu^r, \quad n<\frac{\nu}{2}$$

In particular

$$\beta_1 = \frac{\mu_3^2}{\mu_2^3} = 0$$

$$\beta_2 = \frac{\mu_4}{\mu_3} = \frac{3(\nu-2)}{(\nu-4)}$$

$$= \frac{3\left(1-\frac{2}{\nu}\right)}{\left(1-\frac{4}{\nu}\right)}$$

As $\nu\rightarrow\infty$, we get

$$\beta_2 = \underset{\nu\rightarrow\infty}{\mathrm{Lt}}\frac{3\left(1-\frac{2}{\nu}\right)}{\left(1-\frac{4}{\nu}\right)} = 3$$

10.3 PROPERTIES OF PROBABILITY CURVE

1. Equation of t-probability contains only even powers of t, therefore the curve is symmetrical about the line $t=0$.
2. t-Distribution extends to infinity on either side and is asymptotic to t-axis at each end.
3. t-Distribution has a greater spread than the normal distribution and its graph is similar to that of normal distribution.
4. Sampling distribution of t is independent of the population parameters μ and σ^2, but it depends only on the size of the sample.

5. The curve has a maximum ordinate at $t=0$, the mean and mode coincide at $t=0$.
6. As $\nu \to \infty$, $y \to e^{-\frac{1}{2}t\nu}$ hence t is distributed normally.

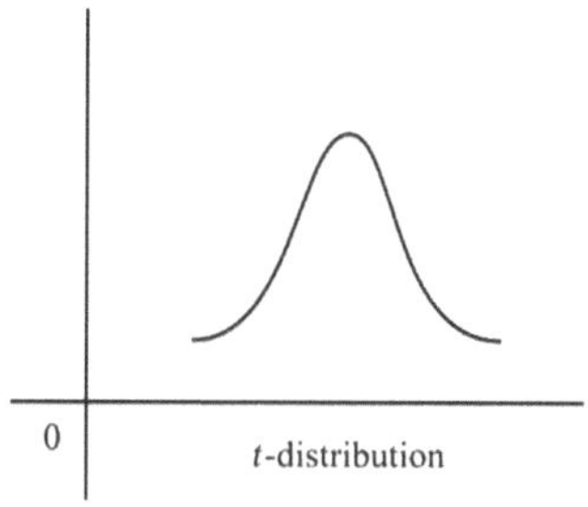

10.4 ASSUMPTIONS FOR *t*-TEST

The t-test is applied under the following assumptions:

1. The samples are drawn from normal population and are random.
2. The population SD may not be known.
3. For testing the equality of two population means, the population variances are regarded as equal.

10.5 USES OF *t*-Distribution

The t-distribution is used to test the significance of:

1. a mean for small sample when the population variance is not known.
2. the difference between two means for small samples.

The values of t for different degrees of freedom are given at the end of this book.

10.6 INTERVAL ESTIMATE OF POPULATION MEAN

Let $\overline{x}$ denote the sample mean and n denote the size of the sample. Then the interval estimate of the population mean μ is given by $\overline{x} \pm t_{0.05}(S/\sqrt{n})$

$$\overline{x} \pm t_{\alpha}\frac{S}{\sqrt{n-1}} \quad \left(\text{where}\frac{S^2}{n} = \frac{s^2}{n-1}\right)$$

$$S = \sqrt{\frac{\sum (x_i - \overline{x})^2}{n}} \text{ and } \frac{S}{n-1} \text{ is the standard error mean.}$$

10.7 TYPES OF *t*-TEST

There are two types of *t*-tests:

1. Unpaired *t*-test.
2. Paired *t*-test.

10.8 SIGNIFICANT VALUES OF *t*

The significant value of t at level of significance (Fig. 10.1) and degrees of freedom ν, for two-tailed test is given by

$$P\{|t| > t_\nu(\alpha)\} = \alpha$$

$$P\{|t| \leq t_\nu(\alpha)\} = 1 - \alpha$$

The significant value of t at level of significance for a one-tailed test can be obtained from those of two-tailed test by referring to the values Z_α. The significant values are given α in the table known as the t-table.

If the calculated value of t exceeds the table value at 5% level of significance then the null hypothesis is rejected at 5%.

Similarly for 1% level of significance, we can reject the null hypothesis at 1% if the calculated value of t exceeds the table value.

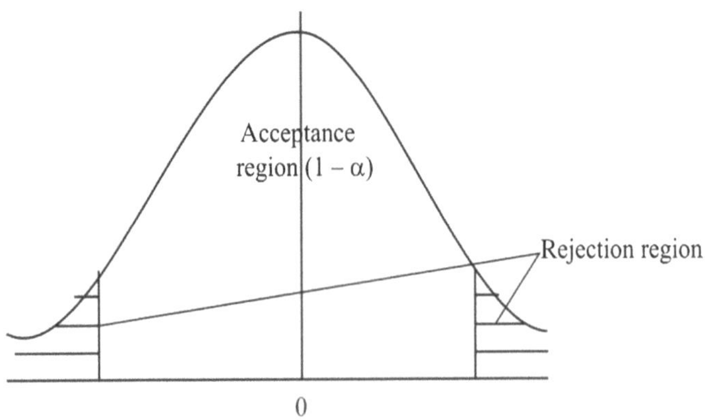

Figure 10.1 Level of significance.

10.9 TEST OF SIGNIFICANCE OF A SINGLE MEAN

To test the significance of a mean of a small sample, the test statistic is

$$t = \frac{\overline{x} - \mu}{\left(\frac{S}{\sqrt{n-1}}\right)}$$

where $\overline{x} = \sum x_i/n$; $S/(n-1)$ is the standard error of mean; $n =$ size of the sample ($n \leq 30$); $S =$ standard deviation $= \sqrt{\frac{\sum (x_i - \overline{x})^2}{n}}$.

If $\overline{x}$ is a fraction, we compute S by using the formulae $\overline{x} = A + \left(\sum d/n\right)$, $d = x - A$

$$S^2 = \frac{1}{n-1}\left[\sum d^2 - \frac{\sum (d)^2}{n}\right]$$

where A is the assumed mean.

If the calculated value of $|t|$ is less than the table value of t_α ($\alpha =$ level of significance), we accept null hypothesis for the degrees of freedom ν.

If the calculated value of $|t|$ is greater than the table value t_α of at the level of significance α, for the degrees of freedom ν, we reject the null hypothesis and accept the alternative hypothesis.

10.9.1 Solved Examples

Example 10.1: A random sample of size 7 from a normal population gave a mean of 977.51 and a SD of 4.42. Find 95% confidence interval for the population mean?

Solution: Here we have $n=7$, $\overline{x}=977.51$, $df=n-1=6$ and $S=4.42$

$$SE = \frac{S}{\sqrt{n-1}} = \frac{4.42}{\sqrt{7-1}} = \frac{4.42}{\sqrt{6}} = \frac{4.42}{2.4494} = 1.8045$$

95% confidence limits for μ are

$$\overline{x} - t_{0.05}\frac{S}{\sqrt{n-1}} < \mu < \overline{x} + t_{0.05}\frac{S}{\sqrt{n-1}}$$

i.e., $977.51-(2.447)\ (1.8045) < \mu < 977.51 + (2.447)\ (1.8045)$

i.e., $977.51-4.4156 < \mu < 977.51 + 4.4156$

i.e., $973.094 < \mu < 981.9256$

The required confidence interval is (973.094, 981.9256).

Example 10.2: A sample of 10 camshafts intended for use in gasoline engine has an average eccentricity of 0.044 in. The data may be treated as a random sample from a normal population. Determine a 95% confidence interval for actual mean eccentricity of the camshaft.

Solution: It is given that

n = size of the sample = 10

S = Standard deviation = 0.044 in.

$\overline{x} = 1.02$ in.

Degrees of freedom $\nu = n-1 = 10-1 = 9$

$$\frac{S}{\sqrt{n}} = \frac{0.044}{\sqrt{10}} = \frac{0.044}{3.1622} = 0.0139$$

The confidence interval is

$$\overline{x} \pm t_{\alpha/2}\frac{S}{\sqrt{n}} = 1.02 \pm 2.262 \times 0.0139$$

$$= 1.02 \pm 0.0314$$

i.e.,

$$(1.02 - 0.0314,\ 1.02 + 0.0314) = (0.9886, 1.0514)$$

Example 10.3: Find the students t for following variable values in a sample of eight.

$-4, -2, -2, 0, 2, 2, 3, 3$ taking the mean of the universe to be zero.

Solution: Table for computing mean $\overline{x}$ and S

x	$x-\overline{x}$	$(x-\overline{x})^2$
−4	−4.25	18.0625
−2	−2.25	5.0625
−2	−2.25	5.0625
0	−0.25	0.0625
2	1.75	3.0625
2	1.75	3.0625
3	2.75	7.5625
3	2.75	7.5625
$\sum x_i = 2$		49.5080

$$\overline{x} = \frac{\sum x_i}{n} = \frac{2}{8} = \frac{1}{4} = 0.25$$

$$S = \sqrt{\frac{\sum (x_i - \overline{x})^2}{n}} = \sqrt{\frac{49.5000}{8-1}} = \sqrt{7.071428}$$

$$= 2.6592$$

Test statistic:

$$t = \frac{\overline{x} - \mu}{\left(\frac{S}{\sqrt{n}}\right)} = \frac{0.25 - 0}{2.6592} \times \sqrt{8}$$

$$= \frac{(0.25)(2.8284)}{2.6592}$$

$$= 0.2659$$

Example 10.4: A certain stimulus administered to each of 12 patients resulted in the following increases of blood pressures:

5, 2, 8, −1, 3, 0, 6, −2, 1, 5, 0, 4.

Can it be calculated that the stimulus will be in general accomplished by an increase in blood pressure given that for 11 degrees of freedom the value of is 2.201?

Solution: For 11 df = 2.201 (given)

$$\text{Mean of the sample } \overline{x} = \frac{\sum x_i}{n} = \frac{5+2+8-1+3+0+6-2+1+5+0+4}{12}$$

$$= \frac{31}{12} = 2.581$$

Calculation of S

x	$x - \overline{x}$	$(x - \overline{x})^2$
5	2.42	5.8564
2	− 0.58	0.3364
8	5.42	29.3764
−1	− 3.58	12.8164
3	0.42	0.1764
0	− 2.58	6.6564
6	3.42	11.6964
−2	− 4.58	20.9764
1	− 1.58	2.4964
5	2.42	5.8564
0	− 2.58	6.6564
4	1.42	2.0614
		104.9268

$$S^2 = \frac{\sum (x_i - \overline{x})^2}{n} = \frac{104.9268}{12 - 1} = 9.5389$$

$$\therefore \qquad S = \sqrt{9.5389} = 3.0885$$

Assuming that mean of the universe is zero, we get

$$t = \frac{\overline{x} - \mu}{\left(\frac{S}{\sqrt{n}}\right)} = \frac{2.58 - 0}{3.0885} \times \sqrt{12}$$

$$= \frac{(2.58)(3.464)}{3.0885} = 2.8936$$

Since the calculated value of t for 11 df at 5% level of significance is greater than the table value of t, we reject null hypothesis and conclude that stimulus will be in general accompanied by an increase in blood pressure.

Example 10.5: A random blood sample to test fasting sugar for 10 taxi drivers gave the following data (in mg/dL):

70, 120, 110, 101, 88, 83, 95, 107, 100, 98.

Do this support, the assumption of population mean of 100 mg/dL. Find a reasonable range in which the most of the mean fasting sugar test of 10 taxi drivers lie?

Solution: We have $n = 10$

$$\sum x_i = 70 + 120 + 110 + 101 + 88 + 83 + 95 + 107 + 100 + 91$$
$$= 972$$

$$\overline{x} = \frac{\sum x_i}{n} = \frac{972}{10} = 97.2$$

Null Hypothesis H_0: $\mu = 100$

Alternative Hypothesis H_1: $\mu \neq 100$ (two-tailed test)

Degrees of freedom $\nu = n - 1 = 10 - 1 = 9$

Level of significance $\alpha = 0.05$

Table value of t for 9 df at 5% level of significance = 2.262

x	$x - \overline{x}$	$(x - \overline{x})^2$
70	−27.2	739.84
120	22.8	519.84
110	12.8	163.84
101	3.8	14.44
88	−9.2	84.64
83	−14.2	201.64

(*Continued*)

x	$x-\overline{x}$	$(x-\overline{x})^2$
95	−2.2	4.84
107	9.8	96.04
100	2.8	7.84
98	0.8	0.64
		$\sum(x_i-\overline{x})^2 = 1833.60$

$$S^2 = \frac{\sum(x_i-\overline{x})^2}{n} = \frac{1833.60}{10-1} = \frac{1833.60}{9} = 203.74$$

$$\text{Standard error of mean} = \frac{S}{\sqrt{n}} = \frac{\sqrt{203.74}}{\sqrt{10}} = \sqrt{20.374}$$

Test statistic:

$$t = \frac{\overline{x}-\mu}{\left(\frac{S}{\sqrt{n}}\right)} = \frac{-2.8}{\sqrt{20.374}} = \frac{-2.8}{4.5137} = -0.6203$$

$$\therefore \quad |t| = 0.6203 < 2.262$$

i.e., calculated value of t at 5% level of significance for 9 df is less than the table value.

Hence we accept null hypothesis.

The 95% confidence limits for μ are

$$\overline{x} \pm t_{0.05}\frac{S}{\sqrt{n}} = 97.2 \pm 2.262\,(4.51)$$

i.e., [97.2−10.20, 97.2 + 10.20]

i.e., [87.0, 107.40]

i.e., The 95% confidence interval for is [87.0, 107.40].

Example 10.6: A sample of 26 bulbs gives a mean life of 990 hours with a SD of 20 hours. The manufacturer claims that the mean life of the bulbs is 1000 hours. Is the sample not up to the standard?

Solution: We have

$$n = 26$$
$$\overline{x} = 990$$
$$\mu = 1000$$
$$SD = 20$$
$$df = n-1 = 26-1 = 25$$

Null Hypothesis H_0: The sample is up to standard
Alternative Hypothesis H_1: The sample is not up to the standard

$$t = \frac{\bar{x} - \mu}{\left(\frac{S}{\sqrt{n-1}}\right)} = \frac{990 - 1000}{\frac{20}{\sqrt{26-1}}} = \frac{-10}{\left(\frac{20}{5}\right)}$$

$$= \frac{-50}{20} = -2.5$$

$$\therefore \quad |t| = 2.5$$

Table value of t for 25 df at 5% level is 1.708. Since the calculated value of t is greater than the table value of t, we reject null hypothesis. Therefore the sample is not up to the standard.

Example 10.7: The average breaking strength of the steel rods is specified to be 18.5 thousand pounds. To test this, a sample of 14 rods was tested. The mean and SDs obtained were 17.85 and 1.955, respectively. Is the result of the experiment significant?

Solution: Here we have $n = 14$, $\bar{x} = 17.85$, $\mu = 18.5$

$$S = SD = 1.955$$

Degrees of freedom $= n-1 = 14-1 = 13$

Table value of t for 13 df at 5% level of significance = 2–16 (i.e., critical value)

Null Hypothesis H_0: The result of the experiment is not significant.
Alternative Hypothesis H_1: $\mu \neq 18.5$
The test statistic

$$t = \frac{\bar{x} - \mu}{\left(\frac{S}{\sqrt{n-1}}\right)} = \frac{17.85 - 18.5}{\frac{1.955}{\sqrt{13}}} = \frac{-0.65}{0.542} = -1.199$$

Therefore

$$|t| = 1.199 < 2.166$$

Calculated value of t is less than the table value of t. Hence we accept null hypothesis, i.e., the result of the experiment is not significant.

Example 10.8: A machine is designed to produce insulating washers for electrical devices of average thickness of 0.025 cm. A random sample of 10 washers was found to have a thickness of 0.024 cm, with SD of

0.002 cm. Test the significance of the deviation (value of t for 9 degrees of freedom at 5% level of significance is 2.262).

Solution: Size of the given sample = 10 (small sample)

Sample mean = $\overline{x} = 0.024$ cm

Population mean $\mu = 0.025$ cm

S = Standard deviation = 0.002 cm

Degrees of freedom = $n - 1 = 10 - 1 = 9$

Table value of t for 9 df at 5% level of significance = 2.262

Null hypothesis H_0: The difference between $\overline{x}$ and μ is not significant.

Alternative hypothesis H_1: $\mu \neq 0.025$

The test statistic

$$t = \frac{\overline{x} - \mu}{\left(\frac{S}{\sqrt{n-1}}\right)} = \frac{0.024 - 0.025}{\frac{0.002}{\sqrt{10-1}}}$$

$$= \frac{-0.01}{0.002} \times 3 = -1.5$$

$$\therefore \quad |t| = 1.5 < 2.262$$

Calculated value of t is less than the table value of t at 5% level of significance and 9 df. We accept null hypothesis and conclude that the difference between $\overline{x}$ and μ is not significant.

Example 10.9: A sample of 20 items has a mean 42 units and SD 5 units. Test the hypothesis that it is a random sample from a normal population with mean 46 units.

Solution: We have $n = 20$ (small sample), $\overline{x} = 46$, $\mu = 46$, $S = 5$

Null Hypothesis H_0: $\mu = 46$

Alternative Hypothesis H_1: $\mu \neq 46$

Degrees of freedom = $n - 1 = 20 - 1 = 19$

Table value of t for 19 df at 5% level of significance = 2.09

The test statistic

$$t = \frac{\overline{x} - \mu}{\left(\frac{S}{\sqrt{n-1}}\right)} = \frac{42 - 46}{\frac{5}{\sqrt{19}}}$$

$$= \frac{-4\sqrt{19}}{5} = \frac{-4 \times 4.3588}{5}$$

$$= -3.4871$$

Therefore

$$|t| = 3.4871 > 2.09$$

Calculated value of t is greater than the table value at 5% level for 19 df.
Hence we reject null hypothesis H_0: $\mu = 46$
i.e., the difference between sample mean and population mean is significant.

Example 10.10: A machine is designed to produce insulating washers for electrical devices of average thickness of 0.025 cm. A random sample of 10 washers was found to have an average thickness of 0.024 cm with a SD of 0.002 cm. Test the significance of the deviation of mean (Table value of t for df at 5% level is 2.2627).

Solution: We have $n = 10$ (small sample), $\bar{x} = 0.024$, $\mu = 0.025$, $S = 0.002$

Null Hypothesis H_0: $\mu = 0.025$ cm
Alternative Hypothesis H_1: $\mu \neq 0.025$
Degrees of freedom $= n - 1 = 10 - 1 = 9$
Table value of t for 9 df at 5% level of significance $= 2.262$
The test statistic

$$t = \frac{\bar{x} - \mu}{\left(\frac{S}{\sqrt{n-1}}\right)} = \frac{0.024 - 0.025}{\frac{0.002}{3}}$$

$$= \frac{-(0.001) \times 3}{0.002} = -1.5$$

Therefore

$$|t| = 1.5 < 2.262$$

Calculated value of t is less than the table value at 5% level for 9df.
Hence we accept null hypothesis. We conclude that there is no significant deviation between the sample mean and population mean.

10.10 STUDENT'S *t*-TEST FOR DIFFERENCE OF MEANS

Suppose we want to test if two independent samples $x_1, x_2, x_3, \ldots, x_{n_1}$ and $y_1, y_2, y_3, \ldots, y_{n_2}$ of sizes n_1 and n_2 have been drawn from the normal population with means μ_1 and μ_2, respectively.

Assume that the population variances are equal

The statistic t

$$t = \frac{(\overline{x} - \overline{y}) - (\mu_1 - \mu_2)}{S\sqrt{\dfrac{1}{n_1} + \dfrac{1}{n_2}}}$$

where

$$\overline{x} = \frac{\sum x_i}{n_1}, \quad \overline{y} = \frac{\sum y_i}{n_2}$$

$$S^2 = \frac{1}{n_1 + n_2 - 2}\left[\sum_1^n (x_i - \overline{x})^2 + \sum_1^n (y_i - \overline{y})^2\right]$$

is an unbiased estimate of the population variance σ^2.

t follows t-distribution with degrees of freedom $n_1 + n_2 - 2$

Under the null hypothesis H_0, that

1. Samples have been drawn from the populations with the same means, i.e., $\mu_1 = \mu_2$
2. Sample means $\overline{x}$ and $\overline{y}$ do not differ significantly.

We compute the test statistic

$$t = \frac{(\overline{x} - \overline{y})}{S\sqrt{\dfrac{1}{n_1} + \dfrac{1}{n_2}}}$$

The degrees of freedom is $n_1 + n_2 - 2$

If $|t|$ is less than the table value of t_α for a given level of significance α, and df $= n_1 + n_2 - 2$

We accept null hypothesis H_0, otherwise reject the null hypothesis.

The confidence interval for $\mu_1 - \mu_2$ is a $(100, 1 - \alpha)\%$ interval given by

$$\overline{x} - \overline{y} \pm t_{\alpha/2}\sqrt{\frac{\sum (x_i - \overline{x})^2 + \sum (y_i - \overline{y})^2}{n_1 + n_2 - 2}\left[\frac{1}{n_1} + \frac{1}{n_2}\right]}$$

i.e.,

$$\overline{x} - \overline{y} \pm t_{\alpha/2} S\sqrt{\frac{1}{n_1} + \frac{1}{n_2}}$$

where the degrees of freedom is $\nu = n_1 + n_2 - 2$.

Example 10.11: The nicotine content in milligrams of the samples of tobacco were found as follows:

Sample A	24	27	26	21	25	
Sample B	27	30	28	31	22	36

Can it be said that the two samples come from normal populations with the sample mean.

Solution: We have

$$\overline{x} = \frac{24 + 27 + 26 + 21 + 25}{5} = \frac{123}{5} = 24.6$$

$$\overline{y} = \frac{27 + 30 + 28 + 31 + 22 + 36}{6} = \frac{174}{6} = 29$$

Calculation of sample SD's

x	$x - \overline{x}$	$(x - \overline{x})^2$	y	$y - \overline{y}$	$(y - \overline{y})^2$
24	−0.6	0.36	27	−2	4
27	2.4	5.76	30	1	1
26	1.4	1.96	28	−1	1
21	−3.6	12.96	31	2	4
25	0.4	0.16	22	−7	49
—	—	—	36	7	49
123	0	21.2	174	0	108

$$\therefore \quad \sum (x_i - \overline{x})^2 = 21.2, \sum (y_i - \overline{y})^2 = 108$$

$$S^2 = \frac{1}{n_1 + n_2 - 2}\left[\sum_1^n (x_i - \overline{x})^2 + \sum_1^n (y_i - \overline{y})^2\right]$$

$$= \frac{1}{5 + 6 + 1}[21.2 + 108] = 14.35$$

$$\therefore \quad S = \sqrt{14.35} = 3.78$$

Null Hypothesis H_0: $\mu_1 = \mu_2$, i.e., the two samples have been drawn from normal populations with the same mean

Alternative Hypothesis H_1: $\mu_1 \neq \mu_2$

Level of significance = 5%,

Degrees of freedom = $n_1 + n_2 - 2 = 5 + 6 - 2 = 9$

$t_{0.05}$ for 9 df = 2.262

Test statistic is

$$t = \frac{(\overline{x} - \overline{y})}{S\sqrt{\frac{1}{n_1} + \frac{1}{n_2}}} = \frac{24.6 - 29}{3.78\sqrt{\frac{1}{5} + \frac{1}{6}}} = -1.92$$

$$\therefore \qquad |t| = 1.92 < 2.262$$

The calculated value of t is less than the table value of t at 5% level of significance for 9 df. Therefore we accept H_0.

The samples are drawn from normal populations with the same mean.

Example 10.12: A group of 7-week-old chicken reared on a high-protein diet weigh 12, 15, 11, 16, 14, 14, and 16 ounces. A second group of 5 chickens similarly treated except that they receive a low-protein diet weigh 8, 10, 14, 10, and 13 ounces. Test whether there is evidence that additional protein has increased the weight of the chicken (The table value of t for $\nu = 10$ at 5% level of significance is 2.33).

Solution: Let the weights of chicken reared on high-protein diet be denoted by x_i and the weights of chicken reared on low-protein diet be denoted by y_i

We have

$$\overline{x} = \frac{\sum x_i}{n_1} = \frac{12 + 15 + 11 + 16 + 14 + 14 + 16}{7} = \frac{98}{7} = 14,$$

$$\overline{y} = \frac{\sum y_i}{n_2} = \frac{8 + 10 + 14 + 10 + 13}{5} = \frac{55}{5} = 11$$

Null Hypothesis H_0: $\mu_1 = \mu_2$

i.e., additional protein has not increased the weight of the chickens

Alternative Hypothesis H_1: $\mu_1 > \mu_2$ (one-tailed test)

Computation of SD.

x	$x - \overline{x}$	$(x - \overline{x})^2$	y	$y - \overline{y}$	$(y - \overline{y})^2$
12	−2	4			
15	1	1	8	−3	9
11	−3	9	10	−1	1
16	2	4	14	3	9
14	0	0	10	−1	1

(*Continued*)

x	$x-\bar{x}$	$(x-\bar{x})^2$	y	$y-\bar{y}$	$(y-\bar{y})^2$
14	0	0	13	2	4
16	2	4			
98		22	55		24

$$S=\sqrt{\frac{\sum(x_i-\bar{x})^2+\sum(y_i-\bar{y})^2}{n_1+n_2-2}}=\sqrt{\frac{22+24}{7+5-2}}=\sqrt{\frac{46}{10}}=\sqrt{4.6}=2.1447$$

Level of significance = 5% ($\alpha = 0.05$)
Critical value, i.e., table value of t for 10 df at 5% level = 1.81
Test statistic is

$$t=\frac{(\bar{x}-\bar{y})}{S\sqrt{\frac{1}{n_1}+\frac{1}{n_2}}}=\frac{14-11}{2.1447\sqrt{\frac{1}{7}+\frac{1}{5}}}$$

$$=\frac{3}{2.1447}\times\frac{\sqrt{25}}{\sqrt{5+7}}=\frac{3\times5.916}{2.1447\times3.4641}$$

$$=\frac{17.748}{7.4294}=2.3888$$

Since the calculated value of t is greater than the table value of t, at 5% level of significance for 10 degrees of freedom. We reject H_0 and conclude that additional protein has increased the weight of the chicken.

Example 10.13: Two salesmen A and B working in a certain district from a sampling survey conducted by the head office. The following results were obtained. State whether there is any significant difference in the average sales between the two salesmen?

	A	B
No. of sales	20	18
Average sales (in Rs.)	170	203
Standard deviation (in Rs.)	20	25

Solution: We have

$$n_1=20,\quad n_2=18$$
$$\bar{x}_1=170,\quad \bar{x}_2=205$$
$$S_1=20,\quad S_2=25$$

Degrees of freedom = $n_1+n_2-2=20+16-2$
Level of significance $\alpha=0.05$

Table value of $t = 1.9$ (for 36 df at 5%)

Standard error of difference of means $= SE(\overline{x}_1 - \overline{x}_2)$

$$= S\sqrt{\frac{1}{n_1} + \frac{1}{n_2}}$$

where

$$S = \sqrt{\frac{n_1 S_1^2 + n_2 S_2^2}{n_1 + n_2 - 2}} = \sqrt{\frac{(20)(20)^2 + (18)(25)^2}{20 + 18 - 2}}$$

$$= \sqrt{\frac{800 + 11250}{36}} = 23.1240$$

Therefore

$$SE(\overline{x}_1 - \overline{x}_2) = 23.1240\sqrt{\frac{1}{20} + \frac{1}{18}} = 23.1240\sqrt{\frac{9 + 10}{180}}$$

$$= 23.1420\sqrt{\frac{19}{180}} = 23.1420 \times 0.3248$$

$$= 7.5128$$

Test statistic is

$$t = \frac{(\overline{x}_1 - \overline{x}_2)}{S\sqrt{\frac{1}{n_1} + \frac{1}{n_2}}} = \frac{170 - 205}{7.5128} = \frac{-35}{7.5128} = -\,4.6587$$

$$\therefore \quad |t| = 4.6587 > 1.96$$

i.e., the calculated value of t is greater than the table value of t at 5% level of significance, for 36 df.

Hence we reject null hypothesis and conclude that there is a significant difference in the average sales between two salesmen.

Example 10.14: A group of 5 patients treated with medicine A weigh 42, 39, 48, 60, and 41 kg. A second group of 7 patients from the same hospital with medicine B weigh 38, 42, 56, 64, 68, 69, and 62 kg. Do you agree with the claim that the medicine B increases the weight significantly (The value of t at 5% level of significance for 10 df is 2.228).

Solution: Let x_i denote the weights of patients treated with medicine A, and y_i denote the weights of patients treated with medicine B.

We have $n_1 = 5$, $n_2 = 7$, $\text{df} = n_1 + n_2 - 2 = 5 + 7 - 2 = 10$
Null hypothesis H_0: $\mu_A = \mu_B$
i.e., there is no significant difference between the medicines A and B regards their effect on increase in weight.
Alternative hypothesis H_1: $\mu_A \neq \mu_B$ (Two-tailed test)
Table value of t at 5% level of significance for 10 df = 2.228

$$\overline{x} = \frac{\sum x_i}{n_1} = \frac{42 + 39 + 48 + 60 + 41}{5} = \frac{230}{5} = 46,$$

$$\overline{y} = \frac{\sum y_i}{n_2} = \frac{38 + 42 + 56 + 64 + 68 + 69 + 62}{7} = 59$$

x	$x - \overline{x}$	$(x - \overline{x})^2$	y	$y - \overline{y}$	$(y - \overline{y})^2$
42	−4	16	38	−19	361
39	−7	49	42	−15	225
48	2	4	56	−1	1
60	14	196	64	7	49
41	−5	25	68	11	121
–	–	–	69	12	144
–	–	–	62	5	25
230	–	290	399	–	926

$$S = \sqrt{\frac{\sum (x_i - \overline{x})^2 + \sum (y_1 - \overline{y})^2}{n_1 + n_2 - 2}} = \sqrt{\frac{290 + 926}{5 + 7 - 2}}$$

$$= \sqrt{\frac{121.6}{10}} = 11.027$$

Test statistic is

$$t = \frac{(\overline{x} - \overline{y})}{S\sqrt{\frac{1}{n_1} + \frac{1}{n_2}}}$$

$$= \frac{46 - 57}{11.027\sqrt{\frac{1}{5} + \frac{1}{7}}} = \frac{-11 \times \sqrt{35}}{(11.027)\sqrt{12}}$$

$$= \frac{-11 \times 5.916}{(11.027) \times 3.464}$$

$$= \frac{-65.076}{38.197} = -1.703$$

Therefore

$$|t| = 1.703$$

The calculated value of t is less than the table value of t at 5% level of significance, for 10 df. Therefore we accept null hypothesis H_0.

There is no significant difference between medicines A and B as regards the increase in weights.

Example 10.15: The students of two schools were measured for their heights, one school was in east coast another in west coast, where there is slight difference in weather. The sampling results are as follows:

East coast	43	45	48	49	51	51	–	–	–
West coast	47	49	51	53	54	55	55	56	57

Find whether there is any impact of weather on height taking other variable constant. (given $t_{0.05}$ for 13 df = 2.16)?

Solution: Here we have

$$n_1 = 6, n_2 = 9$$

Let east coast school be denoted by x and west coast student height be denoted by y. Then

$$\overline{x} = \frac{\sum x_i}{n_1} = \frac{43 + 45 + 48 + 49 + 51 + 52}{6} = \frac{288}{6} = 48$$

$$\overline{y} = \frac{\sum y_i}{n_2} = \frac{47 + 49 + 51 + 53 + 54 + 55 + 56 + 57}{9} = \frac{477}{9} = 53$$

Null hypothesis H_0: Weather has no impact on the heights of the students

Alternative hypothesis H_1: Weather has impact on the heights of the students

Degree of freedom = 6 + 9–2 = 13

Level of significance = 5%

Table value of t for 13 df = 2.16

$$S^2 = \frac{1}{n_1 + n_2 - 2}\left[\sum_1^n (x_1 - \overline{x})^2 + \sum_1^n (y_1 - \overline{y})^2\right]$$

$$= \frac{60 + 90}{6 + 9 - 2} = \frac{150}{13} = 11.5$$

$$\therefore \qquad S = \sqrt{11.5} = 3.39$$

Test statistic is

$$t = \frac{(\bar{x} - \bar{y})}{S\sqrt{\frac{1}{n_1} + \frac{1}{n_2}}} = \frac{48 - 53}{3.39\sqrt{\frac{1}{6} + \frac{1}{9}}}$$

$$= \frac{-5}{3.39\sqrt{\frac{5}{18}}} = \frac{-5}{1.76} = -2.84$$

$$\therefore \qquad |t| = 2.84$$

The calculated value of t is greater than the table value of t at 5% level of significance for 13 df.

Hence we reject null hypothesis and accept the alternative hypothesis.

Exercise 10.1

1. Define t-distribution and write the properties of t-probability curve.
2. Ten students are selected at random from a college and their heights are found to be 100, 104, 108, 110, 118, 120, 122, 124, 126, and 118 cm. In the height of these data, discuss the suggestion that the mean height of the students of the college is 110 cm. Use 5% level of significance.

   Ans: Null hypothesis is accepted, i.e., mean height of the students is 110 cm
3. A machine that produces mica insulating washers having a thickness of 10 miles (1 mile = 0.001 in.). A sample of 10 washers has an average thickness of 9.52 miles with a SD of 0.60 miles. Find the value of t?
4. Ten cartons are taken at random from an automatic machine. The mean net weight of the 10 cartons is 11.8 kg and SD is 0.15 kg. Does the sample means differ significantly from the intended height of 12 kg ($t_{0.05}$ for 9 df = 2.26).

   Ans: H_0 rejected. The sample means differ
5. A fertilizer mixing machine is set to give 12 kg of nitrate for every quintal bag of fertilizer. Ten 100 kg bags are examined. The percentages of nitrate are as follows. 11, 14, 13, 12, 13, 12, 1, 3, 14, 11, 12 is there reason to be believe the machine is effective ($t_{0.05}$ for 9 df = 2.262).

   Ans: There is no reason to believe the t the machine is defective

6. Two samples of sizes 6 and 5 gave the following data:

	Sample I	Sample II
Mean	40	50
SD	8	10

Is the difference of the mean significant? The value of t for t degrees of freedom at 5% level of significance is 2.26.

***Ans*:** The difference of means is not significant

7. Samples of two types of sodium vapor were tasted for the length of life and the following data were obtained:

	Type I (h)	Type II (h)
Sample No.	8	7
Sample mean	1234	1036
Sample SD	36	40

Is the difference in the mean sufficient to warrant an inference that type I is superior to type II regarding the length of life (Type I is superior to type II)?

8. Sandal powder is packed into packets by a machine. A random sample of 12 packets is drawn and their weights are found to be (in kg) 0.49, 0.48, 0.47, 0.48, 0.49, 0.50, 0.51, 0.49, 0.50, 0.51, 0.48. Test if the average packing can be taken as 0.5 kg ($t_{0.05}$ for 11 df = 2.20)?

***Ans*:** H_0 rejected

9. The following results are obtained from a sample of 10 boxes of biscuits:

Mean weight of the content = 490 g

SD of the weights = 9 g

Could the sample come from a population having a mean of 500 g ($t_{0.05}$ for 9 df = 2.26)

***Ans*:** H_0 is rejected

10. The wages of 10 workers taken at random from a factory are given as follows:

Wages (in Rs.): 578, 572, 570, 568, 572, 578, 570, 572, 596, 584

It is possible that the mean height of all workers of this factory could be Rs. 580 (given at 9 df $t_{0.05} = 2.26$)?

***Ans*:** H_0 is accepted

11. The 9 items of a sample have the following values 45, 47, 50, 52, 48, 47, 49, 53, 51. Does the mean values differ significantly from the assumed mean 47.5 (for 8 df, $t_{0.05} = 2.31$)?

***Ans*:** H_0 is accepted

12. Write short notes on:
 a. Students t-distribution
 b. Properties of t-distribution

13. Test whether the sample having values 63, 63, 64, 55, 66, 69, 70, 70, and 71 have been chosen from a population with mean of 65 at 5% level of significance ($t_{0.05} = 2.31$ for 8 df)?

Ans: The difference is significant

14. A certain medicine was administered to each of 10 patient's results in the following increases in the blood pressure:

8, 8, 7, 5, 4, 1, 0, 0, −1, −1

Concluded that the medicine was responsible for the increase in blood pressure.

Ans: The difference is significant. The medicine is responsible for the increase in the blood pressure

15. The weight of 10 people of a locality are found to be 70, 67, 62, 68, 61, 68, 70, 64, 66 kg. Is it reasonable to believe that the average weights of the people of locality are greater than 64 kg. Test at 5% level of significance (Given $t_{0.05} = 1.83$ for 9 df).

Ans: H_0 is rejected. Average weight is greater than 64 kg

16. Following table contains the data resulting from a sample of managers trained under different t programs:

Program sampled	Mean sensitivity of the program	Number of managers observed	Estimate SD for sensitivity after the program
Formal	92%	12	15%
Informal	84%	15	19%

Test at 0.05 level of significance whether the sensitivity achieved by formal program is significantly higher than that achieved under informal program?

Ans: H_0 rejected: The sensitivity achieved by formal program is higher than that achieved under the informal program

17. A random sample of 10 tins of oil filled in by an automatic machine gave the following weights (in kg): 2.05, 2.01, 2.04, 1.91, 1.97, 1.97, 2.04, 2.02. Can we accept at 5% level of significance the claim that the average weight of the tin is 2 kg.

Ans: The average weight of the tin is 2 kg is accepted

18. A random sample of size 16 has 53 as its mean. The sum of the squares of the deviations taken from the mean is 1.50. Can this sample be regarded as taken from the population having 56 as its mean? Obtain 95% and 99% confidence limits of the mean population (for $\nu = 15$, $t_{0.01} = 2.95$, $t_{0.05} = 2.131$).

Ans: H_0 is rejected, [50.316, 54.684], [50.67, 55.33]

19. A sample of 20 items has mean 42 units and SD 5 units. Test the hypothesis that it is a random sample from a normal population with mean 45 units?

Ans: The sample could not have come from the given population

20. The means of two random samples of size 9 and 7 are 196.42 and 198.82, respectively. The sum of the squares of the deviations from the mean is 26.94 and 18.73 respectively. Can the sample be considered that they have been drawn from the same normal population?

Ans: The two samples are not drawn from the same population

21. Two horses A and B were tested according to the time (in seconds) to run a particular track with the following results:

Horse A	28	30	32	33	33	29	34
Horse B	29	30	30	24	27	27	-

Test whether you can discriminate between two horses you can use the fact that at 5% level of significance value of t for 11 df is 2.2?

Ans: The medicines A and B do not differ significantly as regards their effect on the increase of weight

22. A group of 5 patients treated with medicine A weighed 42, 39, 48, 60 and 41 kg. Second group of 7 patients from the same hospital treated with medicine B weighed 38, 42, 56, 64, 68, 69, and 62 kg. Do you agree with the claim that medicine B increases the weight significantly?

Ans: The medicines A and B do not differ significantly as regards their effect on the increase of weight

23. Memory capacity of 10 students was tested before and after training. State whether the training was effective or not from the following scores:

Before	12	14	11	8	7	10	3	0	5	6
After	15	16	10	7	5	13	10	2	8	8

Ans: Not significant

24. An IQ test was administered on 5 persons before and after they were trained. The results are given below. Test whether there is any change in IQ after the training program?

Ans: No change

25. Measurements of a sample of weights were determined as
8.3, 10.6, 9.7, 8.8, 10.2, 9.4.
Determine unbiased and efficient estimates of
a. The population mean
b. The population variance
Ans: 9.5 kg, 0.736

26. Ten students were given intensive coaching for a month in mathematics. The scores obtained in tests 1 and 5 are given below:

Serial no. of students:	1	2	3	4	5	6	7	8	9	10	11
Marks in first test:	50	52	53	60	65	67	48	69	72	80	
Marks on second test:	65	55	65	65	60	67	49	82	74	86	

Does the scores from test 1 to test 5 show an improvement? Test at 5% level of significance ($t_{0.05}$ for 9 df = 1.833 and two-tailed test $t_{0.05}$ for 9 df = 2.262)?
Ans: For one-tailed test null hypothesis is rejected

27. A certain stimulus administered to each of 12 patients resulted in the following changes in the blood pressure 5, 2, 8, −1, 3, 0, −2, 1, 5, 0, 4, 6. Can it be concluded that the stimulus in general be accompanied by an increase in blood pressure?
Ans: H_0 rejected. The stimulus in general be accompanied by an increase in blood pressure

28. Experience shows that a fixed dose of certain drug causes an average increase of pulse rate by 10 beats per minute with SD of 4. A group of 9 patients given the same dose showed the following increase: 13, 15, 14, 10, 8, 12, 16, 9, 20. Test 5% level of significance whether this group is different in response to the drug?
Ans: H_0 rejected: The given group is different in response to the drug

10.11 PAIRED *t*-TEST

In this case the size of two samples are equal, i.e., $n_1 = n_2 = n$

Let $x_1, x_2, \ldots, x_n$ and $y_1, y_2, \ldots, y_n$ be two samples. Let $d_1, d_2, \ldots, d_n$ be the differences between the corresponding members of the two samples.

If

$$\overline{d} = \frac{\sum d}{n}$$

and

$$S = \frac{\sum (d-\bar{d})^2}{n-1}$$

then the t-statistic is defined as

$$t = \frac{\bar{d} - 0}{\frac{S}{\sqrt{n}}}$$

t-statistic is applied for $n - 1$ degrees of freedom.

10.11.1 Solved Examples

Example 10.16: A certain diet was newly introduced to each of 12 pigs resulted in the following increase in body weight:

6, 3, 8, −2, 3, 0, −1, 1, 6, 0, 5, and 4.

Can you conclude that the diet was effective in increasing the weight of pigs (given $t_{0.05} = 2.20$ for 11 df)?

Solution: Here we have $n = 12$

$$d_1 = 6, d_2 = 3, d_3 = 8, d_4 = -2, d_5 = 3, d_6 = 0,$$

$$d_7 = -1, d_8 = 1, d_9 = 6, d_{10} = 0, d_{11} = 5, d_{12} = 4$$

where d_i denotes the increase in weights of the pigs.

Null hypothesis H_0: There is no significant difference in the body weights of the pigs before and after introduction of the new diet, i.e., $\mu_x = \mu_y$

Alternative hypothesis H_1: $\mu_x \neq \mu_y$

Degrees of freedom $= n - 1 = 12 - 1 = 11$

Level of significance = 5%, $t_{0.05}$ for 11 df = 2.20

$$\sum d = 6 + 3 + 8 + (-2) + 3 + 0 + (-1) + 1 + 6 + 0 + 5 + 4 = 33$$

$$\bar{d} = \frac{\sum d}{n} = \frac{33}{12} = 2.75$$

$$\sum d^2 = 6^2 + 3^2 + 8^2 + (-2)^2 + 3^2 + 0^2 + (-1)^2 + 1^2 + 6^2 + 0^2 + 5^2 + 4^2 = 201$$

$$\therefore \quad S^2 = \frac{1}{n-1}\left[\sum d^2\left(\frac{\left(\sum d\right)^2}{n}\right)\right] = \frac{1}{12-1}\left[201 - \frac{(33)^2}{12}\right]$$

$$= \frac{1}{11}[201 - 90.75] = 10.0227$$

$$S = \sqrt{10.0227} = 3.1658$$

Test statistic t is

$$t = \frac{\bar{d} - 0}{\frac{S}{\sqrt{n}}} = \frac{2.75}{\frac{3.1658}{\sqrt{12}}} = \frac{2.75 \times 3.464}{3.1658} = 3.0091$$

Since the calculated value of t is greater than the table value of t at 5% level of significance for 11 df, we reject null hypothesis. Therefore the new diet is effective in increasing the body weight of the pigs.

10.12 *F*-DISTRIBUTION

In this section we introduce F-distribution and another statistic F

F-statistic is defined by

$$F = \frac{S_1^2}{S_2^2},$$

where

$$S_1^2 > S_2^2$$

and

$$S_1^2 = \frac{1}{n_1 - 1}\sum (x - \bar{x})^2$$

$$S_2^2 = \frac{1}{n_2 - 1}\sum (y - \bar{y})^2$$

For the degrees of freedom $\nu_1 = n_1 - 1, \nu_2 = n_2 - 1$

The value of F is always greater than 1.

If

$$S_1^2 < S_2^2 (\text{i.e., } S_1^2 < S_2^2)$$

F is given by

$$F = \frac{S_2^2}{S_1^2}$$

So that the value of F is greater than 1.

F-test is based on F-distribution which is defined as the ratio of two independent chi-square variates.

F-distribution with (ν_1, ν_2) degrees of freedom is given by the probability function.

$$\mathrm{d}p = \frac{\nu_1^{\frac{1}{2}\nu_1} \nu_2^{\frac{1}{2}\nu_2} F^{\frac{\nu_1}{2}-2}}{\beta\left(\frac{\nu_1}{2}, \frac{\nu_2}{2}\right)(\nu_1 F + \nu_2)^{\frac{1}{2}(\nu_1+\nu_2)}} \mathrm{d}F, \quad 0 < F < \infty$$

From the definition of F, we have

$$\frac{\nu_1}{\nu_2} F = \frac{(n_1 - 1)\frac{S_1^2}{\sigma^2}}{(n_2 - 1)\frac{S_2^2}{\sigma^2}}$$

The numerator and the denominator of the second member are independent χ^2 variates with ν_1 and ν_2 degrees of freedom respectively. Hence $(\nu_1/\nu_2)F$ is a $\beta_2((\nu_1/2), (\nu_2/2))$ variate so that the probability that a random value of F will fall in the interval $\mathrm{d}F$ is given by

$$\mathrm{d}p = \frac{\nu_1^{\frac{1}{2}\nu_1} \nu_2^{\frac{1}{2}\nu_2} F^{\frac{\nu_1}{2}-2}}{\beta\left(\frac{\nu_1}{2}, \frac{\nu_2}{2}\right)(\nu_1 F + \nu_2)^{\frac{1}{2}(\nu_1+\nu_2)}} \mathrm{d}F, \quad 0 < F < \infty$$

F-Probability curve is J-shaped. If the curve is bell shaped if $\nu_1 > 2$.

The distribution of F is independent of the population variance and depends on ν_1 and ν_2 only.

***F*-test is used to test:**

1. Whether two independent samples have been drawn from the normal populations with same variance.
2. Whether the two independent estimates of the population variances are homogeneous or not.

***F*-test is based on the following assumptions:**

1. The distribution in each group showed to be normally distributed.
2. Error should be independent of each observed value.
3. Variance within each group should be equal for all groups.

When the sample sizes are large, the assumption of normality is not necessary.

10.12.1 Solved Examples

Example 10.17: Two horses A and B were tested according to the time (in seconds) to run on a particular track with the following results:

Horse A	28	30	32	33	33	29	34
Horse B	29	30	30	24	27	29	–

Test whether the two horses have the same running capacity?
Solution: We have $n_1 = 7$, $n_2 = 6$
Then

$$\overline{x} = \frac{\sum x_i}{n_1} = \frac{219}{7} = 31.28, \quad \overline{y} = \frac{\sum y_i}{n_2} = \frac{169}{6} = 28.2$$

Let the time taken by horse A be denoted by x and the time taken by horse B be denoted by y
Degrees of freedom $= n_1 - 1,\ n_2 - 1 = 6,\ 5$
Critical value, i.e., level of significance = 5%

x	$x - \overline{x}$	$(x - \overline{x})^2$	y	$y - \overline{y}$	$(y - \overline{y})^2$
28	− 3.28	10.75	28	0.	0.64
30	− 1.28	1.64	30	1.8	3.24
32	0.72	0.52	30	1.8	3.24
33	1.72	2.96	24	− 4.2	17.64
33	1.72	2.96	27	− 1.2	1.44
29	− 2.28	5.20	29	0.8	0.64
34	2.72	7.40	–	–	–
219		31.43	169		26.84

$$S_1^2 = \frac{1}{n_1 - 1}\sum (x - \overline{x})^2 = \frac{31.43}{6} = 5.224$$

$$S_2^2 = \frac{1}{n_2 - 1}\sum (y - \overline{y})^2 = \frac{26.84}{5} = 5.368$$

Null Hypothesis H_0: The two horses have the same running capacity, i.e., $\sigma_1^2 = \sigma_2^2$
Alternative Hypothesis H_1: $\sigma_1^2 \neq \sigma_2^2$
Table value of F for (5, 6) degrees of freedom at 5% level of significance is 4.39.

Test statistic is $F = \frac{S_2^2}{S_1^2} = \frac{5.368}{5.224} = 1.02$ (since $S_2^2 > S_1^2$)

Since the calculated value of F is less than the table value of F for (5, 6) df at 5% level of significance, we accept H_0.

The two horses have the same running capacity.

Example 10.18: In a sample of 8 observations, the sum of squares of deviations of the sample values from the sample mean was 94.5 and in another sample of 10 observations it was found to be 101.7. Test whether the difference is significant?

Solution: It is given that

$$\sum (x-\overline{x})^2 = 94.5, \ n_1 = 8$$

$$\sum (y-\overline{y})^2 = 101.7, \ n_2 = 10$$

$$S_1^2 = \frac{1}{n_1 - 1}\sum (x-\overline{x})^2 = \frac{94.5}{8-1} = 13.5$$

$$S_2^2 = \frac{1}{n_2 - 1}\sum (y-\overline{y})^2 = \frac{1010.7}{10-1} = 11.3$$

Null Hypothesis H_0: $\sigma_1^2 = \sigma_2^2$, i.e., The difference is not significant

Alternative Hypothesis H_1: $\sigma_1^2 \neq \sigma_2^2$

Level of significance: 5%

Critical value, i.e., table value of F for $(8 - 1, 10 - 1) = (7, 9)$ df is 3.29.

$$F = \frac{S_1^2}{S_2^2} = \frac{13.5}{11.3} = 1.195$$

Since the calculated value of F is less than the table value of F at 5% level of significance for (7, 9) df, we accept Null hypothesis and conclude that the difference is not significant.

Example 10.19: In a test given two groups of students drawn from two normal populations the marks are as follows:

Sample A	18	20	36	50	49	36	34	49	41
Sample B	29	28	26	35	30	44	46		

Examine 5% level whether the two populations have the same variance.

Solution:

Null Hypothesis H_0: $\sigma_A^2 = \sigma_B^2$

Alternative Hypothesis H_1: $\sigma_A^2 \neq \sigma_B^2$

Level of significance: 5%,

We have

$$\bar{x} = \frac{\sum x_i}{n_1} = \frac{18 + 20 + 36 + 50 + 49 + 36 + 34 + 49 + 41}{9} = \frac{333}{9} = 37$$

$$\bar{y} = \frac{\sum y_i}{n_2} = \frac{29 + 28 + 26 + 35 + 30 + 44 + 46}{7} = \frac{238}{7} = 34$$

Calculation of variances

Sample A			Sample B		
x	$x-\bar{x}$	$(x-\bar{x})^2$	y	$y-\bar{y}$	$(y-\bar{y})^2$
18	−19	361			
20	−17	289	29	−5	25
36	−1	1	28	−6	36
50	13	169	26	−8	64
49	12	144	35	−1	1
36	−1	1	30	−4	16
34	−3	9	44	−10	100
49	12	144	46	−12	144
41	4	16			
		1134			386

$$S_A^2 = \frac{1}{n_1 - 1}\sum (x-\bar{x})^2 = \frac{1134}{9-1} = 141.75$$

$$S_B^2 = \frac{1}{n_2 - 1}\sum (y-\bar{y})^2 = \frac{386}{7-1} = 64.33$$

We have

$$S_A^2 > S_B^2$$

$$F = \frac{S_A^2}{S_B^2} = \frac{141.75}{64.33} = 2.203$$

$$\text{Degrees of freedom} = (\nu_1, \nu_2) = (n_1 - 1,\ n_2 - 1) = (9-1, 7-1) = (8, 6)$$

Table value of F at 5% for (8,6) df = 4.15.

Calculated value of F is less than the table value of F at 5% level (8,6) df. Hence we accept H_0 and conclude that the populations from where the samples have been drawn have the same variance.

Example 10.20: Two samples of sizes 9 and 8 gave the sum of the squares of deviations from their respective means equal to 160 and 91, respectively. Can they be regarded as drawn from the same normal population?

Solution: It is given that

$$\sum (x - \overline{x})^2 = 160, n_1 = 9$$

$$\sum (y - \overline{y})^2 = 91, n_2 = 8$$

$$S_1^2 = \frac{1}{n_1 - 1} \sum (x - \overline{x})^2 = \frac{160}{9 - 1} = 20$$

$$S_2^2 = \frac{1}{n_2 - 1} \sum (y - \overline{y})^2 = \frac{91}{8 - 1} = 13$$

Null Hypothesis H_0: Both the samples have come from the same normal population.

Alternative Hypothesis H_1: The samples are not from the same normal population.

Since $S_1^2 > S_2^2$, we have

Degrees of freedom $(n_1 - 1, n_2 - 1) = (8, 7)$

Level of significance: 5%

Table value of F at 5% level of significance for (8,7) df = 3.73

Test statistic:

$$F = \frac{S_1^2}{S_2^2} = \frac{20}{13} = 1.5385$$

Calculated value of F is less than the table value of F at 5% level (8, 6) df. Hence we accept null hypothesis.

The two samples can be regarded as drawn from the same normal population.

Example 10.21: Two samples are drawn from two normal populations from the following data, test whether the two samples have the same variances at 5% level?

Sample I		60	65	71	74	76	82	85	87	
Sample II	64	66	67	85	78	88	86	85	63	91

Solution: Null hypothesis H_0: $\sigma_1^2 = \sigma_2^2$, i.e., the two samples have same variances.

Alternative hypothesis H_1: $\sigma_1^2 \neq \sigma_2^2$ (Two-tailed test).

$$\overline{x} = \frac{\sum x_i}{n_1} = \frac{60+65+71+74+76+82+85+87}{8} = \frac{600}{8} = 75,$$

$$\overline{y} = \frac{\sum y_i}{n_2} = \frac{64+66+67+85+78+88+86+85+63+91}{10} = \frac{770}{10} = 77$$

Sample A			Sample B		
x	$x-\overline{x}$	$(x-\overline{x})^2$	y	$y-\overline{y}$	$(y-\overline{y})^2$
			61	−16	256
60	−15	225	66	−11	121
65	−10	100	67	−10	100
71	−4	16	85	8	64
74	−1	1	78	1	1
76	1	1	88	11	121
82	7	49	86	9	81
85	10	100	85	8	64
87	12	144	63	−14	196
			91	14	196
600		636	770		1200

$$S_1^2 = \frac{1}{n_1-1}\sum(x-\overline{x})^2 = \frac{636}{8-1} = 90.857$$

$$S_2^2 = \frac{1}{n_2-1}\sum(y-\overline{y})^2 = \frac{1200}{10-1} = 133.33$$

Since $S_2^2 > S_1^2$

$$F = \frac{S_2^2}{S_1^2} = \frac{133.33}{90.851} = 1.467$$

Calculated value of F for $(n_1 - 1, n_2 - 1) = (9, 7)$ df at 5% level = 3.68.
Therefore the calculated value of F is less than table value of F at 5% level of significance for (9, 7) df.

We accept null hypothesis. The samples 1 and 11 have the same variance.

Example 10.22: In a laboratory experiment, two samples of blood gave the following results:

Sample	Size	Sample SD	Sum of squares of deviations from mean
1	10	15	90
2	12	14	108

Test the equality of sample variances at 5% level of significance?

Solution: Here we have $n_1 = 10,\ n_2 = 12$

$$\overline{x} = 15, \overline{y} = 14$$

$$\sum (x - \overline{x})^2 = 90, \quad \sum (y - \overline{y})^2 = 108$$

$$S_1^2 = \frac{1}{n_1 - 1} \sum (x - \overline{x})^2 = \frac{90}{10 - 1} = 10$$

$$S_2^2 = \frac{1}{n_2 - 1} \sum (y - \overline{y})^2 = \frac{108}{12 - 1} = 9.82$$

Null hypothesis H_0: $\sigma_1^2 = \sigma_2^2$
Alternative hypothesis H_1: $\sigma_1^2 \neq \sigma_2^2$
Test statistic:

$$F = \frac{S_1^2}{S_2^2} = \frac{10}{9.82} = 1.018$$

Table value of F at 5% level of significance for $(10 - 1,\ 12 - 1) = (9,\ 11)$ df is 2.90.

Therefore the calculated value of F is less than the table value. Hence we accept null hypothesis. The two programs have the same variance.

Exercise 10.2

1. In one sample of 10 observations, the sum of squares of the deviations of the sample values from the sample mean was 120 and in another sample of 12 observations it was 314. Test whether the difference is significant at 5% level of significance?

   Ans: H_0 accepted, i.e., $\sigma_1^2 = \sigma_2^2$
2. Two random samples gave the following results:

Sample	Size	Sample	Sum of squares of deviations from mean
1	10	15	90
2	12	14	108

   Test whether the samples came from the same population?

   Ans: The samples came from the populations with equal variance. $\sigma_1^2 = \sigma_2^2$, i.e., H_0 accepted
3. Write short notes on F-distribution and F-test.
4. Mention the applications of F-test.
5. Two independent samples are of sizes 8 and 10. The sum of squares of deviations of sample values from their respective sample means were 84.4 and 102.6. Test whether the difference of variances of the populations is significant or not?

   Ans: $\sigma_1^2 = \sigma_2^2$, i.e, Accept H_0

6. Two random samples of sizes 8 and 7 had the following values of variances:

Sample A	9	11	13	11	15	9	12	14
Sample B	10	12	10	14	9	8	10	

Do these eliminates of population variance differ significantly?
***Ans*:** Not significant

7. Two independent samples of sizes 9 and 7 from a normal population has the following values of the variables:

Sample I	18	13	12	15	12	14	16	14	15
Sample II	16	19	13	16	18	13	15		

Do these eliminates of population variance differ significantly?
***Ans*:** H_0 accepted. The difference is not significant

CHAPTER 11

ANOVA (Analysis of Variance)

11.1 INTRODUCTION

In this chapter we briefly introduce Analysis of variance (ANOVA) which is statistical test used to determine if more than two population means are equal. The test uses the *F*-distribution (probability distribution) function and information about the variances of each population (within) and grouping of populations (between) to help decide if variability between and within each population is significantly different. It is based on the comparison of the average value of a common component. The method of ANOVA tests the hypotheses that:

$$H_0 = \mu_1 = \mu_2 = \cdots = \mu_k$$
$$H_1 = \text{Not all the means are equal}$$

The purpose of ANOVA test statistic is

1. To see if more than two population means are equal.
2. To know the difference between the within-sample estimate of the variance and the between the sample estimate of the variance and how to calculate them. The *within-sample* variance is often called the *unexplained variation.*

The *between-sample variance* or error is the average of the square variations of each population mean from the mean or all the data and is an estimate of σ^2 only if the null hypothesis, H_0 is true.

When the null hypothesis is false this variance is relatively large and by comparing it with the within-sample variance we can tell statistically whether H_0 is true or not.

The ANOVA technique helps in performing the test in one go and, therefore, is considered to be important technique of analysis. The basic principle underlying the technique is that the total variation in the dependent variable is broken into two parts—one which can be attributed to some specific causes and the other that may be attributed to chance. The one which is attributed to the specific causes is known as the variation

Statistical Techniques for Transportation Engineering.
DOI: http://dx.doi.org/10.1016/B978-0-12-811555-8.00011-8

between the samples and the one which is attributed to chance is called the variation within the samples. In ANOVA it is assumed that each of the samples drawn from a normal population and each of these populations has an equal variance.

The measure of variability used in the analysis of variance is called a "mean square," i.e.,

$$\text{Mean square} = \frac{\text{Sum of squared deviations from mean}}{\text{Degrees of freedom}}$$

The ANOVA table stands for Analysis of Variance, which is used as a statistical technique or test in detecting the difference in population means or whether or not the means of different groups are all equal when there are more than two populations.

11.2 ASSUMPTIONS

1. The samples are independently drawn.
2. The populations are normally distributed with common variance.
3. They occur at random and independent of each other in the groups.
4. The effects of various components are additive.
5. Variances of populations are equal.

11.3 ONE-WAY ANOVA

The one-way ANOVA is also called a single-factor analysis of variance because there is only one independent variable or factor. The independent population means are equal.

The one-way ANOVA compares the means of the samples or groups in order to make inferences about the population means of the independent variables are equal. It involves single independent variable.

When using one-way analysis of variance, the process of looking up the resulting value of F in an F-distribution table, is proven to be reliable under the following assumptions:

- The values in each of the groups (as a whole) follow the normal curve.
- With possibly different population averages (though the null hypothesis is that all of the group averages are equal) and equal population standard deviations.

The assumption that the groups follow the normal curve is the usual one made in most significance tests, though here it is somewhat stronger in that it is applied to several groups at once. Of course many distributions do not follow the normal curve, so here is one reason that ANOVA may give incorrect results. It would be wise to consider whether it is reasonable to believe that the groups' distributions follow the normal curve.

Of course the different population averages imposes no restriction on the use of ANOVA; the null hypothesis, as usual, allows us to do the computations that yield *F*.

The third assumption, that the populations' standard deviations are equal, is important in principle, and it can only be approximately checked by using as bootstrap estimates the sample standard deviations. In practice statisticians feel safe in using ANOVA if the largest sample SD is not larger than twice the smallest.

Completely randomized design involves the testing of equality of means of two or more groups. In this design there is only one independent variable and one dependent variable. The dependent variable is metric whereas the independent variable is a categorical variable. We briefly explain the steps involved in one-way ANOVA.

Step 1: Set up null hypothesis (H_0) and alternative hypothesis (H_1).

Step 2: Find the total T of all observations in all the samples, i.e.,

$$\sum X_1 + \sum X_2 + \cdots + \sum X_k.$$

Step 3: Find the value of the correction factor $\frac{T^2}{N}$, where

$$N = n_1 + n_2 + \cdots n_k$$

Step 4: Calculate the sum of squares of deviations SST where

$$\text{SST} = \sum X_1^2 + \sum X_2^2 + \cdots + \sum X_k^2 - \frac{T^2}{N}$$

Step 5: Find the sum of squares of deviations between the samples SSB where

$$\text{SSB} = \left[\frac{\sum X_1^2}{n_1} + \frac{\sum X_2^2}{n_2} + \cdots + \frac{\sum X_k^2}{n_k}\right] - \frac{T^2}{N}$$

Step 6: Calculate the mean square deviations within the samples (SSW) where

$$\text{SSW} = \text{SST} - \text{SSB}$$

Step 7: Find degrees of freedom $\nu_1 = df_1 = k - 1$ (k = number of columns) and degrees of freedom

$$\nu_2 = df_2 = N - k$$

Step 8: Find the mean square deviation between the samples (MSB), where

$$\text{MSB} = \frac{\text{SSB}}{\nu_1}$$

and the mean square deviation within the samples (MSW), where

$$\text{MSW} = \frac{\text{SSW}}{\nu_2}$$

Step 9: In this step calculate the F statistic by applying the formula

$$F = \frac{\text{MSB}}{\text{MSW}} \quad \text{when MSB} > \text{MSW}$$

and

$$F = \frac{\text{MSW}}{\text{MSB}} \quad \text{when MSW} > \text{MSB}$$

Step 10: In this Final step, compare calculated value of F with the tabulated value. If the calculated value is less than the tabulated value of F, accept the null hypothesis and if the calculated value is greater than table value of F, reject the null hypothesis.
Table value of F, reject the null hypothesis.

ANOVA Table

Source of variation	Sum of scores	Degrees of freedom	Mean square	Statistic—F
Between varieties	SSB	$df_1 = \nu_1$	MSB	$F = \frac{\text{MSB}}{\text{MSW}}$
(column means)	SSW	$df_2 = \nu_2$	MSW	

The method is explained with the help of an example given below:

Example 11.1: Three varieties of A, B, C wheat were shown in 4 plots each and the following yields in tonnes per acre were obtained:

A	B	C
7	3	4
4	5	5
6	5	4
8	7	2

(The table value of $F=$ at 5% level of significance for (2, 9) degrees of freedom is 2.165).

Test the significance of difference between yields of the varieties.

Solution: Null Hypothesis H_0: Varieties of wheat are not significantly different from each other in their yielding capacities.

Alternative Hypothesis H_1: Varieties of wheat are significantly different from each other in their yielding capacities.

We have the following table:

Sample A		Sample B		Sample C	
X_1	X_1^2	X_2	X_2^2	X_3	X_3^2
7	49	3	9	3	9
4	16	5	25	5	25
6	36	5	25	4	16
8	64	7	49	2	4

$n_1=4,\ \ n_2=4,\ \ n_3=4,\ \ N=n_1+n_2+n_3=4+4+4=12$ and $k=3$

$$\sum X_1 = 25, \quad \sum X_2 = 20, \quad \sum X_3 = 15$$

$$\sum X_1^2 = 165, \quad \sum X_2^2 = 108, \quad \sum X_3^2 = 61$$

$$T = \sum X_1 + \sum X_2 + \sum X_3 = 25 + 20 + 15 = 60$$

$$T^2 = (60)^2 = 3600$$

$$\text{Correction factor} = \frac{T^2}{N} = \frac{3600}{12} = 300$$

$$\text{Total sum of squares (SST)} = \sum X_1^2 + \sum X_2^2 + \sum X_3^2 - \frac{T^2}{N}$$

$$= 165 + 108 + 61 - 300 = 334 - 300 = 34$$

$$\text{Degrees of freedom} = \text{df}_1 = \nu_1 = k - 1 = 3 - 1 = 2$$

$$\text{Degrees of freedom} = \text{df}_2 = \nu_2 = N - k = 12 - 3 = 9$$

Sum of squares between the samples (SSB)

$$= \left[\frac{\sum X_1^2}{n_1} + \frac{\sum X_2^2}{n_2} + \ldots + \frac{\sum X_k^2}{n_k}\right] - \frac{T^2}{N}$$

$$= \left[\frac{(25)^2}{4} + \frac{(20)^2}{4} + \ldots + \frac{(15)^2}{4}\right] - 300$$

$$= \frac{1250}{4} - 300 = 312.5 - 300 = 12.5$$

Sum of squares = within the varieties (SSW) = SSW − SSB = 34 − 12.5 = 21.5

Mean square between varieties = MSB $= \frac{SSB}{\nu_1} = \frac{12.5}{2} = 6.25$

Mean square within varieties MSW $= \frac{SSW}{\nu_2} = \frac{21.5}{2} = 2.39$ (approx.)

Test statistic $= F = \frac{MSB}{MSW} = \frac{6.25}{2.39} = 2.615$

ANOVA Table

Source of variation	Sum of scores	Degrees of freedom	Mean square	Statistic—*F*
Between varieties (column means)	SSB = 12.5	$df_1 = v_1 = 2$	MSB = 6.25	$F = \frac{MSB}{MSW}$ = 2.615
Within samples (errors)	SSW = 21.5	$df_2 = v_2 = 9$	MSW = 2.39	
Total	34.0			

Since the calculated value of *F* is less than the table value of *F* for (2, 9) df at 5% level of significance we accept Null hypothesis.

11.3.1 Two-Way ANOVA

A one-way analysis is used to compare the populations for one variable or factor. The one-way ANOVA measures the significant effect of one independent variable. The two-way analysis of variance (ANOVA) test is an extension of the one-way ANOVA test that examines the influence of different categorical independent variables on one dependent variable. It involves two independent variables. The "two-way" comes because each item is classified in two ways, as opposed to one way. The test is useful when we desire to compare the effect of multiple levels of two factors and we have multiple observations at each level.

An important advantage of this design is it is more efficient than its one-way counterpart. There are two assignable sources of variation—age and gender in our example—and this helps to reduce error variation thereby making this design more efficient. Unlike one-way ANOVA it enables us to test the effect of two factors at the same time. The assumptions in both versions remain the same—normality, independence, and equality of variance. One can also test for independence of the factors provided there are more than one observation in each cell. The only restriction is that the number of observations in each cell has to be equal (there is no such restriction in case of one-way ANOVA).

11.3.1.1 Assumptions

The assumptions in both versions remain the same—normality, independence, and equality of variance.

The populations from which the samples were obtained must be normally or approximately normally distributed.

- The samples must be independent.
- The variances of the populations must be equal.
- The groups must have the same sample size.

11.3.1.2 Hypothesis

There are three sets of hypothesis with the two-way ANOVA.

The null hypothesis for each of the sets are given below:

1. The population means of the first factor are equal. This is like the one-way ANOVA for the row factor.
2. The population means of the second factor are equal. This is like the one-way ANOVA for the column factor.
3. There is no interaction between the two factors. This is similar to performing a test for independence with contingency tables.

The two independent variables in a two-way ANOVA are called factors. The idea is that there are two variables, factors, which affect the dependent variable. Each factor will have two or more levels within it, and the degrees of freedom for each factor is one less than the number of levels.

11.4 WORKING RULE

The following steps are involved in the short-cut method for ANOVA:

Step 1: Set up null hypothesis (H_0) and alternative hypothesis (H_1)

Step 2: Find the sum of the values of all items of all samples and denote it by T

i.e.,

$$T = \sum X_1 + \sum X_2 + \cdots + \sum X_k$$

Step 3: Find the value of correction factor $= \dfrac{T^2}{N}$

Step 4: Find the sum of squares of all items of all samples

Step 5: Find the total sum of squares SST, where

$$\text{SST} = \sum X_1^2 + \sum X_2^2 + \cdots + \sum X_k^2 - \frac{T^2}{N}$$

Step 6: Find the total sum of squares between the samples SSC

$$= \left[\frac{\sum X_1^2}{n_1} + \frac{\sum X_2^2}{n_2} + \cdots + \frac{\sum X_k^2}{n_k}\right] - \frac{T^2}{N}$$

Step 7: Find the sum of squares within the samples SSE where

$$\text{SSE} = \text{SST} - \text{SSC}$$

Step 8: Set up ANOVA table and calculate F, which is the test statistic.

Source of variation	Sum of scores	Degrees of freedom	Variance ratio	F
Between columns	SSC	$c-1$	MSC	$F_C = \dfrac{\text{MSC}}{\text{MSE}}$
Between rows	SSR	$r-1$	MSR	$F_R = \dfrac{\text{MSR}}{\text{MSE}}$
Error	SSE	$(c-1)\ (r-1)$	MSE	

***Example 11.2*:**

Plot of land	Yield A	Yield B	Yield C	Yield D
I	3	4	6	6
II	6	4	5	3
III	6	6	4	7

(For (3,9) df $F_{0.05} = 4.76$ and for (2, 6) df $F_{0.05} = 5.14$).

Solution:

Null hypothesis

1. There is no significant difference in the yield of four varieties of wheat.
2. There is no significant difference in the plots of land with regard to yield.

X_1	X_1^2	X_2	X_2^2	X_3	X_3^2	X_4	X_4^2
3	16	4	16	6	36	6	36
6	36	4	16	5	25	3	09
6	36	6	36	4	16	7	49
$\sum X_1=15$	$\sum X_1^2=15$	$\sum X_2=14$	$\sum X_2^2=68$	$\sum X_3=15$	$\sum X_3^2=77$	$\sum X_4=16$	$\sum X_4^2=94$

We have

$$\sum X_1 = 15, \quad \sum X_2 = 14, \quad \sum X_3 = 15, \quad \sum X_4 = 16,$$

$$n_1 = 3, \ n_2 = 3. \ n_3 = 3, \ n_4 = 3, \ N = 3 + 3 + 3 + 3 = 12$$

$$T = \sum X_1 + \sum X_2 + \sum X_3 + \sum X_4 = 15 + 14 + 15 + 16 = 60$$

$$\text{Correction factor} = \frac{T^2}{N} = \frac{(60)^2}{12} = \frac{3600}{12} = 300$$

Total sum of squares

$$= SST = \sum X_1^2 + \sum X_2^2 + \cdots + \sum X_k^2 - \frac{T^2}{N}$$

$$= 88 + 68 + 77 + 94 - 300$$

$$= 320 - 300 = 20$$

Sum of squares between the samples (i.e., between columns) (SSC)

$$= \left[\frac{\sum X_1^2}{n_1} + \frac{\sum X_2^2}{n_2} + \cdots + \frac{\sum X_k^2}{n_k}\right] - \frac{T^2}{N}$$

$$= \left[\frac{(15)^2}{3} + \frac{(14)^2}{3} + \frac{(15)^2}{3} + \frac{(16)^2}{3}\right] - 300$$

$$= \frac{902}{3} - 300 = 300.67 - 300 = 0.67 = 303.5 - 300 = 3.5$$

Sum of squares within the samples (between rows) SSR

$$= \left[\frac{(19)^2}{4} + \frac{(18)^2}{4} + \frac{(23)^2}{4}\right] - 300$$

Error = SST − (SSC + SSR) = 20 − (0.67 + 3.5) = 15.83

Source of variation	Sum of squares	Degrees of freedom (v)	Variance ratio	F
Between columns (yields)	0.67	$v_1 = c - 1 = 3$	$MSC = \frac{MSC}{v_1} = 0.223$	$F_C = \frac{MSC}{MSE} = 0.076$
Between rows	3.50	$v_2 = 2$	$MSR = \frac{SSR}{R-1} = 1.75$	$F_R = \frac{MSR}{MSE} = 0.597$
Error	15.83	6	$MSE = \frac{15.83}{6} = 2.930$	

For (3, 6) df, the calculated value of F is less than the table value of F, hence we accept null hypothesis, i.e., there is no significant difference in the yields of four varieties of wheat.

For (3, 6) df, the calculated value of F is less than the table value of F, hence we accept null hypothesis, i.e., there is no significant difference in the yields of four varieties of wheat.

For (2, 6) df, the calculated value of F is less than the table value of F, hence we accept null hypothesis, i.e., there is no significant difference in the y plots of land with regard to yield.

Exercise 11.1

1. A farmer applied three types of fertilizers on four separate plots. The figure on yields per acre are tabulated below:

	Yield				
Fertilizers/plots	***A***	***B***	***C***	***D***	**Total**
Nitrogen	6	4	8	6	24
Potash	7	6	6	9	28
Phosphates	8	5	10	9	32
Total	21	15	24	24	84

Find out if the plots are materially different in fertility, as also, if the three fertilizers make any material difference in yields?

2. The varieties of A, B, C wheat were sown in four plots each and the following yields in tonnes per acre were obtained:

A	*B*	*C*
8	7	2
4	5	5
6	5	4
7	3	4

Test the significance of difference between the yield of the varieties.

(Table value of F at 5% level of significance for (2, 9) is (4, 26)).

3. The three samples below have been obtained from normal populations with equal variances. Test the hypothesis at 5% level that the population means are equal.

X	*Y*	*Z*
8	7	12
10	5	9
7	10	13
14	9	12
11	9	14

Test the significance of difference between the yield of the varieties.

(Table value of F at 5% level of significance for (2, 12) is 3.88).

4. The varieties of A, B, C wheat were sown in four plots each and the following yields in tonnes per acre were obtained:

A	*B*	*C*
20	18	25
21	20	28
	17	22
23	17	28
16	25	32
20	15	

Test at 5% level of significance whether the average yields of land under different varieties seed show significance difference.

(Table value of at 5% level for (2, 12) is 3.88).

5. The following data represents the number of units of a product produced by 3 different workers using three different types of machines:

Workers	Machine A	Machine B	Machine C
X	8	32	20
Y	28	36	38
Z	6	28	14

Test (a) whether the mean productivity is the same for the different machine types, and (b) whether the three workers differ with respect to mean productivity. Table value of F_C at 5% level for (2, 4) df is 6.95 (And the table value of $F_R = 6.95$).

CHAPTER 12

Analysis of Time Series

12.1 INTRODUCTION

There are many factors that change with the passage of time. For example, consumer price index, rate of inflation, the yearly demands of a commodity, balance of trade, annual profit of a firm, etc. In this chapter we consider data that are collected sequentially over time: i.e., time series data. Numerical variables that are calculated, measured, or observed sequentially on a regular chronological basis are called Time Series. Many economists and statisticians have defined time series in different words. Some of them are quoted below:

A time series is a set of statistical observations arranged in chronological order

—Morris Hamburg

A time series is a sequence of values of the same variate corresponding to successive points of time.

—Werner Z. Hirsch

A time series is a set of statistical data which are collected, recorded or observed over successive increments.

—Patterson

The importance of time series is obvious once it is realized that the main problem of business is making forecasts for the future, which cannot be done unless data representing change over a period of time is analyzed.

The time series data are subjected to two kinds of analysis: Descriptive and Inferential. The descriptive analysis uses graphical and numerical techniques. It provides clear understanding of the time series. Forecasts and their measures of reliability are examples of inferential techniques in time series analysis. They are generally focused on the problems of forecasting future values of the time series.

12.2 PURPOSE OF TIME SERIES STUDY

The analysis of time series is useful to economists, scientists, business that persons, etc. It is also found useful in the study of seismology,

Statistical Techniques for Transportation Engineering.
DOI: http://dx.doi.org/10.1016/B978-0-12-811555-8.00012-X

oceanography, meteorology, etc. The purpose of time series study is to measure chronologically particular variations. Time series analysis helps in understanding the cream in phenomena.

1. It describes the past movements and fluctuations.
2. It helps us to compare the present accomplishments with past performances.
3. It helps in predicting the future behavior like price, production, demand, etc.
4. It helps us to compare two or more series at a time.
5. Time series analysis helps to evaluate the achievements.

12.3 EDITING OF DATA

The time series data are always given over specified periods and one value is generally compared with the other. So before the analysis of time series, the data have to be critically examined and adjusted for various factors. Otherwise discrepancies are likely to arise leading to wrong conclusions. Mainly one or more types of adjustments are needed. Normally the types of adjustments incorporated in a time series data are:

1. *Calendar Variation*: In some cases monthly consumption of an item is compared within a year. The number of days are different in different months, e.g., the number of days in March is 31 and the number of days in April is 30. Therefore some adjustments are made, therefore are should be worry of the type of variable dealing with before implementing the adjustments for calendar variation. Wages should not be adjusted as the salary in India is paid on monthly basis.
2. *Price Variation*: Production or sales are to be adjusted for price variation by the formula.

   $$q = v/p$$

   where q = Quantity of production or sales in a specified period; v = Sales in terms of money; and p = Price per unit in the reference period.

   If this adjustment is not done, rising of prices will lead to the conclusion that production or sales have increased, though in reality it may not be so.
3. *Population Changes*: Consumption of sugar, food grain, etc., depends on the number of consumers. If there is an increase in population, then there will be increase of consumption. Similarly demand is also directly related to the population.

If the demand doubles and population also doubles, it should not be that demand has increased.

4. *Miscellaneous Changes*: We observe that many changes occur with the lapse of time. Synthetic fiber cloth that is more durable is now produced in our country. Some years ago synthetic fiber was not available in our country. We know that units of measurement have also changed. Earlier it was yards and miles, now it is meter and kilometer. Therefore for comparison of two time series, they should be converted to the same unit of measurement.

12.4 COMPONENTS OF TIME SERIES

There are four components of a time series, namely

1. Secular trend;
2. Seasonal variation;
3. Cyclic variation;
4. Irregular variation.

The components are also called the elements of a time series.

1. *Secular trend (or Trend)*(T_t): The secular trend also known as long-term trend means long-term movement. It is denoted by T_t (or by T). The population, production, sales, etc., have an increasing or decreasing tendency over periods of time, secular trend measures long-term changes occurring in a time series without bothering about short-term fluctuations occurring in between. If you want to characterize the secular trend of the production of two-wheelers since 1970. You would show T_t as an upward-moving time series over the period from 1970 to the present. This does not imply that the two-wheeler production series has always moved upward from month to month and from year to year, but it does mean the long-term trend has been an increasing one over that period of time. Thus the secular trend is concerned with regular growth or decline.
2. *Seasonal variation* (S_t): The seasonal variation describes the fluctuations in the time series that recur during specific time periods. It is denoted by S_t (or S) For instance certain items have tremendous sale during festivals only in a particular month; rain coats are sold in rainy season; soft drinks and ice cream in summer. All such variations in time series comes under seasonal variation.
3. *Cyclic variation* (C_t): Cyclic variation relates to periodic changes, cycles related to business are termed as business cycles (or trade cycles). A business

normally exceeds a year in length. The changes that occur for periods more than 1 year come under the category of cyclic variations. These cycles are never regular in periodicity and amplitude. Hardly any time series has strict cycles. A large number of factors are responsible for the occurrence of cycles. The changes in social customs, like and dislikes of people, new scientific and technological developments, etc., are some factors responsible for creation of cycles. Cyclic variation is denoted by the symbol C_t (or by C).

4. *Irregular variations* (I_t): These variations are due to famines, strikes, wars, droughts, earthquakes, and other calamities. These variations are also called residual variations.

12.5 MATHEMATICAL MODEL FOR A TIME SERIES

A Mathematical model is used to describe a time series. The objective of a mathematical model is to produce accurate forecasts of future values of the time series. Many different algebraic representations of time series models have been proposed. If Y_t denotes any particular observation at time t, then the multiplicative model is of the form:

$$Y_t = T_t \times S_t \times C_t \times I_t$$

or

$$(Y = T \times S \times C \times I)$$

and the additive model is of the form

$$Y_t = T_t + S_t + C_t + I_t$$

or

$$(Y = T + S + C + I)$$

The multiplicative model does not assume the independence of the four components T, S, C, and I of the time series, but the additive model is based on the assumption that four components are independent of each other. We can also use mixed models. Some of the examples of mixed models are given below:

$$Y = T \times C + S \times I$$

$$Y = T + S \times C \times I$$

12.6 METHODS OF MEASURING TREND

The idea of measuring the trend is to estimate the average growth or decline. The precondition for the measurement of secular trend is that the data must be available for a long period. Otherwise it will not be possible to isolate and estimate the growth or decline in the trend. The following methods are used for measuring the trend:

1. Free-hand method or Graphic method
2. Semiaverage method
3. Moving average method
4. Least square method

12.6.1 Free-Hand Method

The free-hand method is the simplest method of ascertaining trend. On the graph paper, time t is measured horizontally, whereas the values of the variable y are measured vertically. The points (t_1, y_1), (t_2, y_2), ..., (t_n, y_n) are plotted on a graph paper by taking t_i on the x-axis and y_i on the y-axis. The plotted points are then joined by straight lines and the trend line is fitted with the help of a transparent ruler or a smooth curve by hand. The trend line is drawn in such way that the following are satisfied as far as possible:

1. The algebraic sum of the deviations of actual values from the trend values are zero.
2. The sum of the squares of the deviations of actual values from the trend values is least.
3. The area above the trend is equal to the area below it.

A free-hand curve removes the short-term variations and exhibits a general trend. Free-hand method is one of the simplest and most flexible methods and sometimes yield good results. It is too subjective. In this method the trend line varies from person to person, as it depends on individual's judgment. Hence it cannot be used as a basis of prediction.

Example: Fit a straight line trend to the following data by using free-hand graph method:

Years	1970	1971	1972	1973	1974	1975	1976
Profit of the firm X20 (in lakhs of Rs.)	20	30	25	40	45	30	55

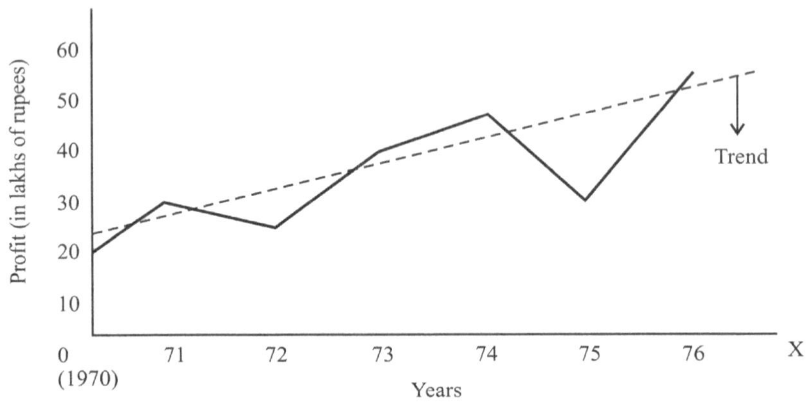

12.6.2 Semiaverage Method

In this method the series is divided into two halves. Then the average is found out for each half of the series. The average values are plotted on the graph paper against the mid periods of the corresponding each half series. The line joining these two plotted points gives the trend line. The direction of the line indicates about rising, falling, or constant trend of business movements.

If the number of years in a series is odd, the middle year is excluded at the time of dividing the series into two halves, and then either excluded in both series or excluded totally depending on whether the leftover series contains an even or odd number of years respectively.

The semiaverage method is not subjective and for one series, there is only one trend line. But it is affected by extreme values. This method does not ensure elimination of short-term and cyclic variations. Semiaverage method is superior to free-hand method. It may be used when the data are given for a longer period.

Example 12.1: Fit a straight line trend by using the following data:

Years	1981	1982	1983	1984	1985	1986	1987
Profit (in lakhs of Rs.)	20	22	27	26	30	29	40

Use semiaverage method. Also estimate the profit for the year 1988.

Solution: Trend line by semiaverage method number of years is 7, i.e., odd. We leave the middle most value, i.e., the value corresponding to the year 1984.

First part: (First half)		Ordinate to be plotted	Second part: (Last half)		Ordinate to be plotted
Year	Profit		Year	profit	
1981	20	Middle year = 32	1985	30	Middle year = 33
1982	22		1986	29	
1983	27		1986	40	
1984	26				

Mean of the first half of the series = Mean

$$= \frac{20 + 22 + 27}{3} = \frac{69}{3} = 23$$

Mean of the last half of the series = Mean

$$= \frac{30 + 29 + 40}{3} = \frac{99}{3} = 33$$

Plot 23 against 1982 and 33 against 1986. Join these points as shown in figure below.

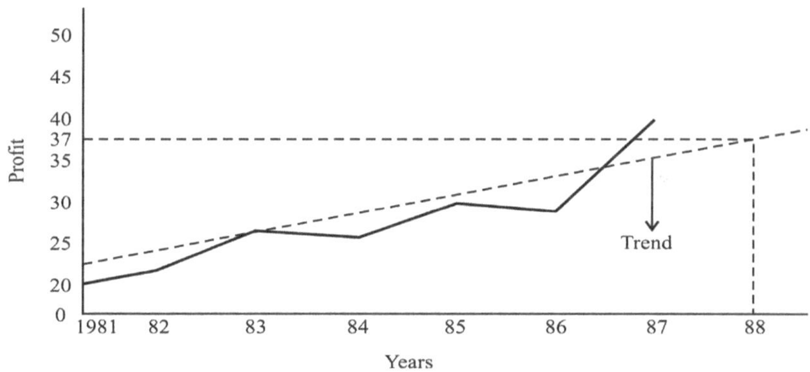

The profit in the year 1988 is Rs. 37 lakhs.

Example 12.2: Draw the trend by semiaverage method from the following data:

Year	1800	1810	1820	1830	1840	1850	1860
Price (Index No.)	129	131	106	91	95	84	93
Year	1870	1880	1890	1900	1910	1920	1930
Price (Index No.)	135	100	82	82	103	226	126

Solution: The data correspond to 14 years

$$\text{No. of years in each half} = \frac{14}{2} = 7$$

We break the data into two equal parts of 7 years each.

First part (first half)		Ordinate to be plotted	Second part (last half)		Ordinate to be plotted
Year	Index No.		Year	Index No.	
1800	129		1870	135	
1810	131		1880	100	
1820	106	Middle year = 104	1890	82	Middle year = 122
1830	91		1900	82	
1840	95		1910	103	
1850	84		1920	226	
1860	93		1930	126	
	729			854	

$$\text{Average of the first half series} = \frac{729}{7} = 104.14 = 104 \text{ (approx.)}$$

$$\text{Average of the half of the series} = \frac{854}{7} = 122$$

Plot 104 against 1830 and 122 against 1900. Join these points (Fig. 12.1).

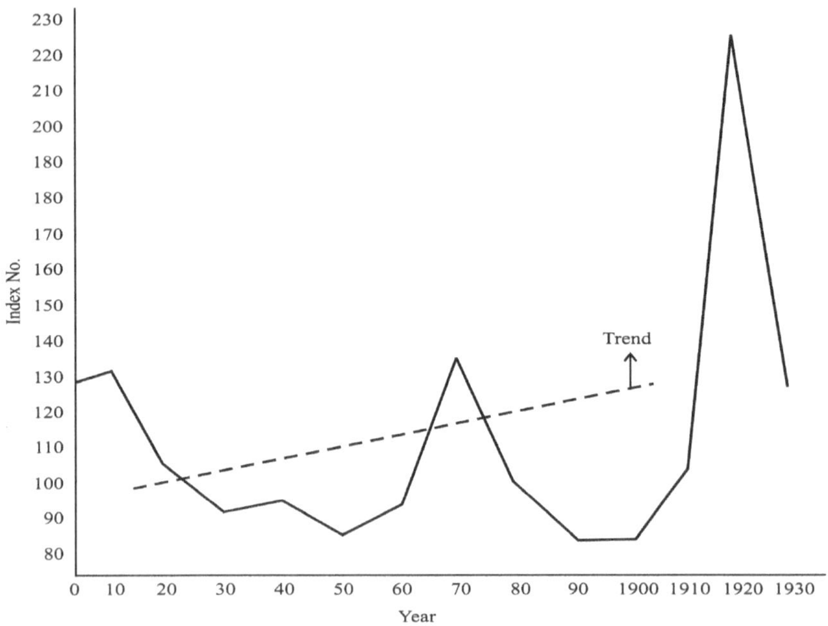

Figure 12.1 Semiaverage method.

12.6.3 Moving Average Method

The moving average method is an improvement over the semiaverage method and short-term fluctuations are eliminated by it. A moving average is defined as an average of fixed number of items in the time series which move through the series by dropping the top items of the previous averaged group and adding the next in each successive average.

Let $(t_1, y_1), (t_2, y_2), \ldots, (t_n, y_n)$ denote given time series $y_1, y_2, \ldots, y_n$ are the values of the variable y; corresponding to time periods $t_1, t_2, \ldots, t_n$, respectively.

The moving averages of order m are defined as

$$\frac{y_1 + y_2 + \cdots + y_m}{m}; \quad \frac{y_2 + y_3 + \cdots + y_{m+1}}{m};$$

Here $y_1 + y_2 + \cdots + y_m, y_2 + y_3 + \cdots + y_{m+1}, \ldots$ are called moving totals of m.

In using moving averages in estimating the trend, we shall have to decide as what should be the order of the moving averages. The order of the moving average should be equal to the length of the cycles in the time series. In case the order of the moving averages is given in the problem itself, then we shall use that order for computing the moving average. The order of the moving averages may either be odd or even.

The moving averages of order 3 are

$$\frac{y_1 + y_2 + y_3}{3}; \quad \frac{y_2 + y_3 + y_4}{3}; \ldots; \quad \frac{y_{n-2} + y_{n-1} + y_n}{3}$$

These moving averages are called the trend values. They are considered to correspond 2nd, 3rd, ..., $(n-1)$th years, respectively. Calculation of trend values by using moving averages of even order is slightly complicated. The following steps are involved in the method:

Step 1: In the first step, a group of beginning years (periods), which constitute cycle is chosen for calculating the average. This average is placed in front of the mid-year of the group.

Step 2: Now delete the first year value from the group and add a succeeding year value in the group. Find the average of the reconstituted group and place it in front of this group.

Step 3: If the number of years in a group is odd, middle year is located without any problem. But if the number of years in the group is even, the average of the averages in pairs is calculated and placed against the mid-year of the two.

Step 4: Repeat the Step 2 till all years of the data are exhausted.
Step 5: The moving averages calculated are considered as an artificially constructed time series.
Step 6: Plot the moving averages on a graph paper taking years along x-axis and moving averages along y-axis by choosing a proper scale.
Step 7: Join the plotted point in the sequence of time periods. The resulting graph provides the trend.

12.6.3.1 Merits and Demerits of the Moving Average Method

Merits

1. The moving average method eliminates the short-term fluctuations.
2. It reduces the effect of extreme values.
3. As the free-hand method, this method is not subject to personal prejudice and bias of the estimator.
4. This method is a flexible method.

Demerits

1. Moving average method is not fully mathematical.
2. If the series given is a very large one, then the calculation of moving average is cumbersome.
3. The choice of the period of moving average needs a great amount of care. If an inappropriate period is selected, a true picture of the trend cannot be obtained.
4. It is very much affected by extreme values.

12.6.3.1.1 Solved Examples

Example 12.3: Estimate the trend values using the data given below by taking a 3-yearly moving averages.

Year	1980	1981	1982	1983	1984	1985	1986	1987	1988
Sales (in lakhs of units)	65	95	80	115	105	135	125	150	140

Solution: Trend by 3-yearly moving averages

Year	Sales (in lakhs of units)	3-year moving totals	3-year moving averages
1980	65	–	–
1981	95	65 + 95 + 80 = 240	240/3 = 80
1982	80	95 + 80 + 115 = 290	290/3 = 96.67
1983	115	80 + 115 + 105 = 300	300/3 = 100

(Continued)

Year	Sales (in lakhs of units)	3-year moving totals	3-year moving averages
1984	105	115 + 105 + 135 = 355	355/3 = 118
1985	135	105 + 135 + 125 = 365	365/3 = 121.67
1986	125	135 + 125 + 150 = 410	410/3 = 136.67
1987	150	125 + 150 + 140 = 415	415/3 = 138.33
1988	140	–	–

Example 12.4: Estimate the trend values using the data given below by taking a 4-yearly moving averages:

Year	Value	Year	Values
1969	4	1975	24
1970	7	1976	36
1971	20	1977	25
1972	15	1978	40
1973	30	1979	42
1974	28	1980	45

Solution: We use a process called "centering of moving averages." In this method we first calculate 4-yearly moving averages. The first and second 4-year moving averages are added and the total is divided by two and written opposite the gap between the first two 4-year moving averages. The second and third 4-year moving averages are added and the total is divided by two. In short, 2-year moving averages of the 4-year moving averages are calculated. The 2-year moving averages of the 4-year moving averages are called the central 4-year moving averages.

The following table given the centered 4-year moving averages:

Year	Value	4-yearly moving totals	4-yearly moving averages	4-yearly centered moving averages
1969	4			
1970	7	46	11.5	14.75
1971	20	72	18.0	20.625
1972	15	93	23.25	23.75
1973	30	97	24.25	26.875
1974	28	118	29.5	28.875
1975	24	113	28.25	29.75
1976	36	125	31.25	33.50
1977	25	143	35.75	36.875
1978	40	152	38	
1979	42			
1980	45			

Example 12.5: Compute 7 years' moving averages.

Year	Values	Year	Values	Year	Values
1954	496	1959	1081	1964	1442
1955	615	1960	1132	1965	1617
1956	686	1961	1139	1966	1678
1957	835	1962	1320		
1958	888	1963	1389		

Solution: See Table 12.1.

Table 12.1 Seven-year moving averages

Year	Value	7-year moving totals	7-year moving averages
1954	496	–	–
1955	615	–	–
1956	686	5733	–
1957	835	6376	819.00
1958	888	7081	910.85
1959	1081	7784	1011.60
1960	1132	8391	1112.00
1961	1139	9120	1198.70
1962	1320	9717	1303
1963	1389	–	1388
1964	1442	–	–
1965	1617	–	–
1966	1678	–	–

12.6.4 Method of Least Square

The method of least squares is a widely used method of fitting curve for a given data. It is the most popular method used to determine the position of the trend line of a given time series. The trend line is technically called the best fit. In this method a mathematical relationship is established between the time factor and the variable given. Let (t_1, y_1), (t_2, y_2), ..., (t_n, y_n) denote the given time series. In this method the trend value y_c of the variable y are computed so as to satisfy the conditions:

1. The sum of the deviations of y from their corresponding trend values is zero.

   i.e.,

$$\sum(y - y_c) = 0$$

2. The sum of the square of the deviations of the values of y from their corresponding trend values is the least.
i.e.,

$$\sum (y-y_c)^2$$

is least.

The equation of the trend line can be expressed as

$$y_c = a + bx$$

where a and b are constants and the trend line satisfies the conditions:

1. $\sum(y - y_c) = 0$
2. $\sum (y-y_c)^2$ is least.

The values of a and b determined such that they satisfy the equations.

$$\sum y = na + b\sum x \tag{12.1}$$

$$\sum xy = b\sum x + a\sum x^2 \tag{12.2}$$

Eqs. (12.1) and (12.3) are called normal equations.
Solving Eqs. (12.1) and (12.2) we get

$$a = \frac{\sum y \cdot \sum x^2 - \sum x \sum xy}{n\sum x^2 - \left(\sum x\right)^2}$$

and

$$b = \frac{n \cdot \sum xy - \sum x \sum y}{n\sum x^2 - \left(\sum x\right)^2}$$

In the equation, $y_c = a + bx$, of the trend, a represents the trend of the variable when $x = 0$ and b represents the slope of the trend line. If b is positive, the trend line will be upward and if b is negative the trend line will be downward.

When the origin is mentioned and the deviations from the origin is denoted by x, we get

$$a = \frac{\sum y}{n}, \quad b = \frac{\sum xy}{\sum x^2}$$

$\because$ (The sum of derivation from the origin $= \sum x = 0$).

Note: If n is odd, we take the middle value (Middle year) as the origin. If n is even, there will be two middle values. In this case we take the mean of the two middle values as the origin.

Merits

1. The method is mathematically sound.
2. The estimates a and b are unbiased.
3. The least square method gives trend values for all the years and the method is devoid of all kinds of subjectivity.
4. The algebraic sum of deviations of actual values from trend values is zero and the sum of the deviations $\sum(y-y_c)^2$ is minimum.

Demerits

1. The least square method is highly mathematical, therefore, it is difficult for a layman to understand it.
2. The method is not flexible. If certain new values are included in the given, time series, the values of n, $\sum x$, $\sum y$, $\sum x^2$, and $\sum xy$ would change. Which affects the trend values.
3. It has been assumed that y is only a linear function of time period x. Which may not be true in many situations.

12.6.4.1 Solved Examples

Example 12.6: Find the least square line $y = a + bx$ for the data:

x	−2	−1	0	1	2
y	1	2	3	3	4

Solution:

x	y	x^2	xy
−2	1	4	−2
−1	2	1	−2
0	3	0	0
1	3	1	3
2	4	4	8
$\sum x=0$	$\sum y=13$	$\sum x^2=10$	$\sum xy=7$

$\therefore \sum x = 0, \sum y = 13, \sum x^2 = 10, \sum xy = 7$, and $x = 5$

The normal equations are

$$na + b\sum x = \sum y$$

$$a\sum x + b\sum x^2 = \sum xy$$

Putting the values of n, $\sum x, \sum y, \sum x^2, \sum xy$ in the above equation, we get

$$5a + b \cdot 0 = 13 \Rightarrow 5a = 13 \tag{12.3}$$

$$a \cdot 0 + b(10) = 7 \Rightarrow 10b = 7 \tag{12.4}$$

From Eqs. (12.3) and (12.4) we get

$$a = \frac{13}{5} = 2.6, \quad b = \frac{7}{10} = 0.7$$

The required of least square line is

$$y = 2.6 + (0.7)x$$

Example 12.7: Fit a straight line trend by the method of least square from the following data and find the trend values.

Year	1958	1959	1960	1961	1962
Sales (in lakhs of units)	65	95	80	115	105

Solution: We have $n = 5$

$\therefore$ n is odd.

Taking middle year, i.e., 1960 as the origin. We get,

Year	Sales	x	x^2	xy
1958	65	−2	4	−130
1959	95	−1	1	−95
1960	80	0	0	0
1961	115	1	1	115
1962	105	2	4	210
Total	$\sum y = 460$	$\sum x = 0$	$\sum x^2 = 10$	$\sum xy = 100$

$\therefore$ $n = 5, \sum x = 0, \sum x^2 = 10, \sum y = 460$, and $\sum xy = 100$

$$a = \frac{\sum y}{n} \Rightarrow a = \frac{460}{5} = 92$$

$$b = \frac{\sum xy}{\sum x^2} = \frac{100}{10} = 10$$

$\therefore$ The equation of the straight line trend is

$$y_c = a + bx \Rightarrow y_c = 92 + 10x$$

For the year 1958, $x = -2$

$$\Rightarrow y_c = y_{1958} = 92 + 10(-2)$$
$$= 92 - 20 = 72$$

For the year 1959, $x = -1$

$$\Rightarrow y_c = y_{1959} = 92 + 10(-1)$$
$$= 92 - 10 = 82$$

For the year 1960, $x = 0$

$$\Rightarrow y_c = y_{1960} = 92 + 10(0)$$
$$= 92 + 0 = 92$$

For the year 1961, $x = 1$

$$\Rightarrow y_c = y_{1961} = 92 + 10(1)$$
$$= 92 + 10 = 102$$

For the year 1962, $x = 2$

$$\Rightarrow y_c = y_{1962} = 92 + 10(2)$$
$$= 92 + 20 = 112$$

We have

Year	Trend value
1958	72
1959	82
1960	92
1961	102
1962	112

and the straight line trend is $y_c = 92 + 10x$ or simply $y = 92 + 10x$.

Example 12.8: Determine the trend by the method of least squares. Also find the trend values.

Year	1950	1951	1952	1953	1954	1955	1956	1957
Value	346	411	392	512	626	640	611	796

Solution: Here $n = 9$

$\therefore$ n is even.

1953 and 1954 are the middle years.

The origin is $\dfrac{1953 + 1954}{2} = 1953.5$

We take $x = \text{year} - 1953.5$

Year	x	y	x^2	xy
1950	−3.5	346	12.25	−1211.00
1951	−2.5	411	6.25	−1027.50
1952	−1.5	392	2.25	−588.00
1953	−0.5	512	0.25	−256.00
1954	0.5	626	0.25	313.00
1955	1.5	640	2.25	960.00
1956	2.5	611	6.25	1527.50
1957	3.5	796	12.25	2786.00
Total	0	4334	42.00	2504.00

∴ We have $n = 8$, $\sum x = 0$, $\sum y = 4334$, $\sum x^2 = 42$, and $\sum xy = 2504$

$$a = \frac{\sum y}{n} \Rightarrow a = \frac{4334}{8} = 541.75$$

$$b = \frac{\sum xy}{\sum x^2} = \frac{2504}{42} = 59.60$$

∴ The equation of the trend line is

$$y = 541.75 + (59.60)x$$

The trend values are:

Year	x	Trend: $y = 541.75 + (59.60)x$
1950	−3.5	333.15
1951	−2.5	392.75
1952	−1.5	452.35
1953	−0.5	511.95
1954	0.5	571.95
1955	1.5	631.15
1956	2.5	690.75
1957	3.5	750.35

12.6.5 Nonlinear Trend

So far the discussion about trend is confined mostly to the linear trend. There are situations where linear trend is not found suitable. For instance consider the population growth. In the beginning of a period population growth will be slow but will start to multiply at a faster rate, in the later period. For such a type of time series data, a nonlinear trend would depict a better trend than the linear one. Some well-known methods of determining nonlinear trends are:

1. Free-hand method
2. Moving average method and
3. Method of least squares.

The above methods have been discussed in this chapter. If the trend is changing frequently, a curve will give a picture of trend. Which curve is best suited to the data may be guessed by plotting time series data on a graph. When the shape of the curve is known, the mathematical equation for it may be given. Once the equation is decided, it can be fitted to the available time series data. Some of the frequently used curve are:

1. Parabola
2. Exponential curve
3. Compertz curve
4. Pearl-Reed curve and
5. Modified exponential curve.

We shall restrict ourselves only to the study of parabolic curves.

Parabola: The parabolic curve is mathematically represented by a second-degree polynomial of the form

$$y = a + bx + cx^2$$

It can be fitted by the method of least square, using the method of least squares we get the three normal equations.

$$na + b\sum x + c\sum x^2 = \sum y$$

$$a\sum x + b\sum x^2 + c\sum x^3 = \sum xy$$

$$a\sum x^2 + b\sum x^3 + c\sum x^4 = \sum x^2y$$

Solving the above equations we obtain the values for a, b, and c. Substituting the values of a, b, and c is

$$y = a + bx + cx^2$$

We get the parabola. The curve is a quadratic curve and is also known as a response curve. The shape of the curve depends on the values of b and c. Its geometric shapes are as shown in Fig. 12.1 (Fig. 12.2).

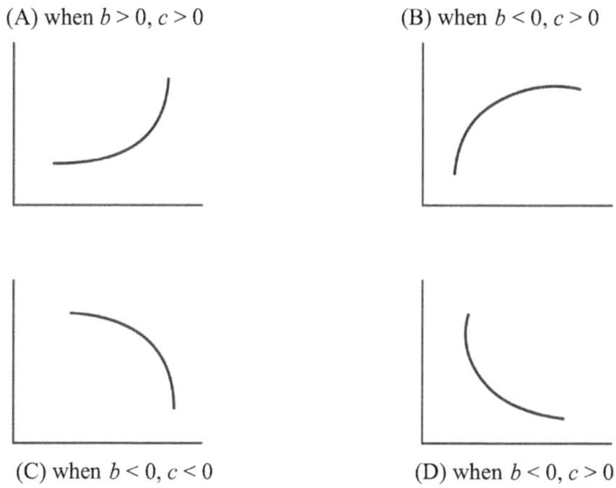

Figure 12.2 Response curve.

Example 12.9: The prices to the commodities during 1978–83 are given below. Fit a parabola $y = a + bx + cx^2$ to the data. Estimate the price of the commodity for the year 1984.

Year	1978	1979	1980	1981	1982	1983
Price	100	107	128	140	181	192

Also find the trend values.

Solution: No. of periods $= n = 6$

We take the origin as $\dfrac{1980 + 1981}{2} = 1980.5$

We define $x = (\text{Year} - 1980.5) \times 2$

Let y denote the price of the commodity.

Trend by least square method.

S. No.	Year	y	x	x^2	x^3	x^4	xy	x^2y
1.	1978	100	−5	25	−125	625	−500	2500
2.	1979	107	−3	9	−27	81	−321	963
3.	1980	128	−1	1	−1	1	−128	128
4.	1981	140	1	1	1	1	140	140
5.	1982	181	3	9	27	81	543	1629
6.	1983	192	5	25	125	625	960	4800

The normal equation are

$$na + b\sum x + c\sum x^2 = \sum y$$

$$a\sum x + b\sum x^2 + c\sum x^3 = \sum xy$$

$$a\sum x^2 + b\sum x^3 + c\sum x^4 = \sum x^2 y$$

i.e.,

$$6a + b.0 + 70c = 848 \Rightarrow 6a + 70c = 848 \quad (12.5)$$

$$a.0 + 70b + c.b = 694 \Rightarrow 70b = 694 \quad (12.6)$$

$$70.a + b.0 + 1414c = 10160 \Rightarrow 70a + 1414c = 10160 \quad (12.7)$$

From Eq. (12.6) we get

$$70b = 694 \Rightarrow b = \frac{694}{70} = 9.914$$

Consider Eqs. (12.5) and (12.7)

$$\begin{array}{r} 35 \times (6a + 70c = 848) \\ 3 \times \underline{(70a + 1414c = 10160)} \\ 210a + 2450c = 29680 \\ \underline{210a + 4242c = 30480} \\ -1792c = -800 \end{array} \quad (12.8)$$

$$\Rightarrow \quad c = \frac{800}{1792} = 0.446$$

$$\therefore \quad c = 0.446$$

From Eq. (12.7) $6a + 70(0.446) = 848$

$$\Rightarrow \quad 6a + 848 - 70(0.446) \Rightarrow a = 136.13$$

$$\therefore \quad a = 136.13, \quad b = 9.14x + 0.4462x^2$$

where 1980.5 is the origin and x unit is 1/2 years.

Trend values

Year	x	Trend value
1978	−5	136.13 + 9.914 (−5) + 0.446 (25) = 97.710
1976	−3	136.13 + 9.914 (−3) + 0.446 (9) = 110.402
1980	−1	136.13 + 9.914 (−1) + 0.446 (1) = 126.662

(Continued)

Year	x	Trend value
1981	1	$136.13 + 9.914\ (1) + 0.446\ (1) = 146.990$
1982	3	$136.13 + 9.914\ (3) + 0.446\ (9) = 169.886$
1983	5	$136.13 + 9.914\ (5) + 0.446\ (25) = 196.85$

Estimated value of price for the year 1984:
We have $x = (1984 - 1980.5) \times 2 = 7$
Estimated price for the year 1984:

$$y_c = y_{1984} = 136.13 + (9.914)(7) + (0.446)(49)$$
$$= 227.382 \text{ units}$$

12.6.6 Conversions of Trend Equations

A time series has three characteristics namely

1. Origin of time
2. Unit of time and
3. Unit of variable.

We can shift the origin and change the trend equation or we can change unit of variable to find the new trend equation. In this section we discuss the following conversions:

1. Conversion of the origin and
2. Conversion of trend values.

1. *Conversion of the origin*: Let $y_c = f(x)$ the equation of the trend where x denotes the deviations of time periods from the origin and y denotes the variables of the time series. If the origin is shifted by k periods forward, then the new trend equation is of the form

$$y_c = f(x + k)$$

If the origin is shifted back by k periods then the new trend equation is of the form

$$y_c = f(x - k)$$

(where $k > 0$)

Example 12.10: The parabolic trend of the equation of a time series is given by

$$y_c = 1200 + 4x - 2x^2$$

with origin = 1980, and x unit = 1 year shift the origin to (a) 1984, (b) 1979 and find the new trend equations.

Solution: We have
$y_c = 1200 + 4x - 2x^2$
The origin under reference = 1980

a. When the origin is shifted to 1984, x is to be replaced by $x+(1984-1980) = x+4$ in the given equation.
i.e.,

$$k = 1984 - 1980 = 4$$

The new trend equation with new origin at 1984 is

$$y_c = 1200 + 4(x+4) - 2(x+4)^2$$

$$\Rightarrow \quad y_c = 1200 + 4x + 16 - 2x^2 - 16x - 32$$

$$\Rightarrow \quad y_c = 1184 - 12x - 2x^2$$

b. When the origin is shifted to 1975, then

$$k = 1979 - 1980 = -1$$

The new trend equation with new origin at 1984 is

$$y_c = 1200 + 4(x-1) - 2(x-1)^2$$

$$\Rightarrow \quad y_c = 1200 + 4x - 4 - 2(x^2 - 2x + 1)$$

$$\Rightarrow \quad y_c = 1194 + 8x - 2x^2$$

2. *Conversion of trend values*: Usually trend is computed from figures. If it is required to compute monthly trend, it is more convenient to compute the trend equation from the annual data and then convert it to a monthly trend. To convert a trend equation operative on annual level to a monthly level, we divide the constant a by 12 and the constant b by 144 (i.e., b is divided twice by 12).
Let

$$y_c = a + bx \text{ is the trend line.}$$

with

$$x \text{ unit} = 1 \text{ year}$$
$$y \text{ unit} = \text{annual total}$$

If we convert the trend equation, so that x unit is 1 month and y unit is monthly average then the equation of new trend line will be

$$y_c = \frac{a}{12} + \frac{b}{12 \times 12}x$$

i.e.,

$$y_c = \frac{a}{12} + \frac{b}{144}x$$

If the annual trend equation $y_c = a + bx$ is converted to half yearly trend equation, then the new trend equation will be of the form

$$y_c = \frac{a}{\left(\frac{12}{6}\right)} + \frac{b}{\left(\frac{144}{6}\right)}x = \frac{a}{2} + \frac{b}{24}x$$

If the annual trend equation is converted into quarterly trend equation, then it is of the form

$$y_c = \frac{a}{\left(\frac{12}{3}\right)} + \frac{b}{\left(\frac{144}{3}\right)}x = \frac{a}{4} + \frac{b}{48}x$$

where x unit: one quarter; y unit: quarterly values.

Example 12.11: Convert the following annual trend equation for cloth production in a factory to a monthly trend equation.

$$y = 108 + 7.2x$$

(origin 1976, time limit 1 year, y = cloth production in '000 m).

Solution: Annual trend equation is $y = 108 + 7.2x$

$\therefore$ Monthly trend equation is

$$y = \frac{108}{12} + \frac{7.2}{144}x$$

$$y = 9 + 0.5x$$

Note: When the data are given monthly averages per year, the value of a remains unchanged in the conversion process, in which case we take new trend lines as

$$y = a + \frac{b}{12}x$$

Example 12.12: Convert the following annual trend equation of ABC corporation to a monthly level: $y = 41 + 7.2x$ (Origin: 1976, x units are year, y = average monthly sales).

Solution: y is average monthly sales.

∴ The monthly trend equation can be written as

$$y = 41 + \frac{7.2}{12}x = 41 + (0.6)x$$

(Origin July 1, 1976, time unit 1 month, y = monthly cloth production in ′000 m).

Example 12.13: The parabolic trend equation of a time series is given by $y_c = 5 + x - x^2$ with origin 1980, and x limit = 1 year, y= annual production. Convert the trend equation so that

1. x unit = 1 month, y = average monthly production.
2. x unit = 1 month, y = average monthly production.

Solution: The trend equation is $y_c = 5 + x - x^2$ with origin 1980, x unit = 1 year, and y unit = annual production.

1. When x limit = 1 year and y unit = average monthly production.
   The new trend equation is

$$y_c = \frac{15}{12} + \frac{x}{12} - \frac{x^2}{12}$$

2. When x unit = 1 month; y unit = average monthly production.
   The new trend equation is

$$y_c = \frac{15}{12} + \frac{x}{12^2} - \frac{x^2}{12^3}$$

i.e.,

$$y_c = \frac{15}{12} + \frac{x}{144} - \frac{x^2}{1728}$$

Exercise 12.1

1. Define
   a. Time series
   b. Secular trend
   c. Seasonal variation
2. What is meant by cyclic variation?
3. Give various adjustments usually practiced during editing of data meant for analysis of time series.
4. Discuss irregular variation in the context of time series.
5. What is semiaverage method?
6. What are the merits and demerits of semiaverage method?

7. Draw free-hand trend from the following time series:

Year	1957–58	1958–59	1959–60	1960–61	1961–62	1962–63	1963–64
Reserves	612	719	820	907	1001	1106	1231

8. Draw a free-hand trend for the following series:

Year	1956	1957	1958	1959	1960	1961	1962	1963	1964	1965
Yield of wheat (Million tons)	12.8	13.9	12.8	13.9	13.4	6.5	29	14.8	14.9	15.9

9. Fit a straight line trend to the following data, using free-hand graph method:

Year	1980	1981	1982	1983	1984	1985	1986
Profit (in lakhs of Rs.)	20	30	25	40	42	30	50

10. Calculate 3-year moving averages for the following data:

Year	1970	1971	1972	1973	1974	1975	1976	1977	1978
Profit (in Lakhs of Rs.)	45	80	70	110	100	125	115	140	130

11. Compute trend by moving average method assuming "a four year cycle."

Year	Sales	Year	Sales
1984	75	1990	70
1985	60	1991	75
1986	55	1992	85
1987	60	1993	100
1988	65	1994	70
1989	70		

12. The following table gives the number of workers employed in a small industry during 1980–89. Calculate the trend values by using 3-yearly moving averages.

Year	No. of workers	Year	No. of workers
1980	20	1984	27
1981	24	1985	26
1982	25	1986	28
1983	18	1987	30

13. Below are given figures of production (in thousand pounds) of a sugar factory.

Year	1981	1982	1983	1984	1985	1986	1987
Production	80	90	92	83	94	99	92

Find the trend by least squares method?

Ans: $Y_c = 84 + 2x$

14. Find out the straight line trend and trend values by the method of least squares for the following data. Also find out the expected value of the year 1938.

Year	1929	1930	1931	1932	1933	1934
No. of industrial failures	23	26	28	32	20	12

Ans: $y_c = 21.857 - 2.46x$, y1938 = 7.097.

15. Draw trend by semiaverage method.

Year	1944	1945	1946	1947	1948	1949
Value	16	18	25.3	35.3	46.6	53.2
Year	1950	1951	1952	1953	1954	1955
Value	4.6	50.9	53.6	84.5	70	79
Year	1956	1957	1958	1959	1960	1961
Value	89.5	97.5	105.92	119	119.62	114.5

16. The production of pig iron and ferro-alloys in thousand metric tons in India is given below:

Year (x)	Production ('000 M tons)	Year (x)	Production ('000 M tons)
1974	620	1979	745
1975	713	1980	726
1976	833	1981	806
1977	835	1982	861
1978	810		

Find the trend time $y = a + bx$, by the method of least squares?

17. Fit a straight line of the form $y = mx + c$ for the following time series showing production of a commodity over a period of 8 years.

Year	Production (in lakhs)	Year	Production (in lakhs)
1961	8	1965	20
1962	12	1966	23
1963	15	1967	27
1964	18	1968	30

Ans: $y = 3.03x + 19.125$

18. Compute 5-year moving averages.

Year	1950	1951	1952	1953	1954	1955
Value	901	95.3	99.7	98.2	104.8	96.1
Year	1956	1957	1958	1959	1960	1961
Value	99.8	113.1	113.9	126.0	128.4	141.4

Year	1962	1963	1964	1965	1966	1967
Value	148.0	154.0	172.1	204.3	203.4	231.2

Ans: 87.6, 98.8, 99.7, 102.4, 105.5, 109.8, 116.2, 124.6, 131.5, 139.6, 148.8, 164.0, 176.4, 193.0

19. Compute 5-years moving average.

Year	1	2	3	4	5	6	7
Value	328	317	357	392	402	405	410
Year	8	9	10	11	12	13	14
Value	427	405	438	445	447	480	482

Ans: 359.2, 374.6, 393.2, 407.2, 409.8, 417, 425, 432.4, 443, 458.4

20. Compute the trend from the following data by the method of least squares:

Year	1970	1971	1972	1973	1974
Population (in lakhs)	830	920	710	900	1690

21. Obtain the trend of bank clearances by the method of moving averages (assume a 5-yearly cycle).

Year	Bank clearance (in lakhs of Rs.)	Year	Bank clearance (in lakhs of Rs.)
1951	53	1957	105
1952	79	1958	87
1953	76	1959	79
1954	66	1960	104
1955	69	1961	97
1956	74	1962	92

Ans: 68.6, 76.8, 82, 84.2, 86.8, 93.8, 94.4, 91.8

22. Below are the given figures of production (in thousand tons) of a sugar factory:

Year	1966	1967	1968	1969	1970	1971	1972
Production (in thousand tons)	80	90	92	83	94	99	92

Find the average production, rate of growth, and trend ordinates (apply least square method)?

CHAPTER 13

Index Numbers

13.1 INTRODUCTION

The most common technique for characterizing a business or economic time series is to compute Index Numbers. An Index Number is a measure of relative change in the value added by a variable or a group of related variables over time or space. It is a statistical device for comparing the general level of magnitude of a group of distinct but related variables in two or more situations. If we want to compare the price level in 2001 with what it was in 1990, we shall have to consider a group of variables, such as the prices of rice, cloth, oil, vegetables, etc. If all these variables change in exactly the same ratio, and in the same direction, there will be no difficulty in finding out the change in the price level as a whole. But the prices of different commodities change by different ratios and in different directions. Also the prices of different commodities are expressed in different measurement units, e.g., rice and pulses are expressed in kilograms or quintals. The prices of oil, milk, etc., are expressed in liters. To avoid the difficulty in finding the average or relative changes, we use an index number that is an indicator of the change in the magnitude of the prices of different commodities as a whole. The index numbers are intended to show the average percentage of changes in the value of certain product (or products), at a specific time, place, or situation as compared to any other time, place, or situation. The study of index numbers is of great importance to the industry, the business, and to the governments for chalking out policies or fixing prices.

13.2 DEFINITIONS AND CHARACTERISTICS

13.2.1 Definition

Index numbers are specialized averages. They are also known as economic barometers, because they reveal the state of inflation or deflation. Index numbers measure how a time series changes over time. Change is measured relative to preselected time period called the base period.

Statistical Techniques for Transportation Engineering.
DOI: http://dx.doi.org/10.1016/B978-0-12-811555-8.00013-1

Definition: An index number is a number that measures the change in a variable over time relative to the value of the variable during a specific base period.

Many economists and statisticians have defined index numbers in their own way. Some of them are given below:

Clark and Schkade: "An index number is a percentage relative that compares economic measure, in a given period with those some measures at a fixed time period in the past."

John I. Griffin: "An index number is a quantity which by reference to base period, shown by its variations, the changes in the magnitude over a period of time. In general, index numbers are used to measure changes over time in magnitudes which are not capable of direct measurements."

M.M. Blair: "Index numbers are the signs and guide posts along the business highway that indicates to the business how he should drive or manage his affairs."

A.M. Tuttle: "An index number is a single ratio (usually in percentages) which measures the combined (i.e., averaged) change of several variables between two different times, places or situations."

Irving Fisher: "The purpose of index number is that it shall fairly represent, so far as one single figure can, the general of the many diverging ratios from which it is calculated."

13.2.2 Characteristics

Index numbers are not only associated with the study of economic or business data, but are also used to make comparisons in the other branches of social and natural sciences. They play a key role in business planning and drafting of executive policies. Following are the characteristics of index numbers:

1. Index numbers are special type of weighted averages.
2. They are expressed in percentages, which make it feasible to compare any two or more index numbers.
3. Index numbers are of comparable nature any two timings or places or any other situation.
4. Index numbers measure changes not capable of direct measurement.

13.2.3 Uses

The most general and the best known use of index number is to study the changes in price over a period of time. Index numbers are very widely used in dealing with the economic and management problems.

Various uses of index numbers are:

1. *To measure and compare changes*: The main purpose of index numbers is to measure the relative temporal or cross-sectional changes at a point of time over some previous time.
2. *To measure purchasing power of money*: The purchasing power of money keeps constantly changing. The consumer's price index helps in computing the real wages of a person. To arrive at real income of people, the income or wages for the time period are deflated by dividing them by equivalent cost of living index or wholesale price index. This helps in adjusting the wages and salaries of the employees.
3. *To find trend*: We use index numbers to measure the changes from time to time, which enable us to study the general trend 3/2 of the economic activity under consideration. As a measure of average change in an specified group, the index number may be used for forecasting.
4. *An aid of framing policies*: Economic policies like, volume of trade fixing of wholesale and retail prices are guided by index number. Fixing of wages and dearness allowance is mainly based on consumer price index.

13.3 TYPES OF INDEX NUMBERS

Index numbers are the devices to measure the relative movements in variables. Which are incapable of measuring directly? The usefulness of any index number lies in the types of questions it can answer. Each index number is designed for a particular purpose and it is the purpose that determines its method of construction. Different kinds of usually constructed index numbers are:

1. *Price Index Numbers*: These index numbers measure changes in prices between two points of time. They measure the general change in the retail or wholesale prices of a commodity or group of commodities at current period as compared to some previous period known as reference period or base period.
2. *Quantity Index Number*: A quantity index number, measure the changes occurring in the quantity of goods demanded, consumed, produced, etc.
3. *Value Index*: A value index compares total value in some period with total value in the base period.
4. *Consumer Price Index*: This is a special kind of index that is constructed for the prices of only the essential items.

13.4 PROBLEMS IN THE CONSTRUCTION OF INDEX NUMBERS

In the process of constructing index numbers (i.e., price index numbers) one encounters several problems, which are to be tackled and solved very carefully. In this section we shall discuss some of these problems. The following are the important problems faced in the construction of index numbers:

1. *Definition of the purpose*: The first and foremost objective is to clearly delineate the purpose of index number for which it is going to be constructed. All the other factors involved in the construction of index numbers mostly depend on the purpose. Choice of items, collection price quotations, selection of sources of data, and base period chiefly depend on the purpose for which the index number is required. For example, while constructing consumer price index numbers, consumer goods with retail prices will be considered, wholesale prices or exfactory prices will not be useful. Therefore it is necessary to define clearly the purpose of index number before it is constructed.
2. *Selection of base period*: Every index number must have a base. The base period should be recent and normal as far as possible.

   According to Marris Hamburg, "It is desirable that the base period be not too far away in time from the present. The further away we move from the base period the dimmer are our recollections of economic conditions prevailing at that time. Consequently, comparisons with these remote periods tend to loose significance and to become rather tenuous in meaning."

   The base period should be clearly related also to the time period when the patterns of spending habits do not change materially. The index for base period is always taken to be 100. The base period should be fairly normal one free from fluctuations and disturbances.
3. *Selection of items to be included*: The items to be selected should be relevant, representative, reliable, and comparable. They should be adequate in number and importance so as to cover and reflect the overall picture and should be unaffected by violent movements. The items selected should be of standard quantity. The number to be selected will depend on the technique of collection and processing of data no. definite limit can laid for this. The larger is the number of items, the lesser will be the chances of error in the average. The list of items included in the construction of index number is called "Regimen" or "Basket."

4. *Collection of data*: Data are collected on the items, which are to be included in the construction of index numbers. The choice of items totally depends on the purpose of index numbers. The sources of data should be selected with discrimination. These should be reliable and representative.
5. *Choice of an average*: Any measure of central tendency may be used for constructing an index number. Median and mode are erratic; hence they are not suitable for constructing index numbers. Harmonic mean is difficult to calculate, hence it is not used. Geometric mean (GM) is the best measure for measuring the relative changes and give more importance to small items and less importance to bigger items. Therefore GM is most preferred average. For the construction of general index numbers, Arithmetic mean (AM) is duly affected by extreme values and it is still widely used because of its easy computations.
6. *System of weighting*: In order to allow each commodity to have a reasonable influence on the index, it is advisable to use a suitable weighting system. The system of weighting depends on the purpose of index. But they ought to reflect the relative importance of the commodities in the basket in the relevant sense. The system of weighting may be either arbitrary or rational. In the case of arbitrary weighting, the statistician is free to assign weights according to his judgment and in the case of rational weighing, the statistician has some fixed criteria for assigning weights. Weights may be either implicit or explicit. Weights should be allowed to vary from period to period. Index number will give better results if weights are allowed to vary.

13.5 METHOD OF CONSTRUCTING INDEX NUMBERS

There are two general methods for constructing index number:

1. Aggregative method;
2. Average of price relative method.

They are further be classified as follows:

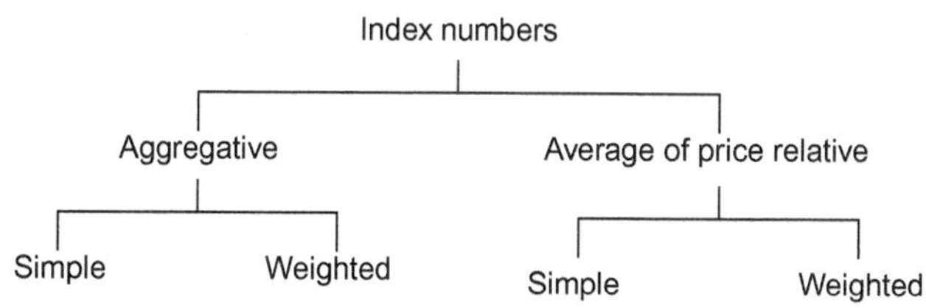

1. *Simple aggregative method*: This is a simple method. In this method, the prices of different commodities of the current year are added and the sum is divided by the sum of the prices of those commodities in the base year and the quotient thus obtained is multiplied by 100.
Symbolically:

$$P_{01} = \frac{\sum p_1}{\sum p_0} \times 100$$

where 0 and 1 stand for the base period and the current period, respectively. P_{01} = The required price index number for the current period; Σp_1 = The sum of the prices of commodities per unit in the current period; Σp_0 = The sum of the prices of commodities per unit in the base period.

13.5.1 Solved Examples

Example 13.1: Calculate the index numbers from the following by simple aggregate method:

Index numbers

Item	Price in base period (in Rs.)	Price in current period (in Rs.)
A	5	7
B	6	8
C	8	14
D	16	27
E	30	28
F	96	67

Solution:

Items	Price in base period (p_0)	Price in current period (p_1)
A	5	7
B	6	8
C	8	14
D	16	27
E	26	28
F	100	67
Total	$\Sigma p_0 = 161$	$\Sigma p_1 = 151$

Index number for the current period

$$P_{01} = \frac{\sum p_1}{\sum p_0} \times 100 = \frac{151}{161} \times 100$$

$$= 93.79$$

Note: The price of every item in the current period shows that they are increased, except the item F. But the index number shows that there is a fall in the commodities. The extent of fall = 100 − 93.7% = 6.21%.

Due to presence of item F, the index number is declaring a decrease in the prices of the commodities on an average. Therefore the presence of extreme items will give misleading results. This is a demerit of this method.

Example 13.2: Calculate the index numbers from the following data by simple aggregative method, taking 1980 as base:

Commodity	Prices per unit (in Rs.)				
	1980	1981	1982	1983	1984
A	0.30	0.33	0.36	0.36	0.39
B	0.25	0.24	0.30	0.32	0.30
C	0.20	0.25	0.28	0.32	0.30
D	2.00	2.40	2.50	2.50	2.60

Solution:

Commodity	Prices per unit (in Rs.)				
	1980	1981	1982	1983	1984
A	0.30	0.32	0.36	0.36	0.39
B	0.25	0.24	0.30	0.32	0.30
C	0.20	0.25	0.28	0.32	0.30
D	2.00	2.40	2.50	2.50	2.60

Index number for 1981 (1980 as base) is

$$\frac{\sum p_1}{\sum p_0} \times 100 = \frac{3.22}{2.75} \times 100 = 117.1$$

Index number for 1982 (1980 as base) is

$$\frac{\sum p_1}{\sum p_0} \times 100 = \frac{3.44}{2.75} \times 100 = 125.1$$

Index number for 1983 (1980 as base) is

$$\frac{\sum p_1}{\sum p_0} \times 100 = \frac{3.50}{2.75} \times 100 = 127.3$$

Index number for 1984 (1980 as base) is

$$\frac{\sum p_1}{\sum p_0} \times 100 = \frac{3.59}{2.75} \times 100 = 130.5$$

2. *Weighted aggregative method*: This method is known as Laspayre's method. In this method the base year quantities (consumption, demand, production, etc.) corresponding to the prices of the items are taken as weights. The weighted aggregative index number is constructed by using the formula:

$$P_{01} = \text{Index number } P_{01} \frac{\sum p_1 q_0}{\sum p_0 q_0} \times 100$$

where q_0 = base period quantities used as weights; p_0 = prices of the base year; p_1 = prices of the current year; $\sum p_0 q_0$ = sum of products of prices of the commodities in the base period with their corresponding quantities used in the base period; $\sum p_1 q_0$ = sum of products of prices of the commodities in the current period with their corresponding quantities used in the base period.

Example 13.3: Calculated weighted aggregative index number by taking 1980 as base from the following data:

Commodity	Quantity	Prices per unit (in Rs.)				
		1980	1981	1982	1983	1984
A	12 units	0.30	0.33	0.36	0.36	0.39
B	10 units	0.25	0.24	0.30	0.32	0.30
C	20 units	0.20	0.25	0.28	0.32	0.30
D	1 unit	2.00	2.40	2.50	2.50	2.60

Solution:

Commodity	Quantity (q_0)	1980		1981		1982		1983		1984	
		p_0	p_0q_0	p_1	p_1q_0	p_1	p_1q_0	p_1	p_1q_0	p_1	p_1q_0
A	12	0.30	3.60	0.33	3.96	0.36	4.32	0.36	4.32	0.39	4.68
B	10	0.25	2.50	0.24	2.40	0.30	3.00	0.32	3.20	0.30	3.00
C	20	0.20	4.00	0.25	5.00	0.28	5.60	0.36	6.40	0.30	6.00
D	1	2.00	2.00	2.40	2.40	2.50	2.50	2.50	2.50	2.60	2.60
Total			12.10		13.76		15.42		16.42		16.28

Weighted index numbers taking 1980 as base:
(Laspeyre's price index numbers)
Weighted index number for 1981

$$(1980 \text{ as base period}) = \frac{\sum p_1 q_0}{\sum p_0 q_0} \times 100$$

$$= \frac{13.76}{12.10} \times 100 = 113.7$$

Weighted index number for 1982

$$(1980 \text{ as base period}) = \frac{\sum p_1 q_0}{\sum p_0 q_0} \times 100$$

$$= \frac{15.42}{12.10} \times 100 = 127.4$$

Weighted index number for 1983

$$(1980 \text{ as base period}) = \frac{\sum p_1 q_0}{\sum p_0 q_0} \times 100$$

$$= \frac{16.42}{12.10} \times 100 = 135.7$$

Weighted index number for 1981

$$(1980 \text{ as base period}) = \frac{\sum p_1 q_0}{\sum p_0 q_0} \times 100$$

$$= \frac{16.25}{12.10} \times 100 = 134.5$$

Example 13.4: Construct index numbers of price, by applying Laspeyre's from the data:

Commodity	**1993**		**1994**	
	Price	**Quantity**	**Price**	**Quantity**
A	2	8	4	6
B	5	10	6	5
C	4	14	5	10
D	2	19	2	13

Solution:

Base year is 1993

Commodity	p_0	q_0	p_1	q_1	p_0q_0	p_1q_0
A	2	8	4	6	16	32
B	5	10	6	5	50	60
C	4	14	5	10	56	70
D	2	19	2	13	38	38
Total					$\Sigma p_0q_0 = 160$	$\Sigma p_1q_0 = 200$

Laspeyre's price index number

$$\text{(base period 1993)} = \frac{\sum p_1 q_0}{\sum p_0 q_0} \times 100$$

$$= \frac{200}{160} \times 100 = 125$$

3. *Simple average price of relative method*: Before introducing the method, we shall first explain the concept of "Price Relative."

   Price Relative of a Commodity: The price relative of a commodity in the current period with respect to base period is defined as the price of the commodity in the current period expressed as a percentage of the price in the base period.

   If p_0 denotes the price of commodity per unit in the base period, and

   p_1 denotes the price of commodity in the current period, then

$$\text{Price Relative} = P = \frac{p_1}{p_0} \times 100$$

4. *Index number by simple average price relative method*: In this method the simple average of price relatives of all the items in the data is taken as the index number.

   If the AM is used as the average then the index number is

$$\text{Price index number} = P_{01} = \frac{(\Sigma p_1/p_0) \times 100}{n}$$

where n denotes the number of items, i.e., commodities.

If GM is used for computing price relatives, then the price index number is

$$P_{01} = \text{Antilog}\left[\frac{\Sigma \log P}{n}\right]$$

Example 13.5: Compute index numbers by simple AM of price relative method.

	Wheat	Cotton	Oil
1st year	4	2	2
2nd year	3	1.5	1.25
3rd year	2.5	1	0.75

The price quotations are given in quantity terms converting the money prices, i.e., computing price of the commodity per quintal we get,

	Price of wheat per quintal	Price of cotton per quintal	Price of oil per quintal
1st year	100/4 = 25	100/2 = 50	100/2 = 50
2nd year	100/3 = 33.3	100/1.5 = 66.7	100/1.25 = 80
3rd year	100/2.5 = 40.0	100/1 = 100	100/0.75 = 133.3

Index numbers by simple AM of price relative method:

Year	Prices per Quintal			Price relatives			Total
	Wheat	Cotton	Oil	Wheat	Cotton	Oil	
I	25	50.0	50.0	$\left(\frac{25}{32.8}\right)\times100 = 76.2$	$\left(\frac{50}{72.2}\right)\times100 = 69.3$	$\left(\frac{50}{87.8}\right)\times100 = 56.9$	202.4
II	33.3	66.7	80.0	$\left(\frac{33.3}{32.8}\right)\times100 = 101.5$	$\left(\frac{66.7}{72.2}\right)\times100 = 92.4$	$\left(\frac{80}{87.8}\right)\times100 = 91.1$	285
III	40.0	100.0	133.3	$\left(\frac{40}{32.8}\right)\times100 = 138.5$	$\left(\frac{100}{72.2}\right)\times100 = 138.5$	$\left(\frac{133}{87.8}\right)\times100 = 151.8$	412.3
Total	98.3	216.7	263.3				
Average	32.8	72.2	87.8				

We have $n = 3$

$$\text{Average index numbers} = \frac{\Sigma P_1/P}{n}$$

$$\text{For I year} = \frac{202}{3} = 67.3$$

$$\text{For II year} = \frac{285}{3} = 95$$

$$\text{For III year} = \frac{412.3}{3} = 137.4$$

Example 13.6: From the following details, construct an index for 1982, taking 1975 as the base by the price relative method using (1) AM, (2) GM.

Commodities	Prices (1975)	Prices (1982)
A	10	13
B	20	17
C	30	60
D	40	70

Solution:

1.

Commodities	Prices (1975) P_0	Prices (1982) P_1	Price relative (P) $(P_1/p_0) \times 100$
A	10	13	$(13/10) \times 100 = 130$
B	20	17	$(17/20) \times 100 = 85$
C	30	60	$(60/30) \times 100 = 200$
D	40	70	$(70/40) \times 100 = 175$
Total			590

We have $n = 4$

$$\text{Index number (by Arithmetic mean)} = \frac{(\Sigma p_1/p_0) \times 100}{n}$$

$$= \frac{590}{4} = 147.5$$

2. Index number (by GM)

Commodities	Prices (1975), P_0	Prices (1982), P_1	Price relatives, P	log P
A	10	13	130	2.1139
B	20	17	85	1.9294
C	30	60	200	2.3010
D	40	70	175	2.2430
Total				8.5873

We have $n = 4$

$$\text{Index number (by Geometric mean)} = \text{Antilog}\left[\frac{\Sigma \log P}{n}\right]$$

$$= \text{Antilog}\frac{8.5873}{4}$$

$$= \text{Antilog}\,(2.1468) = 140.3$$

5. *Weighted index number by price relative method*: This is a method for computing weighted index numbers. We use value weights. The values of weights may correspond to either base period or current period or any other period. We use the formula:

$$\text{Index number (by Arithmethic mean)} = P_{01} = \frac{\Sigma Pw}{\Sigma w} \left(P = \left(\frac{p_1}{p_0} \right) \times 100 \right)$$

i.e.,

$$P_{01} = \frac{\Sigma \frac{P_1}{P_0} \times w}{\Sigma w}$$

where p_0 and p_1 have their usual meanings.

When GM is used, the formula is

$$P_{01} = \frac{\Sigma w \log P}{\Sigma w} \quad \left(P = \left(\frac{p_1}{p_0} \right) \times 100 \right)$$

Example 13.7: Construct index numbers from the following data for 1986 and 1987 taking prices of 1985 as base:

Commodity	Price (1985)	Price (1986)	Price (1987)
A	2.00	2.25	2.12
B	5.00	8.00	8.00
C	1.25	1.50	1.00
D	20.00	24.00	21.00

Solution: Given that the weights of the four commodities are 1, 2, 3, 4, respectively

Commodity	Weight,	1985			1986			1987		
	w	p_0	R	$R.W.$	P_1	R	$R.W.$	p_1	R	$R.W.$
A	1	2.00	100	100	2.25	112.5	112.5	2.12	106	106
B	2	5.00	100	300	1.50	120	360	1.00	80	240
C	3	1.25	100	300	1.50	120	360	1.00	80	240
D	4	20.00	100	400	24.00	120	480	21.00	105	420
Total	10			1000			1272.5			1086

We have $\Sigma w = 10$

$$\text{Index number for 1986 (1985 as base)} = \frac{\Sigma \frac{p_1}{p_0} \times w}{\Sigma w}$$

$$= \frac{\Sigma Rw}{\Sigma w} = \frac{1272.5}{10} = 127.25$$

$$\text{Index number for 1987 (1985 as base)} = \frac{\Sigma Rw}{\Sigma w} = \frac{1086}{10} = 108.6$$

Exercise 13.1

1. What is an Index Number? Examine the various problems involved in the construction of an index number. Discuss in brief the use of an index number.
2. Explain the use of index numbers. Describe the procedure followed in the preparation of general and cost of living index numbers.
3. What main points should be taken into consideration while constructing simple index numbers? Explain the procedure of construction of simple index numbers taking example of 10 commodities.
4. Define an "Index Number." Distinguish between the fixed base and chain base methods of constructing the index numbers and discuss their relative methods.
5. "Index Numbers are Economic Barometers." Explain this statement, and mention the precautions that should be taken in making in use of any published index numbers.
6. Discuss the problems of (a) Selection of base and (b) Selection of weights in the construction of index numbers.
7. What are the uses of an index number? Discuss the role of the weight in the construction of an index for the general price level.
8. Explain how "cost of living index number" are constructed? Describe briefly the problems involved and suggest their solution.
9. Laspeyre's price index generally shows an upward trend in the price changes while Paasche's method shows a downward trend in them elucidate the statement.
10. What is Fishers Ideal Index? Why it is called ideal? Show that it satisfies both the time reversal test as well as the factor reversal test.

11. From the following data calculate price index by using:

(a) Laspeyre's method

(b) Paasche's method

Commodities	Base year		Current year	
	Quantity	Price	Quantity	Price
A	20	4	30	6
B	40	5	60	7
C	60	3	70	4
D	30	2	50	3

Ans: (a) 140.38; (b) 141.140

12. Calculate price index numbers for the year 1990 by using the following methods:

(a) Laspeyre's method

(b) Paasche's method

(c) Bowley's method

(d) Fisher's method

(e) Marshall's method

Ans: (a)124.699; (b) 121.769; (c) 123.234; (d) 123.225; (e) 123.323

13. Apply Fisher's method and calculate the price index number from the following data:

Commodities	1994		1995	
	p_0	q_0	p_1	q_1
A	10	4	12	3
B	15	6	20	5
C	2	5	5	6
D	4	4	4	4

Ans: 135.4

14. Compute Fisher's ideal price index number for the following data:

Commodity	1993		1994	
	Price/unit	Expenditure	Price/unit	Expenditure
A	5	125	6	180
B	10	50	15	90
C	2	30	3	60
D	3	36	5	75

Ans: 137.11

15. From the following data construct a price index number of the group of four commodities by using an appropriate formula:

Commodity	Base year		Current year	
	Price/unit	Expenditure (in $)	Price/unit	Expenditure (in $)
A	2	40	5	75
B	4	16	8	40
C	1	10	2	24
D	5	25	10	60

Ans: 219.12

16. Calculate weighted aggregative price index number taking 1972 as base from the following data:

Commodity	Quantity consumed in 1972 (Quintals)	Units	Price in base year 1972	Price in current year 1987
Wheat	4	Per Quintal	80	100
Rice	1	Per Quintal	120	250
Gram	1	Per Quintal	100	150
Pulses	2	Per Quintal	200	300

Ans: 148.94

17. Prepare index numbers of prices for 3 years with average price as base:

(Rate per Rs.)

	Wheat (kg)	Cotton (kg)	Oil (kg)
1st year	10	4	3
2nd year	9	3.5	3
3rd year	9	3	2.5

(Hint: The prices have been given in terms of quantity, convert them to money prices)

Ans: 91, 98, 110

18. Construct index number of agricultural production for 1968–69 with 1949–50 as base.

Agricultural production in India ('000 tonnes)

Crop	1949–50	1968–69
Rice	25,100	39,761
Jowar	6958	9809
Bajra	3190	3802
Maize	2780	5701
Ragi	1510	1648

(*Continued*)

Crop	1949–50	1968–69
Small naillets	1943	1804
Wheat	6757	18,652
Barley	2390	2424

***Ans*:** 150.4

19. Construct wholesale price index for the year 1968–69 with 1952–53 = 100

Commodity	1952–53		1968–69	
	Production ('000 tonnes)	Price per tonne (Rs.)	Production ('000 tonnes)	Price per tonne (Rs.)
Rice	23,420	680	39761	1300
Jowar	6040	390	9809	560
Bajra	2920	370	3802	611
Maize	2816	350	5701	611
Wheat	6760	440	18,652	1100
Barley	2660	375	2424	625
Gur	5260	420	12,003	900
Gram	3800	445	4310	1100

Using

(a) Pasche's method

(b) Laspeyre's method

(c) Marshall-Edgeworth method, and

(d) Construct Fishers ideal index method

***Ans*:** (a) 199.6; (b) 196; (c) 198; (d) 197.8

20. Construct chain index numbers for the year 1965–66, 1966–67, and 1967–68.

Employees consumers price index (1960–61–100)

Year	Bombay	Ahmedabad	Calcutta	Delhi	Kanpur
1965–66	130	130	131	136	446
1966–67	147	148	148	152	153
1967–68	162	168	163	172	174

***Ans*:** 134.6, 149.8, 167.9

21. Construct new index numbers using chain base method.

Year	Rice	Milk	Coffee	Tea	Total	Average
1961	81	77	119	55	332	83.0
1962	82	54	128	82	346	86.0
1963	104	87	111	100	402	100.5
1964	93	75	154	96	418	104.5

(Continued)

Year	Rice	Milk	Coffee	Tea	Total	Average
1965	60	43	165	88	356	89.0
1966	60	44	159	89	352	88.0
1967	62	47	139	84	332	83.0

***Ans*:** 100, 107, 124.15, 102.6, 80.15, 99.95, 97.98

6. *Paausche's method*: This is a method of finding weighted index number. In this method current period quantities (q_1) are used as weights. Paasche's index is also known as current year method index (or given year method index).

   If p_{01} is the required index number for the current period then

$$P_{01} = \frac{\Sigma p_1 q_1}{\Sigma p_0 q_1} \times 100$$

where p_{01}, p_1 represent price per unit of commodities in the base period and current period, respectively. Paasche's indices posses a downward bias.

Both Laspeyre's and Paasche's formulae stand at their sound logic and anyone cannot be out rightly rejected at the cost of the other. All the more, if the base year and given year (current year) are not much distant, both the formulae almost lead to the same index number and it makes little difference which one is chosen.

Example 13.8: Calculate the Paasche's index number from the following data:

Commodity	Base year		Current year	
	Quantity	Price	Quantity	Price
A	12	10	15	12
B	15	7	20	5
C	24	5	20	9
D	5	16	5	14

Solution: Construction of Paasche's index

Commodity	Base year		Current year		p_1q_0	p_0q_0	p_1q_1	p_0q_1
	q_0	p_0	q_1	p_1				
A	12	10	15	12	144	120	180	150
B	15	7	20	5	75	105	100	140
C	24	5	20	9	216	120	180	100
D	5	16	5	14	70	80	70	80
					$\Sigma p_1q_0=505$	$\Sigma p_0q_0=425$	$\Sigma p_1q_1=530$	$\Sigma p_0q_1=470$

We have

$$\Sigma p_1 q_0 = 505, \quad \Sigma p_0 q_0 = 422, \quad \Sigma p_1 q_1 = 530, \quad \text{and } \Sigma p_0 q_1 = 470$$

$$\text{Paasche's index} = p_{01} = \frac{\Sigma p_1 q_1}{\Sigma p_0 q_1} \times 100$$

$$= \frac{530}{470} \times 100 = 112.8$$

Example 13.9: Construct index number from the data by applying Paasche's method.

Commodity	1995		1996	
	Price	Quantity	Price	Quantity
A	2	8	4	6
B	5	10	6	5
C	4	14	5	10
D	2	19	2	13

Solution: Calculation of index number (1995 = 100)

Commodity	p_0	q_0	p_1	q_1	p_0q_0	p_1q_1	p_0q_1	p_1q_0
A	2	8	4	6	16	24	12	32
B	5	10	6	5	50	30	25	60
C	4	14	5	10	56	50	40	70
D	2	19	2	13	38	26	26	38
Total					160	130	103	200
					Σp_0q_0	Σp_1q_1	Σp_0q_1	Σp_1q_0

We have $\Sigma p_1 q_1 = 130$, $\Sigma p_0 q_1 = 103$

$$\text{Paasche's price index number} = \frac{\Sigma p_1 q_1}{\Sigma p_0 q_1} \times 100$$

$$= \frac{130}{103} \times 100 = 126.21$$

Example 13.10: Given the data

	Commodity	
	A	B
p_0	1	1
q_0	10	5
p_1	2	x
q_1	5	2

where p and q respectively stand for price and quantity and subscripts 0 and 1 stand for time period. Find x if the ratio between Laspeyre's (L) and Paasche's (P) index number is:

L : P: : 28:27

Solution:

Commodity	p_0	q_0	p_1	q_1	p_1q_0	p_0q_0	p_1q_1	p_0q_1
A	1	10	2	5	20	10	10	5
B	1	5	X	2	5x	5	2x	2
Total					$\Sigma p_1q_0=20+5x$	$\Sigma p_0q_0=15$	$\Sigma p_1q_1=10+2x$	$\Sigma p_0q_1=7$

$$\text{Laspeyre's index} = P_{01} = \frac{\Sigma p_1q_0}{\Sigma p_0q_0} = \frac{20+5x}{15}$$

$$\text{Paasche's index} = P_{01} = \frac{\Sigma p_1q_1}{\Sigma p_0q_1} = \frac{10+2x}{7}$$

It is given that

L:P::28:27

i.e.,

$$\frac{L}{P} = \frac{28}{27}$$

$$\Rightarrow \quad 27L = 28P$$

$$\Rightarrow \quad 27\left[\frac{20+5x}{15}\right] = 28\left[\frac{10+2x}{7}\right]$$

$$\Rightarrow \quad 27\frac{5(4\times x)}{15} = 4(10+2x)$$

$$\Rightarrow \quad 9(4+x) = 4(10+2x)$$

$$\Rightarrow \quad 36+9x = 40+8x$$

$$\Rightarrow \quad 9x - 8x = 40 - 36$$

$$\therefore \quad x = 4$$

Example 13.11: Calculate Paasche's index number for 1985 from the following data:

Commodity	Price 1975	Quantity 1975	Price 1985	Quantity 1985
A	4	50	10	40
B	3	10	9	2
C	2	5	4	2

Solution:

Commodity	1975		1985					
	Price, p_0	Quantity, q_0	Price, p_1	Quantity, q_1	p_0q_0	p_0q_1	p_1q_0	p_1q_1
A	4	50	10	40	200	160	500	400
B	3	10	9	2	30	6	90	18
C	2	5	4	2	10	4	20	8
Total					240	170	610	426

We have

$$\sum p_0q_0 = 240, \quad \sum p_0q_1 = 170$$
$$\sum p_1q_0 = 610, \quad \sum p_1q_1 = 426,$$

$$\text{Paasche's index number} = \frac{\Sigma p_1q_1}{\Sigma p_0q_1} \times 100$$
$$= \frac{426}{170} \times 100 = 250.6$$

7. *Dorbish and Bowley method*: This method is similar to Fisher's method, but instead of taking GM, arithmetic average is calculated.

If P_{01} is the required index number for the current period, then

$$P_{01} = \frac{\left[\frac{\sum p_1q_0}{\sum p_0q_0} + \frac{\sum p_1q_1}{\sum p_0q_1}\right]}{2} \times 100 \qquad (13.1)$$

where p_0, p_1 represents per unit of commodities in the base and current period, respectively q_0, q_1 represents number of units in the base period and current period, respectively. The Dorbish and Bowley index formula is the AM of Lasypeyre's Paasche's formula.

From Eq. (13.1) we have

$$P_{01} = \frac{\left[\frac{\sum p_1q_0}{\sum p_0q_0} + \frac{\sum p_1q_1}{\sum p_0q_1}\right]}{2} \times 100$$

$$= \frac{\frac{\sum p_1q_0}{\sum p_0q_0} \times 100 + \frac{\sum p_1q_1}{\sum p_0q_1} \times 100}{2}$$

$$= \frac{\text{Laspeyre's index number} + \text{Paasche's index number}}{2}$$

$$= \text{Arithmetic mean of Laspeyr's and Paasche's index number}$$

Example 13.12: Calculate index number from the following data by Dorbish and Bowley method:

Commodity	Base year		Current year	
	Price	Quantity	Price	Quantity
A	6	50	10	56
B	2	100	2	120
C	4	60	6	60
D	10	30	12	24
E	8	40	12	36

Solution:

Commodity	Base year		Current year		P_0q_0	P_0q_1	P_1q_0	P_1q_1
	p_0	q_0	p_1	q_1				
A	6	50	10	56	300	336	500	560
B	2	100	2	120	200	240	200	240
C	4	60	6	60	240	240	360	360
D	10	30	12	24	300	240	360	288
E	8	40	12	36	320	288	480	422
Total					1360	1344	1900	1880

$$\therefore \quad \sum p_0q_1 = 1360, \quad \sum p_0q_1 = 1900$$

$$\sum p_0q_1 = 1344, \quad \sum p_1q_1 = 1880,$$

$$\text{Dorbish and Bowley index number} = \frac{\left[\dfrac{\sum p_1q_0}{\sum p_0q_0} + \dfrac{\sum p_1q_1}{\sum p_0q_1}\right]}{2} \times 100$$

$$= \frac{1}{2}\left[\frac{1900}{1360} + \frac{1880}{1344}\right] \times 100$$

$$= 140 \text{ (approximately)}$$

8. *Fisher's method*: In this method two index numbers with a different set of weights are constructed and a GM is found out. Symbolically it is expressed as

$$\text{Index number} = P_{01} = \sqrt{\frac{\sum p_1q_0}{\sum p_0q_0} \times \frac{\sum p_1q_1}{\sum p_0q_1}} \times 100$$

where p_0 = price of the base year; p_1 = price of the current year; q_0 = quantity of the base; q_1 = quantity of the current year.

This method is also called weight formula.

9. *Marshall-Edgeworth's method*: In this method the sum of base period quantities and current period quantities are used as weights. The formula for computing index numbers by this method is given as

$$P_{01} = \frac{\Sigma p_1(q_0 + q_1)}{\Sigma p_0(q_0 + q_1)} \times 100$$

where p_0, p_1 represent the price per unit of commodities in the base period and current period respectively. q_0, q_1 represent the number of units in the base period and current period, respectively.

The above formula can also be written as

$$P_{01} = \frac{\Sigma p_1 q_0 + \Sigma p_1 q_1}{\Sigma p_0 q_0 + \Sigma p_0 q_1} \times 100$$

Example 13.13: Construct index number by Marshall-Edgeworth's method.

Commodity	**1993**		**1994**	
	Price	**Quantity**	**Price**	**Quantity**
A	2	8	4	6
B	5	10	6	5
C	4	14	5	10
D	2	16	2	13

Solution: Construct of index number (1993–100)

Commodity	p_0	q_0	p_1	q_1	p_0q_0	p_1q_1	p_0q_1	p_1q_0
A	2	8	4	6	16	24	12	32
B	5	10	6	5	50	30	25	60
C	4	14	5	10	56	50	40	70
D	2	16	2	13	38	26	26	38
Total					160 Σp_0q_0	130 Σp_1q_1	103 Σp_0q_1	200 Σp_1q_0

$$\text{Marshall Edgeworth's index number} = \frac{\Sigma p_0 q_0 + \Sigma p_1 q_1}{\Sigma p_0 q_0 + \Sigma p_0 q_1} \times 100$$

$$= \frac{200 + 130}{160 + 103} \times 100$$

$$= \frac{67404.54}{24816} \times 100$$

$$= 125.47$$

10. *Kelly's method*: In this method we compute weighted index numbers. The quantities (q) corresponding to any period can be used as weights in this method; we can also use the average quantities for two more periods as weights. Kelly's method is also called Kelly's fixed weight aggregative method. If p_{01} denotes the index number for the current period, then

$$P_{01} = \frac{\Sigma p_1 q}{\Sigma p_0 q} \times 100$$

where p_0 and p_1 have their usual meanings and q represents the quantities which are used as weights.

The greatest advantage of Kelly's method is that the change in base year does not required to determine the new weights.

Example 13.14: Construct weighted index number by Kelly's method from the data:

Commodity	Base year – 1993		Current year – 1994	
	Price	Quantity	Price	Quantity
A	2	20	4	45
B	4	22	5	30
C	6	30	8	40
D	8	40	10	60

Solution:

Commodity	Base year		Current year		$q = \frac{q_0 + q_1}{2}$	$p_1 q$	$p_0 q$
	p_0	q_0	p_1	q_1			
A	2	20	4	45	32.5	130	65
B	4	22	5	30	26.0	130	104
C	6	30	8	40	35.0	280	210
D	8	40	10	60	50.0	500	400
Total						$\Sigma p_1 q = 1040$	$\Sigma p_0 q = 799$

We have $\Sigma p_1 q = 1040$, $\Sigma p_0 q = 779$

$$\text{Kelly's index number} = \frac{\Sigma p_1 q}{\Sigma p_0 q} \times 100$$

$$= \frac{1040}{779} \times 100$$

$$= 133.5$$

Example 13.15: Calculate price index number for 1945 by

1. Laspeyre's method
2. Paasche's method
3. Bowley's method
4. Marshall's and Edgeworth's method
5. Fisher's method

From following data:

Commodity	Price 1935	Quantity 1935	Price 1945	Quantity 1945
A	4	50	10	40
B	3	10	9	2
C	2	5	4	2

Solution:

Commodity	1935		1945		p_0q_0	p_0q_1	p_1q_0	p_1q_1
	p_0	q_0	p_1	q_1				
A	4	50	10	40	200	160	500	400
B	3	10	9	2	30	6	90	18
C	2	5	4	2	10	4	20	8
Total					240	170	610	426

$$\sum p_0q_0 = 240, \quad \sum p_0q_1 = 170$$

$$\sum p_1q_0 = 610, \quad \sum p_1q_1 = 426$$

Index number for 1945 by

1. Laspeyre's method $= \dfrac{\Sigma p_0 q_0}{\Sigma p_1 q_1} \times 100$

$$= \frac{610}{240} \times 100 = 254.2$$

2. Paasche's method $= \dfrac{\Sigma p_1 q_1}{\Sigma p_0 q_1} \times 100$

$$= \frac{426}{170} \times 100 = 250.6$$

3. Bowley method $= \dfrac{\left[\dfrac{\Sigma p_1 q_0}{\Sigma p_0 q_0} + \dfrac{\Sigma p_1 q_1}{\Sigma p_0 q_1}\right]}{2} \times 100$

$$= \frac{2.542 + 2.506}{2} \times 100 = 252.4$$

4. Marshall Edgeworth's index number $= \dfrac{\Sigma p_1 q_0 + \Sigma p_1 q_1}{\Sigma p_0 q_0 + \Sigma p_0 q_1} \times 100$

$$= \frac{610 + 426}{240 + 170} \times 100 = 252.7$$

5. Fisher's index number $= p_{01} = \sqrt{\dfrac{\Sigma p_1 q_0}{\Sigma p_0 q_0} \times \dfrac{\Sigma p_1 q_1}{\Sigma p_0 q_1}} \times 100$

$$= \sqrt{\frac{610}{240} \times \frac{426}{170}} \times 100$$

$$= \sqrt{6.369} \times 100 = 252.4$$

13.6 TESTS FOR CONSISTENCY OF INDEX NUMBERS

A number of theoretical criteria have been developed to test the consistency of the index numbers and to evaluate various formulae. The following are the tests developed by statisticians for judging the adequacy of a particular index number method:

1. Time reversal test
2. Factor reversal test
3. Unit test
4. Circular test.

13.6.1 Time Reversal Test

An index number method is said to satisfy time reversal test, if

$$P_{01} \times P_{10} = 1$$

where P_{01} and P_{10} are the index numbers for two periods with base period and current period reverser.

In the words of A.M. Tuttle "If relative for a single series of any type (Price, quantity, etc.) are computed for the two periods 'o' and 'n', with period 'o' as a base, then recomputed for the two periods with period 'n' as a base, the two sets of relatives will always be proportional."

Thus an ideal index number should work both ways, i.e., forward and backward. Fisher's formula satisfies, the time reversal test:

According Fisher's formula

$$P_{01} = \sqrt{\frac{\Sigma p_1 q_0}{\Sigma p_0 q_0} \cdot \frac{\Sigma p_1 q_1}{\Sigma p_0 q_1}}$$

And where time is reversed, we get

$$P_{10} = \sqrt{\frac{\Sigma p_0 q_1}{\Sigma p_1 q_0} \cdot \frac{\Sigma p_0 q_1}{\Sigma p_1 q_1}}$$

where $P_{01} \times P_{10} = 1$

where P_{01} = Price index for current year on the base year; P_{10} = Price index for base year on the basis of current year.

The following methods of constructing index numbers also satisfy this test:

1. Simple aggregative method
2. Marshall's Edgeworth's method
3. Kelly's method.

Example 13.16: Given the sum of the products of prices and quantities for the current year 1 and base year 0 for five items as:

$$\Sigma p_0 q_0 = 782, \quad \Sigma p_0 q_1 = 1008, \quad \Sigma p_1 q_0 = 1084, \text{ and } \Sigma p_1 q_1 = 1329$$

On the basis of the given information show that the data satisfies time reversal test.

Solution:

$$P_{01} = \sqrt{\frac{\Sigma p_1 q_0}{\Sigma p_0 q_0} \cdot \frac{\Sigma p_1 q_1}{\Sigma p_0 q_1}}$$

$$= \sqrt{\frac{1084}{782} \times \frac{1329}{1008}} \times 100 = \sqrt{135.19}$$

and

$$P_{01} = \sqrt{\frac{\Sigma p_1 q_0}{\Sigma p_0 q_0} \cdot \frac{\Sigma p_1 q_1}{\Sigma p_0 q_1}} \times 100 = 73.97$$

$$P_{01} \times P_{10} = \frac{135.19 \times 73.97}{100 \times 100} = 1.00$$

∴ The given data satisfies time reversal test.

Example 13.17:

Commodity	Average price 1981 (Base)	Average price 1982
A	16.1	14.2
B	9.2	8.7
C	15.1	12.5
D	5.6	4.8
E	11.7	13.4
F	100.0	117.0

Now reverse the process taking 1982 as base year and 1981 as current year, and show that the two results are strictly consistent.

Commodity	Price 1981	Price 1982	Price relatives for 1982 with 1981 as base R_1	Price relatives for 1981 with 1982 as base R_0	Log R_1	Log R_0
A	16.1	14.2	88.20	113.45	1.9455	2.0551
B	9.2	8.7	94.56	105.78	1.9757	2.0244
C	15.1	12.5	82.77	120.82	1.9179	2.0820
D	5.6	4.8	85.77	120.82	1.9179	2.0820
E	11.7	13.4	114.50	87.32	2.0589	1.9411
F	100.0	117.0	117.00	85.46	2.0682	1.9318
Total					11.8992	12.1015

Solution:

We have $n = 6$, $\Sigma \log R_1 = 11.8992$, and $\Sigma \log R_0 = 12.1015$

$$\text{Index number for 1982 (1981 as base)} = \text{Antilog}\ \frac{\Sigma \log R_1}{n}$$

$$= \text{Antilog}\frac{11.8992}{6}$$

$$= \text{Antilog } 1.9832 = 96.20$$

$$\text{Index number for 1982 (1981 as base)} = \text{Antilog}\ \frac{\Sigma \log R_0}{n}$$

$$= \text{Antilog}\frac{12.1015}{6}$$

$$= \text{Antilog } 2.0169 = 104$$

$$\therefore \quad P_{01} \times P_{10} = \frac{96.20 \times 104}{100 \times 100} = 1.00$$

The results are strictly consistent.

13.6.2 Factor Reversal Test

This rest was originated by Iring Fisher with the logic that a formula that correctly reflects price changes would also reflect quantity changes.

If P_{01} and Q_{01} are the price index number and quantity index number for the period t_1 corresponding to basic period t_0, then we must have

$$P_{01} \times Q_{10} = V_{01} = \frac{\Sigma p_1 q_1}{\Sigma p_0 q_0}$$

In other words an index number method is said to satisfy factor reversal test if the product of price index number and quantity index number, as calculated by the same method is equal to the value of the index number.

Fisher's index number method is the only method which satisfies this test:

i.e.,

$$P_{01} \times Q_{10} = \sqrt{\frac{\Sigma p_1 q_1}{\Sigma p_0 q_0} \cdot \frac{\Sigma p_1 q_1}{\Sigma p_0 q 1}} \times \sqrt{\frac{\Sigma q_1 p_0}{\Sigma q_0 p_0} \cdot \frac{\Sigma q_1 p_1}{\Sigma q_0 p_1}}$$

$$= \sqrt{\frac{\Sigma p_1 q_0 \times \Sigma p_1 q_1 \times \Sigma q_1 p_0 \times \Sigma q_1 p_1}{\Sigma p_0 q_0 \times \Sigma p_0 q_1 \times \Sigma q_0 p_0 \times \Sigma q_0 p_1}}$$

$$= \sqrt{\frac{\Sigma p_1 q_0 \times \Sigma p_1 q_1 \times \Sigma p_0 q_1 \times \Sigma p_1 q_1}{\Sigma p_0 q_0 \times \Sigma p_0 q_1 \times \Sigma q_0 p_0 \times \Sigma p_1 q_0}}$$

$$= \sqrt{\frac{\Sigma p_1 q_1 \times \Sigma p_1 q_1}{\Sigma p_0 q_0 \times \Sigma p_0 q_0}} = \frac{\Sigma p_1 q_1}{\Sigma p_0 q_0}$$

$$= \text{True value ratio}$$

∴ Fisher's method satisfies the factor reversal test.

Example 13.18: Compute Fischer's ideal number from the following data and show that it satisfies factor reversal test:

Year	Article A		Article B		Article C	
	Price	Quantity	Price	Quantity	Price	Quantity
1975	16	4	4	4	2	2
1982	30	3.5	14	1.5	6	2.5

***Solution*:**
Calculation of Fisher's index number

Article	p_0	q_0	p_1	q_1	p_0q_0	p_1q_1	p_1q_0	p_0q_1
A	16	4	30	3.5	64	105	120	56
B	4	4	14	1.5	16	21	56	6
C	2	2	6	2.5	4	15	12	5
					84	141	188	67

$$\therefore \quad \sum p_0q_0 = 84, \quad \sum p_1q_1 = 141$$
$$\sum p_1q_0 = 188, \quad \sum p_0q_1 = 67$$

$$\text{Fisher's Ideal index number} = P_{01} = \sqrt{\frac{\Sigma p_1q_0}{\Sigma p_0q_0} \times \frac{\Sigma p_1q_1}{\Sigma p_0q_1}} \times 100$$

$$= \sqrt{\frac{188}{84} \times \frac{141}{67}} \times 100$$

$\therefore$ Fisher's price index number for 1982 with base 1975 = 217.03
Fisher's quantity index number for 1982 with base 1975

$$Q_{01} = \sqrt{\frac{\Sigma q_1p_0 + \Sigma q_1p_1}{\Sigma q_0p_0 + \Sigma q_0p_1}} \times 100$$

$$= \sqrt{\frac{67}{84} \times \frac{141}{186}} \times 100 = 77.34$$

$$P_{01} \times Q_{01} = \frac{271.03}{100} \times \frac{77.34}{100} \tag{13.2}$$

$$= 2.1703 \times 0.7734 = 1.67865$$

$$V_{01} = \frac{\Sigma p_1q_1}{\Sigma p_0q_0} = \frac{141}{84} = 1.6786 \tag{13.3}$$

From Eqs. (13.2) and (13.3) $P_{01} \times Q_{01} = \dfrac{\Sigma p_1q_1}{\Sigma p_0q_0}$

$\therefore$ Factor reversal test is satisfied.

Example 13.19: Show that Fisher's ideal index satisfies both the time reversal test as well as the Factor reversal test using the data given below:

Commodity	Base year		Current year	
	Price	Quantity	Price	Quantity
A	6	50	10	56
B	2	100	2	120
C	4	60	6	60
D	10	30	12	24
E	8	40	12	36

Solution:

Commodity	p_0	q_0	p_1	q_1	p_0q_0	p_0q_1	p_1q_0	p_1q_1
A	6	50	10	56	300	336	500	560
B	2	100	2	120	200	240	200	240
C	4	60	6	60	240	240	360	360
D	10	30	12	24	300	240	360	288
E	8	40	12	36	320	288	480	432
Total					1360	1344	1900	1880

We have

$$\sum p_1q_0 = 1900, \quad \sum p_0q_0 = 1360$$
$$\sum p_1q_1 = 1880, \quad \sum p_0q_1 = 1344$$

Time Reversal Test:

$$P_{01} = \sqrt{\frac{\Sigma p_1q_0}{\Sigma p_0q_0} \times \frac{\Sigma p_1q_1}{\Sigma p_0q_1}}$$
$$= \sqrt{\frac{1900}{1360} \times \frac{1880}{1344}}$$

$$P_{10} = \sqrt{\frac{\Sigma p_0q_1}{\Sigma p_1q_1} \times \frac{\Sigma p_0q_0}{\Sigma p_1q_0}}$$
$$= \sqrt{\frac{1344}{1880} \times \frac{1360}{1900}}$$

$$\therefore\ P_{01} \times P_{10} = \sqrt{\frac{1900}{1360} \times \frac{1880}{1344} \times \frac{1344}{1880} \times \frac{1360}{1900}}$$

$$\therefore\ P_{01} \times P_{10} = 1$$

Hence Time reversal test is satisfied.

Factor Reversal Test:

$$Q_{01} = \sqrt{\frac{\Sigma q_1 p_0}{\Sigma q_0 p_0} \times \frac{\Sigma q_1 p_1}{\Sigma q_0 p_1}}$$

$$= \sqrt{\frac{1344}{1360} \times \frac{1880}{1900}}$$

$$\therefore\ P_{01} \times Q_{01} = \sqrt{\frac{1900}{1360} \times \frac{1880}{1344} \times \frac{1344}{1360} \times \frac{1800}{1900}}$$

$$= \sqrt{\left(\frac{1880}{1360}\right)^2}$$

$$= \frac{1880}{1360}$$

and

$$\frac{\Sigma p_1 q_1}{\Sigma p_0 q_0} = \frac{1880}{1360}$$

$$\therefore\ P_{01} \times Q_{01} = \frac{\Sigma p_1 q_1}{\Sigma p_0 q_0}$$

∴ Factor reversal test is satisfied.

∴ Fisher's Ideal Index satisfies, Time reversal test as well as Factor reversal test.

Unit Test:

An index number method is said to satisfy unit test if it is not changed by a change in the measuring units of some items under consideration. All index number method, except simple aggregative method satisfies this test.

13.6.3 Circular Test

This test is an extension of the time reversal test and is applicable when the indexes for more than 2 years are given. Suppose an index number is computed for the period 1, on the base period 0, another index number is computer for period 2 on the base period 1, and another index number is computer for the period 3 on the base period 2, and so on, their product should be equal to 1.

Symbolically:

$$P_{01} \cdot P_{12} \cdot P_{23} \ldots P_{(n-1)n} \cdot P_n = 1$$

The following methods satisfy circular test:

1. Simple aggregative method
2. Simple GM of price relative method
3. Kelly's method.

Fisher's Ideal formula does not satisfy circular test.

13.7 QUANTITY INDEX NUMBERS

They are used to the average change in the quantities of related goods with respect to time. Quantity index numbers are also used to measure the level of production. In computing quantity index numbers, either prices or values are used as weights.

Let Q_{01} denote the quantity index number for the current period. The formulae for calculating quantity index numbers are obtained by interchanging the role of "p" and "q" in the formulae for computing price index numbers. Various methods for computing quantity index numbers are as follows:

1. Simple aggregative method

$$Q_{01} = \frac{\Sigma q_1}{\Sigma q_0} \times 100$$

2. Simple average of quantity relative method

$$Q_{01} = \frac{\Sigma Q}{n} \quad \text{(using AM)}$$

or

$$Q_{01} = \text{Antilog}\left[\frac{\Sigma \log Q}{n}\right] \quad \text{(using GM)}$$

where Q = Quantity relative

$$Q = \frac{q_1}{q_0} \times 100$$

3. Laspeyre's method

$$Q_{01} = \frac{\Sigma q_1 p_0}{\Sigma q_0 q_0} \times 100$$

4. Paasche's method

$$Q_{01} = \frac{\Sigma q_1 p_1}{\Sigma q_0 p_1} \times 100$$

5. Dorbish and Bowley's method

$$Q_{01} = \frac{\left[\frac{\Sigma q_1 p_0}{\Sigma q_0 p_0} + \frac{\Sigma q_1 p_1}{\Sigma q_0 p_1}\right]}{2} \times 100$$

6. Fisher's Ideal method

$$Q_{01} = \sqrt{\frac{\Sigma q_1 p_0}{\Sigma q_0 p_0}} \times \sqrt{\frac{\Sigma q_1 p_1}{\Sigma q_0 p_1}} \times 100$$

7. Marshall-Edgeworth's method

$$Q_{01} = \frac{\Sigma q_1 p}{\Sigma q_0 p} \times 100$$

8. Weighted average of quantity relative method

$$Q_{01} = \frac{\Sigma WQ}{\Sigma W} \quad \text{(using AM)}$$

or

$$Q_{01} = \text{Antilog}\left[\frac{\Sigma W \log Q}{\Sigma W}\right] \quad \text{(using GM)}$$

13.8 CONSUMER PRICE INDEX NUMBER

Consumer price index number is a measure of average percentage change in prices at a particular time as compared to a base period for a section of the population for which it is referred.

Formerly consumer price index was known as cost of living index number. It is an index of prices paid by consumers but it does not indicate how much must actually be spent by families to maintain a specified level of living.

According to John I. Griffin, "The index measures only changes in prices it tells nothing about changes in the kinds and amount of goods and services families buy; the total amount families spend for living or the difference in living costs in different places."

The sixth international conference of labor statisticians recommended that the term "cost of living index" should be replaced in appropriate circumstances by the term "consumer price index" or "price of living index" or "cost of living price index." The term retail price index can also be used.

13.8.1 Utility of Consumer Price Index Number

1. Consumer price index is useful in evaluating purchasing power of money.
2. The consumer price index numbers are used in wage fixation and automatic increase in wages.
3. The consumer price index numbers are used by planning commission for framing rent policy, taxation policy, etc.
4. Consumer price index helps to fix dearness allowance to compensate the rise (or fall) in prices of commodities of common utility.

13.8.1.1 Various Steps Involved in the Construction of Consumer Price Index

Consumer price index is a special purpose index. Construction of consumer price index involves consideration of the following:

1. The first step in computing consumer price index number is to decide the category of people for whom the index is to computed. A particular consumer price index number relates to a particular class of people having similar consumption habits and pattern and to a definite region with more or less economic homogeneity.
2. *Selection of Base Period*: Generally the base period for consumer price index is the year declared by the government. A period of comparative economic stability can be selected as the base period, so that the consumption pattern that is reflected in the index number remains practically the same over a fairly long period.

3. *Conducting Family Budget Inquire*: The commodities which are to be included in the index will have to be selected from the standard or average family that will be obtained from the family budget enquiry. A family budget is the detailed statement of expenditure of the family on various commodities. The commodities are generally classified in the following heads:
   - **(a)** Food
   - **(b)** Clothing
   - **(c)** Fuel and lighting
   - **(d)** House rent
   - **(e)** Miscellaneous

13.8.2 Formulas for Constructing Consumer Price Index

There are two methods for constructing consumer price index numbers.

1. Aggregate expenditure method and
2. Family budget method

Aggregate expenditure method: In this method base period quantities are used as weights. The formula is given by:

$$\text{Consumer price index number} = \frac{\Sigma p_1 q_0}{\Sigma p_0 q_0} \times 100$$

where "0" and "1" suffixes stand for base period and current period, respectively.

Example 13.20: Compute the cost of living index from the following data using aggregate expenditure method:

Item	Consumer quantity in the given year	Price in base year	Price in given year
Rice	2½ Quintal × 12	12	25
Pulses	3 kg × 12	0.4	0.6
Oil	2 kg × 12	1.5	22
Clothing	6 m × 12	0.75	1
Housing	—	20 P.M.	30 P.M.
Miscellaneous	—	10 P.M.	15 P.M.

Solution:

Item	q_1	p_0	p_1	p_1q_1	p_0q_1
Rice	30	12	25	750	360
Pulses	36	0.4	0.6	21.6	14.4
Oil	24	1.5	2.2	52.6	36

(Continued)

Item	q_1	p_0	p_1	p_1q_1	p_0q_1
Clothing	72	0.75	1.0	72	54
Housing	12	20	30	360	240
Miscellaneous	12	10	15	180	120
Total				1436.2	824.4

$$\therefore\ \Sigma p_1 q_1 = 1436.4,\quad \Sigma p_0 q_1 = 824.4$$

$$\text{Cost of living Index number} = \frac{\Sigma p_1 q_0}{\Sigma p_0 q_0} \times 100$$

$$= \frac{1436.2}{174.24} \times 100 = 174.24$$

Family budget method: In this method the expenditure on different commodities in the base period are used as weights.

According to the method

$$\text{Consumer price index number} = \frac{\Sigma\, Pw}{\Sigma\, w}$$

where P = price relative = $P_{1/p_0} \times 100$; P_0 = price of commodity in the base period; P_1 = price of commodity in the current period; and $w = p_0 q_0$.

Note:

$$\text{We can write consumer price index number} = \frac{\Sigma p_1 q_0}{\Sigma p_0 q_0} \times 100$$

or

$$\text{C.P. index number} = \frac{\Sigma\, IV}{\Sigma\, V}$$

where I = Relatives, V = Weights.

Example 13.21: Construct cost of living index for 1982 based on 1975 from the following data:

Group	Group Index No. For 1982 (based on 1975)	Weight
Food	122	32
Housing	140	10
Clothing	112	10
Fuel and light	116	6
Miscellaneous	106	42

Solution:

Group	Index No.	Weights, I	Weighted relative, IV
Food	122	32	3904
Housing	140	10	1400
Clothing	112	10	1120
Fuel and Light	116	6	696
Miscellaneous	106	42	4452
Total		100	11,572

$$\text{Index number} = \frac{\Sigma\, IV}{\Sigma\, V} = \frac{11572}{100} = 115.72$$

13.9 CHAIN BASE METHOD

If there are m consecutive years data at hand and each time we consider the same n items to be included in the construction of indices, the chain base method consists of calculating the price index for each year taking the preceding year as base. This brings the homogeneity error to zero level.

In the chain base method, we can use any appropriate index number formula. All such year-to-year indices are called link relatives. Mathematically, a link relative can be defined as:

$$\text{Link relative (L.R.)} = \frac{\text{Price in current period}}{\text{Price in preceding period}} \times 100$$

If there are two more commodities under consideration then Average Link Relatives (A.L.R.) calculated for each period. Generally AM is used for averaging link relative. These averages of link relatives for different periods are called chain index numbers.

Various indices of the series by chain base method can be computed by the following relations:

$$P_{01} = P_{01} \quad \text{(as a first link)}$$
$$P_{02} = P_{01}\, P_{12}$$
$$P_{03} = P_{01}\, P_{12}\, P_{23} = P_{02}\, P_{23}$$
$$\vdots$$
$$P_{0m} = P_{01}\, P_{12}\, P_{23} \ldots P_{(m-1)m} = P_{0(m-1)}\, P_{(m-1)m}$$

Using the above relations the general formula can be written as:

$$\text{Chain index} = \frac{\text{Previous year chain index} \times \text{current year link relative}}{100}$$

13.9.1 Advantages of using Chain Base Method

In this method introduction of new items and deletion of old ones can be done without hazards and hassles. Because of this the chain base index (C.B.I.) numbers are used in consumer and wholesale price indices. By using chain base method, comparison is possible between any two successive periods. Index numbers calculated by the chain base method are free from the effect of seasonal variations. The chain base method brings homogeneity error almost to zero.

13.9.2 Limitation

The chain base indices are not suitable for long-range comparisons.

Example 13.22: From the following index numbers prepare new one by (1) taking the year 1982 as base and (2) using chain base method:

Year	Index number
1979	100
1980	110
1981	175
1982	250
1983	300
1984	400

Solution:

1. Index numbers with 1982 as base (base 1982 = 100)

Year	Index number	Base changed to 1982 (1982 = 100)	Index number (1982 = 100)
1979	100	(100/250) × 100	40
1980	110	(110/250) × 100	44
1981	175	(175/250) × 100	70
1982	250	100	100
1983	300	(300/250) × 100	120
1984	400	(400/250) × 100	160

2. Index numbers using chain base method

Year	Index number	Conversion	Chain base Index number
1979	100	100	100.00
1980	110	(110/100) × 100	110.00
1981	175	(175/110) × 100	159.09
1982	250	(250/170) × 100	142.86
1983	300	(300/250) × 100	120.00
1984	400	(400/300) × 100	133.33

13.10 BASE CONVERSION

In this section we shall discuss situations when the base of an index number is desired to be changed we shall denote, fixed base index numbers by F.B.I. and chain base index numbers by C.B.I.

13.10.1 Base Conversion

(a) Conversion of F.B.I. to C.B.I.

To convert F.B.I. numbers to C.B.I. numbers, the following procedure is adopted:

1. The first years' index number is taken as 100.

2. For subsequent years, the index number is obtained by the formula current year's C.B.I. number.

$$= \frac{\text{Current year's F.B.I.}}{\text{Previous year F.B.I.}} \times 100$$

Example 13.23: From the F.B.I. numbers given below compute C.B.I. numbers.

Year	1975	1976	1977	1978	1979	1980
F.B.I.	376	392	408	380	392	400

Solution:

Year	F.B.I.	Conversion	C.B.I.
1975	376	–	100.0
1976	392	(392/376) × 100	104.3
1977	408	(408/392) × 100	104.1
1978	380	(380/408) × 100	93.1
1979	392	(392/380) × 100	103.2
1980	400	(400/392) × 100	102.0

Example 13.24: Compute C.B.I. numbers for the following series of F.B.I. numbers:

Year	1981	1982	1983	1984	1985	1986
F.B.I.	210	220	250	400	300	400

Solution: It is not possible to compute the C.B.I. number for the year 1981 (since the C.B.I. for 1981 is the index number for 1981 with 1980 as the base year).

Year	F.B.I. (1975-100)	Conversion	C.B.I.
1981	210	–	–
1982	220	(220/210) × 100	104.762
1983	250	(250/220) × 100	113.636
1984	400	(400/250) × 100	160
1985	300	(300/400) × 100	75
1986	400	(400/300) × 100	133.333

Conversion of C.B.I. to F.B.I.

The following steps are involved:

1. The first years' index number is taken what the C.B.I. is, but if it is from the base it is taken equal to 100.
2. In the subsequent years, the index is obtained by the formula:

$$\text{Current year's F.B.I.} = \frac{(\text{Current year's C.B.I.}) \times (\text{Preceding year's F.B.I.})}{100}$$

Example 13.25: From the C.B.I. numbers given below, construct F.B.I. numbers with 1970 as base:

Year	1978	1979	1980	1981	1982
C.B.I.	80	140	130	100	110

Solution:

Year	C.B.I.	Conversion	F.B.I.
1978	80	–	80
1979	140	(80/100) × 140	112
1980	130	(80/100) × (140/100) × 100	145.6
1981	100	(80/100) × (140/100) × (130/100) × 100	145.6
1982	110	(80/100) × (140/100) × (130/100) × (300/400) × 100	

13.10.2 Base Shifting

We sometimes shift the base from one period to another period. When the given indices are referred to a long-back period and one wants the indices based on recent base period, we shift the base.

We also shift the base to make two given series comparable. The method is to divide the indices of the other years by the index of the year selected as base and multiplying the quotient by 100.

Symbolically,

$$\text{New base index number} = \frac{\text{Old index number of current year}}{\text{Index number of new base year}}$$

Example 13.26: Following are the wholesale price index numbers from 1975 to 1980 to base 1970:

Year	1975	1976	1977	1978	1979	1980
Index number	175.8	172.4	185.4	185.0	206.2	246.8

Find the wholesale price indices to the base 1977?

Solution: Index number of 1977 is 185.4.

∴ The wholesale index numbers to the base 1977

Year	By formula	Index numbers
1975	$(175.8/185.4) \times 100$	94.82
1976	$(172.4/185.4) \times 100$	92.99
1977	$(185.4/185.4) \times 100$	100
1978	$(185.0/185.4) \times 100$	99.79
1979	$(206.2/185.4) \times 100$	111.22
1980	$(246.8/185.4) \times 100$	133.12

13.11 SPLICING

Splicing is the statistical procedure, which connects an old index number of the series with a revised series in order to make the series continuous.

In economic phenomenon, we sometimes construct a new series of insides banishing the old one. The base period in the new series is always the last year (period) of the old series. The index numbers of the new series are spliced to first series:

1. *Forward splicing*: If an old series is connected with the new one, it is known as forward splicing, we use the formula:

   Forward spliced index number

$$= \frac{(\text{Old Index No. of the new base year}) \times \text{Index No. for the given year}}{100}$$

2. *Backward splicing*: In this we connect new series with the old one in the sense that the indices of the old series are converted to the base of the new series. The formula is:

$$\text{Backward spliced index number} = \frac{\text{Index No. to be spliced}}{\text{Old index No. of the new base year}} \times 100$$

Example 13.27: The following table gives the index numbers of wholesale prices from 1972–73 to 1981–82 to the base 1970–71 in column (2) and from 1982–83 to 1988–89 to the base 1981–82 in column (3).

Solution:

Calculation of spliced index numbers:

Year	Index No. base 1970–71	Index No. 1981–82	Forward spliced Index No.	Backward spliced Index No.
1972–73	111	–	111	$111 \times (100/264) = 42$
1973–74	142	–	142	$142 \times (100/264) = 54$
1974–75	178	–	178	$178 \times (100/264) = 54$
1975–76	166	–	166	$166 \times (100/264) = 63$
1976–77	167	–	167	$167 \times (100/264) = 63$
1977–78		–	184	$184 \times (100/264) = 70$
1978–79	181	–	181	$181 \times (100/264) = 69$
1979–80	207	–	207	$207 \times (100/264) = 78$
1980–81	238	–	238	$238 \times (100/264) = 90$

Year	Index No. base 1970–71	Index No. 1981–82	Forward spliced Index No.	Backward spliced Index No.
1981–82	264	100	$264 \times (100/100) = 264$	100
1982–83	–	107	$264 \times (100/100) = 282$	107
1983–84	–	118	$264 \times (118/100) = 312$	118
1984–85	–	126	$264 \times (126/100) = 333$	126
1985–86	–	126	$264 \times (126/100) = 333$	126
1986–87	–	137	$264 \times (137/100) = 362$	137
1987–88	–	153	$264 \times (153/100) = 404$	153
1988–89	–	160	$264 \times (160/100) = 422$	160

13.12 DEFLATION

The general downward movement in the set of prices is referred to as deflation. Index numbers are used to compute real income from money income.

Deflating the index number means making allowance in indices for the effect of changes in price levels, increase in the price of a commodity reduces the purchasing power of the commodity. If the present price of a commodity is reduced to half. In this way the money value of our earning changes with the rise or fall in prices of the commodities. The real wages, money income index number, and real income can be calculated by the deflation technique. The following formulae are used:

$$\text{Real wage (or real income)} = \frac{\text{Income of the year (money wage)}}{\text{Price index of the current year}} \times 100$$

(Real Income is also known as deflated income)

$$\text{Money income index number} = \frac{\text{Real income}}{\text{Income of the base year}} \times 100$$

$$\text{Real income index number} = \frac{\text{Money income index number}}{\text{Consumer price index number}} \times 100$$

Purchasing Power of money is calculated by the formula:

$$\text{Purchasing power of money} = \frac{1}{\text{Price index}} \times 100$$

Example 13.28: The annual wages (in \$) of a worker are given along with price indices. Find the real wage indices?

Year	1975	1976	1977	1978	1979	1980	1981
Wages	180	220	340	360	370	385	400
Price indices	100	170	300	320	330	350	375

Solution: Construction of real wage index numbers:

Year	Wages (in \$)	Price indices	Real wage = (Wage/CPI) × 100	Real wage	Real wage Index No. (1975 = 100)
1975	180	100	(180/100) × 100	180	100
1976	220	170	(220/170) × 100	129.41	71.89
1977	340	300	(340/300) × 100	113.33	62.96
1978	360	320	(360/320) × 100	112.50	62.50
1979	370	330	(370/330) × 100	112.12	62.28
1980	385	350	(385/350) × 100	110.00	61.11
1981	400	375	(400/375) × 100	106.66	59.25

Exercise 13.2

1. Construct the cost of living index from the following data:

Group	Index for 1992	Expenditure (%)
Food	550	46
Clothing	215	10
Fuel and lighting	220	7
House rent	160	12
Miscellaneous	275	25

2. The cost of living index for the working class families in 1938 was 168.12. The retail prices with base 1934 = 100 and the percentages of family expenditure in 1934 are given below. Find the retail price for the rent, fuel, and light group?

Group	Family expenditure in 1934 (%)	Retail price in 1938 (1934 = 100)
Food	40	132
Rent, fuel, and lighting	18	?
Clothing	9	210
Miscellaneous	33	200

3. An enquiry into the budgets of middle class families in a certain city gave the following information:

Item	Total expenditure (%)	Price in 1993 (in $)	Price in 1994 (in $)
Food	35	150	145
Fuel	10	25	23
Clothing	20	75	65
Rent	15	30	30
Miscellaneous	20	40	45

What is the cost of living index number of 1994 as compared with 1993?

4. Calculate the cost of living index for the following data:

Group	Price in base year	Price in current year	Weight
Food	39	47	4
Fuel	8	12	1
Clothing	14	18	3
Rent	12	15	2
Miscellaneous	25	30	1

5. Construct a cost of living number from the following price relatives for the year 1985 and 1986 with 1982 as base giving weightage to the following groups in the proportion of 30, 8, 6, 4, and 2, respectively.

Group	1982	1985	1986
Food	100	114	116
Rent	100	115	125
Clothing	100	108	110
Fuel	100	105	104
Miscellaneous	100	102	104

6. Calculate C.B.I. numbers for the following series of F.B.I. numbers:

Year	F.B.I. (1975 = 100)
1981	210
1982	220
1983	250
1984	400
1985	300
1986	400

7. From the following series of C.B.I. number, compute F.B.I. numbers with 1980 as the fixed base.

Year	1981	1982	1983	1984	1985	1986
C.B.I.	110	136.364	120	138.889	20	146.667

8. Construct a new series of index numbers by shifting the base from 1980 to 1981.

Year	1980	1981	1982	1983	1984	1985	1986
Index No. (1980 = 100)	100	110	150	180	250	300	440

9. Construct a single series of index numbers (1975 = 100) based on two separate series given below:

Year	Index No. (1975 = 100)	Index No. (1982 = 100)
1975	100	–
1976	110	–
1977	150	–
1978	200	–
1979	200	–
1980	250	–
1981	300	–
1982	350	100

(*Continued*)

Year	Index No. (1975 = 100)	Index No. (1982 = 100)
1983	–	120
1984	–	150
1985	–	180
1986	–	200

10. From the F.B.I. numbers given below, prepare C.B.I. numbers.

Year	1981	1982	1983	1984	1985	1986
F.B.I. (1980 = 100)	110	120	130	141	200	180

11. From the F.B.I. numbers given below, prepare C.B.I. numbers:

Year	1982	1983	1984	1985	1986
F.B.I. (1980 = 100)	267	275	280	290	320

12. Construct a single series of index numbers, based on two separate series with base 1986.

Year	1980	1981	1982	1983	1984	1985	1986
Index No. (1st series)	100	107	119	138	–	–	–
Index No. (2nd series)	–	–	–	100	105	110	111

13. From the chain index numbers given below, prepare F.B.I. numbers with 1978 as base year:

Year	1979	1982	1981	1982	1983	1984
C.B.I.	94	104	104	93	103	102

14. The following data gives the per capita income and the cost of living index number for a particular class of people. Deflate the per capita income by taking into account the charges in the cost of living.

Year	1979	1980	1981	1982	1983	1984	1985	1986
Per capita income (in \$)	300	320	340	350	375	405	425	480
Cost of living index	120	125	150	160	175	220	240	250

15. Find the real incomes from the data given below:

Year	1980	1981	1982	1983	1984	1985	1986
Average monthly income	400	410	450	500	600	750	800
Income cost of living index (1975 = 100)	120	140	141	150	160	180	205

16. Calculate the index numbers of real wages from the following information using 1981 as base year:

Year	1981	1982	1983	1984	1985	1986
Average monthly wage (in $)	1200	1320	1430	1500	1710	2000
C.B.I.	100	120	130	150	190	200

17. The following table gives annual wages and cost of living index numbers. Calculate

(a) Real wages

(b) Index numbers of real wages with 1979 as base

Year	1985	1986	1987	1988	1989	1990
F.B.I.	376	392	408	380	392	400

INDEX

K

L

M

N

O

P

Q

R

S

T

FSC
www.fsc.org
MIX
Papier | Fördert gute Waldnutzung
FSC® C083411

Kolibri 360 GmbH
Ferdinand-Jühlke-Straße 7
99095 Erfurt, Deutschland
produktsicherheit@kolibri360.de